NOTES ON

WALL VEGETATION

by

S. SEGAL

Dr. W. JUNK N.V. - Publishers

13, van Stolkweg

The Hague - The Netherlands

Preface

This study is published with financial support of the Netherlands Organization for the Advancement of Pure Research (Z.W.O.).

The Department of Education and Sciences, The Hague, subsidised part of the field studies. Other travelling grants were received from the French Government, from the Royal Dutch Academy of Sciences and from the Rudolf Lehmann Foundation (Amsterdam).

It is not possible personally to thank everybody who assisted during my studies. An exception must be made for Professor J. Braun-Blanquet, who gave me the opportunity to make field trips with his institute at Montpellier as a base of operations. I thank also Miss R. A. J. Grabandt, Mrs. M. J. A. B. Verhagen née Zuure, mr. M. Snel, mr. A. L. M. Veltman and mr. D. J. van Weers, who all contributed towards the study of wall vegetation as students of biology at the University of Amsterdam. I am much obliged to Professor J. P. Bakker for his kind suggestions and assistance with substrate analyses, and to Ir. J. Heesterman for ash analyses of plant materials.

A special word of appreciation is due to my teacher Professor J. Heimans to whom this publication is dedicated. Like his successor, Professor A. D. J. Meeuse, he offered much constructive criticism.

Professor J.-M. Géhu put the finishing touches to the French summary and Dr. N. V. Pears to the English text.

Amsterdam, March 1969.

Hugo de Vries-laboratorium
Universiteit van Amsterdam

ISBN-13: 978-90-6193-254-3 e-ISBN-13: 978-94-011-6468-9
DOI: 10.1007/978-94-011-6468-9

Contents

1 Introduction

Ecological studies of vegetation on old walls has been a much neglected subject judging by the number of papers on this topic. However, the floristics of mural vegetation has been the subject of a number of publications, especially from Italy and France (see 9). One of the first papers to cover some ecological aspects was RICHARD (1888) on the churches of Poitiers. He posed the question as to how far plants whose principal habitat is the surface of old walls, such as *Corydalis lutea* and *Cheiranthus cheiri,* can be considered to be indigenous. Much later RISHBETH (1948) devoted a study to the walls of Cambridge and WOODELL & ROSSITER (1959) published a paper on walls in Durham. SCHMITT (1950) is the author of a popular booklet on a wall as a biotic community.

From 1951 onward I made a special study of vegetation on the vertical surfaces of old walls and ramparts of ancient cities, castles, ruins, and other buildings, of stone bridges and dams, of the quay-sides of canals and rivers, and of old fencing walls in many localities in the Netherlands, later extended to Great Britain, Belgium, Luxemburg, France, Italy, Switzerland, Austria, Western Germany, Czechoslovakia and Poland. In nearly 1,200 sites the vegetation cover was recorded and, in addition, as far as time and means permitted, a number of ecological data were assembled for several places. In the Netherlands, the study of this biotope commenced in 1947 by members of the Nederlandse Jeugdbond voor Natuurstudie (Netherlands' Association of Young Naturalists), who explored the quay-sides of the Amsterdam canals systematically since 1951 (SEGAL, 1954; VAN KONINGSDAAL et. al., 1956).

Some preliminary results of the investigation have already been published (SEGAL, 1954, 1961, 1962b, 1963a; REYNDERS & SEGAL, 1963).

Outside the Netherlands, ecological studies of wall vegetation are scanty. The descriptions are almost always based on only a few relevés (see e.g. TÜXEN, 1937; OBERDORFER, 1957).

In a later phase of my studies it appeared that for a better understanding of wall vegetation it was necessary tot include vegetation of cracks in paved roads in the investigation, because it has a certain amount of similarity with vegetation on walls. The resemblance is not at all surprising; paved streets can, up to a point, be regarded as 'horizontal' walls. In spite of their ubiquity in the more densely populated parts of Europe, the vegetation types of paved roads have not been thoroughly studied either. Preliminary accounts of the results of my studies on the subject have already been published (SEGAL 1963a, 1964).

A satisfactory interpretation of mural and of paved road vegetation required the treatment of a number of methodological and theoretical problems in phytosociology.

The study of altogether different kinds of vegetation, more particularly of hygro- and of hydroseres and superficially, halo- and psammoseres, gave me a thorough insight into a number of fundamental problems.

The nomenclature of the vascular plants agrees with FLORA EUROPAEA, Vol. I (1964) as far as they have been treated, with the exception of the names for the species of *Polygonum* sectio *Avicularia,* for which SCHOLZ (in ROTHMALER, 1963) is followed, and of *'Parietaria ramiflora Moench',* for which the name *P. judaica* L. is used (MENNEMA & SEGAL, 1967). As far as the remaining species are concerned, the nomenclature adopted is from HEUKELS & VAN OOSTSTROOM (1962) and, if they do not occur in the Netherlands, from CLAPHAM, TUTIN & WARBURG (1962), or else from FOURNIER (1946). There are some exceptions: I prefer the name *Chamaenerion angustifolium* (L.) Scop. to *Epilobium angustifolium* L. and *Carex scandinavica* Davies to *C. serotina* Mérat ssp. *pulchella* (Lönnr.) van Ooststr.; *Plantago major* L. is regarded as a collective species and two taxa are distinguished viz., *P. major* L. (sensu stricto) and *P. intermedia* Gilib. Of the *Poa pratensis* L. aggregate, the taxa *P. pratensis* L. s.s. and *P. angustifolia* L. are fairly common on walls, whereas *P. subcoerulea* Sm. was recorded only occasionally. *Agrostis gigantea* Roth is regarded as a distinct species, clearly separable from *A. stolonifera* L.

The nomenclature of the bryophytes is based on the Index Muscorum by VAN DER WIJK, MARGADANT & FLORSCHÜTZ (1959-1967) as far as the taxa have been treated, on VAN DER WIJK (1962) as far as other indigenous species are concerned, and on GAMS (1957) for all remaining taxa. For the nomenclature of the lichens HILLMAN & GRUMMAN (1957) are followed.

2 Statement of the problem and methods

The present study is restricted to walls and streets that conform to certain prerequisites. A *conditio sine qua non* was always that a stand of Bryophytes, Pteridophytes and/or Spermatophytes was present. Other organisms, principally Thallophytes, have only sporadically been included in the investigation.

2.1 Walls and streets

Walls which were considered to be eligible for inclusion in the investigation had to fulfil the following requirements: built of stones or bricks, jointed with not too hard a type of mortar, of fairly considerable age, and situated in an environment in which no prolonged period of drought prevails. For the development of a more or less characteristic plant cover the walls must be vertical or at least approximately so. Horizontal parts of walls and walls with an inclination less than about 80° were not included in the present study. All these conditions will be amply discussed. The line of demarcation between mural vegetation and vegetation of the surroundings is often not very sharp: for instance at the foot of walls if much humus has been blown against it; or when a wall has been built up against an earth wall and the latter protrudes over the brickwork; in general when a considerable quantity of soil- and humus particles has been deposited on parts of the wall. In such cases gradual transitions between a wall and its surroundings can be observed. However, even under the smallest tufts of moss soil particles are found. Examination of small cushions of *Tortula muralis,* one of the most common wall-dwellers and usually also the pioneer species, reveals that almost invariably a layer of small sharp particles is present under the cushion and that the protonema usually does not extend to the wall surface proper but ends a little above it. The feet of walls and submerged parts of walls were not included in the investigation.

Ecologically, walls are better comparable with rocks than with any other kind of habitat. Vegetation of walls accordingly shows a certain degree of resemblance to that of bare rocks and rock crevices, but walls differ sufficiently from them to necessitate an explanation of the differences in ecology. Of considerable significance is of course that walls are habitats created by man out of materials that usually differ appreciably from natural rocks. Even if the wall is built of natural stone the effect of the mortar substance is often very noticeable and it is precisely the difference in the material of the stone and the mortar used which causes certain types of weathering to occur, for instance by their different reactions to rapid changes in temperature, which may result in shrinkage cracks. It is noteworthy, in this connection, that generally speaking, walls are situated in the immediate vicinity of human settlements and this applies more particularly to walls carrying a well-developed vegetation cover. Walls with a characteristic plant cover are usually found in places far away from exposed rock surfaces, e.g., in the lowlands of western Europe. A number of species considered to be characteristic elements of wall vegetation, such as *Linaria cymbalaria, Corydalis lutea, Cheiranthus cheiri, Centranthus ruber, Parietaria judaica* and *Tortula muralis,* are not only far less often found growing in rock crevices than on walls, but also not so common on walls in montane regions as they are in the lowlands. In montane regions the walls resemble rocks much more from a vegetational point of view and in the mountains, therefore, walls do not, in my opinion, form such an important and singular habitat as they do in the lower regions of western and mediterranean Europe. Nevertheless, in the latter regions wall vegetation has hardly been regarded as equivalent to other vegetation types or formations. Typical examples of mural vegetation also become rarer outside the (sub)atlantic and (sub)mediterranean zones of Europe. In continental Europe the dryer climate and the sometimes considerable fluctuations of the temperature are presumably limiting factors. In atlantic Scandinavia most of the above-mentioned species are rare or completely lacking, most probably as a result of the relatively low summer temperatures. Actually only a rather limited area of Europe provides suitable conditions for the development of characteristic stands of wall vegetation. Even so I could only study a part of that area and up to now could not include relevés from Ireland, the Iberian Peninsula, Yugoslavia and a considerable part of Italy in my study, but it is to be expected that in these regions many interesting observations could be made.

As I started my work in the Netherlands I have a relatively substantial number of data from this country and could also pay a good deal of attention to floristic and phytogeographic aspects of my vegetational researches. This does not imply, however, that this country is better suited for this kind of study than other parts of Europe. Much better developed and more characteristic examples of mural vegetation are found in the more extreme eu-atlantic regions, e.g. in southern England and in western France. Another unfavourable factor is that in Holland in the last decade restoration of buildings has been carried out in many places, so that good specimens of wall vegetation were destroyed.

The study of vegetation in the joints between the bricks or cobbles of paved roads was not so thoroughly pursued, but became more or less an ancillary study for a better interpretation of vegetation types occurring on walls. Nevertheless this provided so many data which appeared to be contrary to reports in the pertaining literature or opened up some new perspectives, that it seemed worth while to discuss these findings in some detail.

2.2 Vegetation studies

The present study was carried out in successive stages as follows:

1. The selection of suitable objects.
2. Preliminary studies: the search for appropriate methods.
3. Field studies: vegetation recordings and relevant ecological observations.
4. First phase of data processing: the setting-up of a uniform and adequate procedure of notation and of corrected readings and estimations, entering of the data on special forms, punching of the data for computer analysis.
5. Further data processing: calculations.
6. Interpretation.

2.2.1 Selection of the representative experimental surface area

The walls from which relevés were taken are principally situated in the following areas: Benelux, Great Britain (Edinburgh, East Anglia, southern England), France (N. France, Normandy, Brittany, Burgundy, the Loire-basin, Carcasonne, Mediterranean area and Provence), Italy (Tuscany, southern Tyrol), Switzerland (Tessino), Austria (Tyrol), Czechoslovakia (Bohemia) and the W. German Republic (Bavaria, Württemberg, Hessia, Rhineland-Palatinate, Westphalia). The total area from which records were obtained was divided into 11 areas, whose boundaries are drawn on the map (Fig. 1). Each area has been given a serial code number, which for the sake of brevity will be used throughout to indicate these areas. The choice of the walls was not quite at random, but not quite subjective either. In the areas I visited, many towns and villages were thoroughly searched for walls and I inspected walls carrying vegetation along the roads. As starting-points for the routes followed, tourist information concerning towns with old city walls, fortifications, moats, churches, castles or other old walls, including quay-sides, usually served a useful purpose. It follows that the accent fell on a number of old cultural centres. When it appeared during my preliminary investigations that vegetation on walls is not so rich and not so characteristically developed in montane to alpine and more

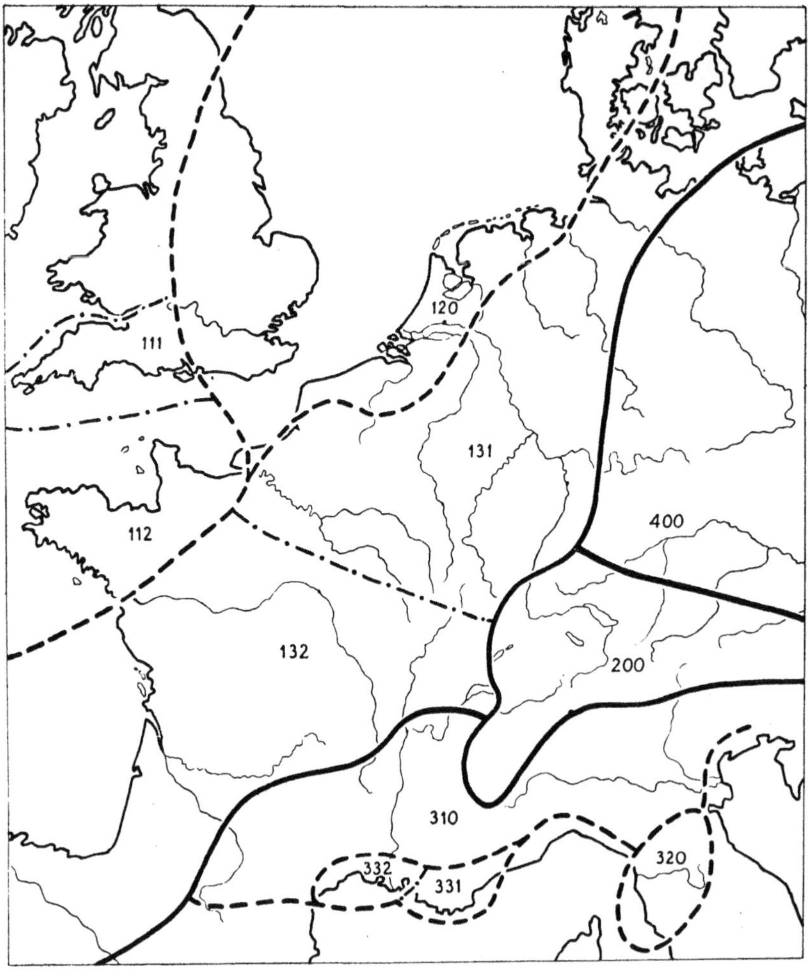

Fig. 1 Geographical subdivision of the area studied.

continental climates, the attention became focussed on the atlantic zone. Vegetation on walls appeared to be particularly well-developed in the eu-atlantic parts of England (Cornwall and Devon) and France (Normandy and Brittany), and in the submediterranean area in places with a fairly high relative humidity. In every town as many walls were studied as the available time permitted (in a few cases not more than a few hours). The vertical walls with the best developed vegetation and not clearly showing a ruderal aspect were always chosen. In a number of towns a fairly large number of relevés could be assembled, e.g. several scores in Amsterdam, Utrecht, Amersfoort, Cambridge, Exeter, Carcasonne, Montpellier and Siena.

When selecting suitable walls I most emphatically attempted to avoid a priori working hypotheses concerning vegetation types. The starting-point was rather the assessment of the variation of wall vegetation in a general sense than a search for special vegetation units, whether or not visualised as discontinuities in a multidimensional system of classification. So-called transitions between supposedly "typical" vegetation types are considered to be of at least the same significance as those "typical" vegetations themselves. It is precisely the continuity in the composition of the vegetation cover in its relation to, among other things, ecological and topographic conditions that is presupposed, and occasional discontinuities are seen as special cases of a general phenomenon.

The size of the experimental surface area is to a large extent dependent on the nature of the vegetation cover and of the relative surface area of the joints. In some instances only vegetation fragments smaller than the representative surface area or even than the qualitative minimum area could be recorded (see under 4.5). If the stand appeared to be well-developed in such cases, the chance was taken to work with such a small fragment. The calculation of the association between at least part of the species could thus be pursued. Principally those walls were included in the investigation which carried a plant cover of at least 5% coverage (of the total surface area of the wall), and contained at least three different species if the coverage was over 10% (as a rule the number of species was much higher: see 4.7). Vegetation of one or of two species was studied, but not used for the mathematical evaluation of the relevés. Experimental areas with the highest possible homogeneity, which was estimated visually, were always looked for. However, like many types of pioneer vegetation, the distribution of the species found growing on walls is not a very homogeneous one. The very regular pattern of the joints in brick walls favours underdispersion. If this pattern is disregarded and the plants growing in the cracks are taken into account as if they were growing on a continuous substratum, the formation of aggregates and hence overdispersion is the rule (overdispersion is more common in nature than random distribution). The limits of the experimental area were always carefully drawn, both ecotones and certain gradients being taken into account (see 4.3.2). If a distinct zonation could be observed on the walls, as is frequently the case with walls of quay-sides of canals and moats and with walls on very wet soil, more relevés were made.

In those cases in which there was only a gradual increase or decrease of all species present (or, in vegetation poor in species), only a single relevé, viz. the one of the densest stand, was used for the mathematical evaluation of the data. The recordings were nearly all made in the season of optimum development of the stands and in which the seasonal changes were at a minimum: in W. and central Europe between June and September, in S. Europe between April and June. In several sites a number of observations were purposely made in other months of the year so as to obtain some idea of the seasonal variation. For the study of bryophyte vegetation the time of the year does not seem to be so important.

2.2.2 Description of the experimental surface area

At least the following data were always noted during every recording (and with each relevé):
Date and locality.
Exposure and inclination.
Nature of the substratum and the relative surface area of the mortar joints.
Size of the experimental surface area.
A complete list of all Spermatophytes, Pteridophytes, Bryophytes and fruticose Lichens.
Visually estimated percentage of coverage of herb layer and bryophyte/lichen layer (and, if present, of more vegetation layers).
Coverage and abundance of each individual species according to the scale of SEGAL & BARKMAN (1960).
Phenological condition of every species according to the symbols proposed by BARKMAN, DOING & SEGAL (1964).

Complementary and ancillary data included, among other things, the presence and relative coverage of the crustaceous and foliaceous Lichens, the average height of certain plants, their sociability, other structural features and patterns of the vegetation cover, vitality and fertility, and of course various data concerning the habitat such as height above the ground- (or water-)level, the vertical extension of vegetation types in a zonation, etc.

2.2.3 Abundance and percentage of coverage

The biomass can be recorded in more than one way e.g. as the fresh and/or the dry weight per unit of surface area, whether or not combined with data concerning the dimensions or the height, or as abundance, percentage of coverage, and density. During the study under discussion a method of estimation was chosen for which abundance and relative coverage serve as the starting-point (BARKMAN, DOING & SEGAL, 1964). The assessments are always related to the quantitative minimum area (see under 4.5). The values reported as corrected or adjusted assessments were calculated as the mean values of a large number of accurately calculated observations (SEGAL, 1965). The adjusted values are expressed as percentages of coverage and were used for the mathematical evaluation of the relevés.

Symbol	Meaning		Readjusted values in per cents
r	of sporadic occurrence in the stand (or ecosystem) as a whole		0
+r	sporadically (1 to 2 individual specimens)		0
+a	,, ,, ,, 3 ,, ,, 20 ,, ,, ,,	1-2%	1.5
+p	not frequent (about 3 to about 20 indiv. spec.), coverage	<1%	0.1
+b	,, ,, ,, 3 ,, ,, 20 ,, ,, ,,	2-5%	3.5
1p	frequent ,, 21 ,, ,, 100 ,, ,, ,,	<1%	0.4
1a	,, ,, 21 ,, ,, 100 ,, ,, ,,	1-2%	1.5
1b	,, ,, 21 ,, ,, 100 ,, ,, ,,	2-5%	3.5
2p	very frequent >100 ,, ,, ,,	<1%	0.4
2m	,, ,, >100 ,, ,, ,,	<5%	2.5
2a	number of individuals various, but ,,	5-12%	8
2b	,, ,, ,, ,, ,, ,,	12-25%	18
3a	,, ,, ,, ,, ,, ,,	25-37%	32
3b	,, ,, ,, ,, ,, ,,	37-50%	44
4a	,, ,, ,, ,, ,, ,,	50-62%	56
4b	,, ,, ,, ,, ,, ,,	62-75%	69
5a	,, ,, ,, ,, ,, ,,	75-87%	81
5b	,, ,, ,, ,, ,, ,,	87-100%	94
5c	,, ,, ,, ,, ,, ,,	>100%	105

The symbol *2p* has not been used in previous papers and is introduced here, because there was a need for it in those cases where small plantlets occur in considerable quantities without attaining a percentage of coverage exceeding 1, as is, for instance, the case when a few cushions of *Bryum argenteum* occur in which up to 400 plantlets per cm^2 may occur. The symbol *2m* would have caused a relatively large over-estimation during the subsequent calculations.

In a few cases the symbols *3, 4* and *5* are used which correspond with the relative values of the scale of BRAUN-BLANQUET (1928) with adjusted values of 37.5, 62.5 and 87.5%, respectively.

For the calculations as much as possible pairs of numbers were used whose sums of readjusted values constitute a round figure (e.g. for the symbols *3b* and *4a, 2b* and *3a* or *1a* and *1b*). The readjusted values obtained in this way are much better suited for a statistical evaluation than those corrected according to TÜXEN & ELLENBERG (1937) for the scale of Braun-Blanquet, because:

1. the scale of Braun-Blanquet is very inaccurate and especially the actual value of his cipher *2* can vary within wide limits;
2. the abundance and the percentage of covering are not sharply distinguished; and
3. a species which only on rare occasions exhibits dominance causes

the value to rise relatively high in a combined table of vegetation recordings.

This last disadvantage has not been eliminated in the method followed during the present study, but it is not so great as it is in the scale of Braun-Blanquet. The advantage of the method proposed by SCHWICKERATH (1931), who expressed the relative predominance of the groups *('Gruppenmächtigkeit')*, is that the abundance can not exert a strong influence on this evaluation. A very serious disadvantage, however, is that the numbers do not represent objective figures. For this reason this method (for which scale-percentage values were given in BARKMAN, DOING & SEGAL, 1964) was not followed. From a number of experiments with mathematical patterns the conclusion could be drawn that, after a person has gained some experience, the error of estimation is usually as follows if enough time is available for the evaluation:

Actual coverage (%)	Maximum error (%)
0- 5	1
5- 12	2
12- 25	3
25- 90	5
90-100	2

For the estimation of the coverage of vascular plants the basis of all calculations was invariably the actual coverage and not, as advocated by Braun-Blanquet, the contour coverage, because this often can not be determined (BARKMAN, DOING & SEGAL, 1964). Of the total herbaceous vegetation layer both values were usually recorded, because the two together provide more information than each does separately. For the bryophytes the percentage of coverage recorded always relates to the damp situation, which, as is generally known, can be appreciably larger than that of the same mosses or liverworts in a completely desiccated condition.

Between the dimensions of the species and the abundance there is a certain relation if there is at least some degree of dominance. This almost always results in a high degree of abundance of certain of the smallest forms, which mostly occur in dense tufts or cushions (see the above-mentioned example of *Bryum argenteum*). This has considerable consequences if the abundance is not related to the minimum area. It is highly probable that in the phytosociological literature numerous mistakes or, at any rate, subjective interpretations of the abundance have been reported, the more so because the abundance of species of small size was hardly ever related to the minimum area of the structural element in question, but to that of the whole vegetation type instead. This usually means that it is related to the minimum area of the largest structural elements (this point is discussed in greater detail under 4.5).

Especially when mosses are concerned the chance of an appreciable error is great if one does not realise that the ratio between the surface areas can be very large. The surface area covered by a single "shoot" of the pleurocarpic moss *Homalothecium sericeum* is about 500 times as large as that of the acrocarpic *Gyroweisa tenuis*! The differences are smaller among herbaceous plants. In practice one apparently tends to assess the abundance of smaller organisms differently from that of larger forms when using the method of Braun-Blanquet.

In proportion to the total surface area, the symbol + or *1* for a moss would have to be much rarer than one can glean from published tables of relevés, which means that the qualification "very frequent" is assessed differently in the case of mosses and in the case of higher plants.

2.2.4 Phenology, vitality and fertility

Phenological observations are especially useful when in a certain site recordings are repeatedly made during at least one vegetation period. However, since the problems were centred around the geographical aspects, phenological records of wall vegetation of necessity were scarce. As we have seen, the vegetation types were studied as much as possible when they were at their peak of seasonal development. Only in the Netherlands was a year-round investigation possible in a number of easily accessible places. In general it can be concluded that in relatively dry habitats winter annuals often occur, most manifestly so in the mediterranean regions. It is noteworthy that certain species which are especially found on south-facing slopes in the dunes of western Europe, such as *Arenaria serpyllifolia, Erophila verna* and species of *Cerastium,* are of common occurrence on north-facing walls in the Mediterranean zone. When winter annuals or other ephemerous forms occur in mural vegetation, there is a distinct seasonal periodicity.

The overall aspect of the plant cover during the summer months may differ appreciably from the spring aspect, but this is not only the result of the dying of the winter annuals, which are not so important in a quantitative sense for that matter. It is striking, however, especially after a prolonged drought or in relatively dry sites, that the mosses and the ferns (which are chiefly species with a high resistance against desiccation such as *Barbula acuta, Ceterach officinarum* and *Asplenium trichomanes)* are often shrivelled up by a contraction of the leaflets and the twisting of the leaflets around the stem (mosses), by a recurving of the margins and the protrusion of the scales on the lower leaf surface *(Ceterach),* or by the enfolding of the leaflets (e.g. *A. trichomanes*). The same phenomena can be observed after frosts (particularly at higher latitudes or at greater orographic altitude). Similar phenomena can be observed in many other species. In vegetation on wall heads, where winter annuals play a considerable role, the seasonal periodicity is usually manifest. The vitality of many wall-dwelling forms is small to moderate. However, this does not apply to the chasmophytes, which prove to be characteristic elements of mural vegetation in many parts of Europe, at least in those habitats where the conditions are otherwise favourable for these species, such as *Tortula muralis,* species of *Asplenium, Ceterach officinarum, Linaria cymbalaria, Corydalis lutea, Parietaria judaica, Cheiranthus cheiri* and *Capparis spinosa.* A number of species frequently occurring on walls in the atlantic zone of western Europe, on the other hand, such as *Phyllitis scolopendrium* and *Dryopteris filix-mas,* are usually far less opulent and not so fertile as when they are found growing in forests where such forms exhibit their optimum development. The same holds for other species in different regions, such as for *Gymnocarpium robertianum* in subatlantic and in continental parts of Europe. Many of the species which are only occasionally recorded from walls have their vitality invariably reduced when they are found growing in these situations. This is reflected, among other things, in the development of their biomass (a lower stature, etc.). Nevertheless such species, more particularly the therophytes of which a relatively large number of species occurs on walls, often attain the flowering stage, albeit with a lower production of flowers and seeds than in more suitable environments.

The fertility of the mosses is usually poor. Capsules were regularly encountered only for *Tortula muralis* and somewhat less frequent for *Ceratodon purpureus.* Many species have exclusively been recorded in the vegetative condition, however. Capsules of *Bryum capillare, B. argenteum, Streblotrichum convolutum, Barbula vinealis* and *Rhynchostegium confertum* were noticed relatively often.

2.2.5 Data processing

The observational data and various records fall into three groups:

1. The vegetation analyses (recordings, relevés, etc.).
2. Lists of species with details concerning each of the species of every recording or relevé.
3. Lists and various relevant details of taxa recorded elsewhere on walls for the purpose of a floristic inventarisation, more particularly in the Netherlands, but also elsewhere if the species concerned occurs in the Netherlands. These data will be treated *in extenso* in a forthcoming publication.

For the arrangement of the relevés of vegetation stands a simple method was employed. The results are not regarded as final and will be compared with the results of other approaches to data processing. The data pertaining to individual taxa are only partly enumerated in this chapter. Structural features, for instance, are discussed in Chapter 4.

2.2.5.1 First phase of processing

All data were entered on mimeographed forms on which, apart from various legenda, the most regularly present species were printed and space was reserved for the notation of additional taxa. Each species was given a code number of 3 ciphers, and the corrected values for the abundance and the percentage of covering were noted behind each species recorded. Code symbols were, furthermore, used for the phytogeographical regions and for the exposure, shaded walls being given a separate symbol (see Appendix I). The exposure was noted in four classes each with a range of 90° and between the limits NNE, ESE, SSW and NNW. Walls rising up from a body of water such as the walls of moats and canals were also given a special code symbol. Each recording or relevé was provided with a reference number, and all data were transferred to punched cards and punched tape for processing on an electronic computing device. On the tapes the following data were also punched:

idiosystematic order (see under 2.3.1)
degree of ploidy (see under 2.3.2)
life form according to Raunkiaer (see under 4.6.2)
life form according to Iversen (see under 4.6.3)
growth form (see under 4.6.4)
sociability (see under 4.3.1)
flower colour (see under 4.6.5)
dissemination type (see under 2.3.3)
formation group (see under 2.3.4)
area of distribution and distribution patterns (see under 2.3.5).

The programmation and the evaluation were partly carried out by the Afdeling Bewerking Waarnemingsuitkomsten of the Centrale Organisatie T.N.O., The Hague, and partly by the Mathematisch Centrum, Amsterdam.

2.2.5.2 Second phase of processing

The data were partly evaluated by means of the heuristic method described by Ellenberg (1956) for the representation in table form, with the purpose of comparing the results of this "orthodox" method with numerical analyses by means of computers. The orthodox method was tried on a table of 177 relevés of vegetation in crevices of paved roads. For the arrangement by means of mathematical or statistical calculations, several methods with a number of modifications have been suggested, such as similarity analysis, association-analysis, and other methods in which factorial analysis plays a role. A critical survey of all these methods, and the respective affinity indices, was given by DAGNELIE (1960) and, later, by WILLIAMS, LAMBERT & LANCE (1966).

For the evaluation of the recorded data two methods were employed to facilitate the grouping of the relevés, viz.,

1. For each of the 11 geographical areas the relevés were arranged in a matrix and of each pair of relevés the relationship was determined according to Poore's equation:

$$A = \frac{200c}{a + b + 2c}$$

in which A = affinity, c = the number of species represented in both relevés, a = the number of species represented in the first of the two relevés and absent from the second, b = the number of species represented in the second of the two relevés and absent from the first.

It follows that the values of A lie between 0 and 100. The calculated values were placed in 5 categories, indicated by appropriate symbols which show an increasing degree of blackening as the value is higher. The relevés were subsequently arranged in such a way that the pair with the highest mutual coefficient of affinity (A) ranks highest, to be followed by the relevé showing the highest degree of affinity with either one of the first pair, etc. In this way clusters were produced with great densities of relatively high values. The same method of data processing was applied to the relevés of paved roads and other paved surfaces [1]).

2. For all species recorded in at least 3% of all stands examined, 2 x 2 convergency tables were calculated according to the following principles: Let *a* (*b*, respectively) be the number of stands at which *A* (*B*, respectively) was observed, *x* the number at which *both A* and *B* were observed, and *n* the total number of stands. The numbers are arranged as follows:

	B	not B	
A	x	a — x	a
not A	b — x	n — a — b + x	n — a
	b	n — b	n

Suppose that this table is tested for independence with a test of size x. The size x = x′ at which the null-hypothesis would just be rejected can be regarded as a measure of association: the smaller x′, the stronger the

[1]) The programmation and execution were both done by the Afdeling Bewerking Waarnemingsuitkomsten T.N.O., The Hague, on an Elliott-503 computer.

association. The association is considered to have the sign (positive or negative) of $x\,(n-a-b+x)-(a-x)\,(b-x)$.
Symbols used in the tables are:

$x' < .01$	X or O
$x \leqslant x' < .05$	x or o
$.1 \leqslant x'$	+ or —

(X, x and + for positive associations,
O, o and — „ negative „).

For all bryophytes represented in all relevés by a degree of presence exceeding 2% a separate matrix was set up. The calculations were carried out on the Electrologica EX 8 Computer of the Mathematisch Centrum, Amsterdam. This procedure had the great advantage of showing both positive and negative associations. In the beginning the arrangement of the taxa was alphabetically. The matrices set up according to these principles are purely based on floristic data.

In this study much importance is attributed to weighted values (in the case of species that are frequently dominant), and that is why sometimes a somewhat different train of thought is followed during the interpretation of the various phytosociological entities than one would have to do if the result of the mathematical evaluation of the matrices alone is taken into account. This weighing was also employed in all cases where the structural features of the stands differ appreciably among a number of relevés of, floristically speaking, closely related communities.

For the mutual comparison of the communities Poore's relation was applied to the percentages of overall presence of each individual species, the correspondence and the difference in the degree of presence being consistently used.

It is not possible to publish all relevés and all matrices in the present report. A selection was made at least of 3 examples of each vegetational entity ("community", "vegetation type", etc.). In addition, the recorded extreme values of the presence of every species are given in the form of the symbols of the scale of visual estimation used, whose ranges were used for the estimations of the biomass (see under 2.2.3), and also the mean of the readjusted values is indicated both for all relevés of each recognised syntaxonomic entity (community, etc.) and for all relevés in which the taxon concerned was represented (these values can be mutually deduced from the degree of presence). All data are available to interested parties upon request and they are available for further mathematical evaluations. If one requires the information, punched cards and tapes can be copied or duplicated.

The originators of the mathematic relations recommend a form of multivariate analysis which is a modification of association-analysis. Such an evaluation is indeed very promising, but its execution is very costly. I was, therefore, forced to start with a form of similarity analysis, which is, as a rule, carried out with simpler coefficients of affinity, such as those of JACCARD (1902) or SØRENSEN (1948). These coefficients do not lead to the inclusion of negative associations in the calculation. The modern, statistical methods have hitherto nearly always been employed for studies in small areas, within a single vegetation type or in only a few, or for studies of a mosaic of vegetation types (of patterns). Similarity analysis with simple coefficients has occasionally been used for the comparison of vegetation units within larger areas. An attempt to apply such techniques to data from a wide geographical range may, at least theoretically, have certain objections. Any statistical evaluation can only be applied to material that is suitable for the purpose, one of the most difficult problems being the prerequisite of random sampling which, as I have explained already, can not be so rigorously adhered to as in an area of smaller size. Data processing along these lines would be completely useless with elementary data collected on the basis of aprioristic working hypotheses. This not being the case in the present study, however, it was decided to try similarity analysis.

Several methods are now being studied and in the future the material, or part of it, may serve as a test object for these methods to ascertain their usefulness.

2.3 Lists of species and spectra

The data for the compilation of the spectra were noted down after the species in the lists by means of code symbols. These symbols and the spectra are represented in the Appendices I and II. The coding was done in such a way that as soon as the combination of certain categories became necessary, this was made as easy as possible. With the aid of an electronic computing device the spectra were calculated for the total amount of recordings or relevés of walls and paved streets, separately for an analysis in different areas, and, finally, separately for the relevés referable to certain categories of vegetation units.

Both unweighted values, (i.e., solely on a qualitative basis or presence) and weighted ones were used, the weighing corresponding with the corrected values of the scale of estimation employed (see under 2.2.3). In this way an idea can be obtained of the quantitative contribution of each of the categories in a certain area or in a specified vegetation unit. The subdivision in areas was done on a floristic basis. Perhaps this could previously have been determined in a more objective fashion by computer analysis, but this would have retarded the data processing appreciably and the chances of a very divergent position of the demarcation lines would, in my opinion, have been almost negligible.

The subdivision used is the following (compare also Fig. 1, Appendix I):

1. Atlantic
   a. Eu-atlantic
      a′. W. England and Scotland
      a″. Normandy and Brittany
   b. N.W. France, W. Belgium and the W. and N.W. parts of the Netherlands
   c. Subatlantic
      c′. N. France (except the N.W. part), E. Belgium, Western Germany and the E. and S.E. parts of the Netherlands
      c″. Central and S.W. France

2. Alpine

3. Mediterranean and submediterranean
   a. Submediterranean France, N. Italy and S. Switzerland
   b. Tuscany
   c. S. France
      c′. Alpes-Maritimes and Var
      c″. Bouches-du-Rhône and Hérault

4. Czechoslovakia and Poland

The tabulated taxa are alphabetically arranged (and numbered in this order), the numbers up to 800 having been used for vascular plants and the higher numbers for bryophytes and lichens. It is to be recommended that all taxa of the European flora would be assigned a code number according to a certain pre-arranged system, so that the same code can be used by various workers. This would certainly facilitate the exchange of data in the form of e.g. punched cards or tapes which could be used for programmes as complementary material, for comparison, or otherwise. Such a universal (and uniform) coding system would also be highly recommendable for a number of invariate characters of the taxa, if necessary combined with a number of alternative systems, to be used by workers in many countries. It is hoped that the Committee for the *Flora Europaea* will act as a co-ordinating body when the time comes.

Carles (1948) advocated the representation of the categories in a biological spectrum (with life forms according to Raunkiaer) according to their quantitative contribution expressed in the units of the combined estimation scale of Braun-Blanquet. In this spectrum he assigned a separate percentage to the open ground and related the proportional contribution of the various life forms to the remaining percentage representing the actually covered area i.e. directly according to their absolute participation in the coverage. Against this rather valuable idea, in principle, some objections can be made, one of which was already foreseen by Carles himself:

1. The spectrum is subject to changes in the course of a year if seasonal periodicity occurs. For special (structural) studies in connection with periodicity this may have a positive meaning.
2. In open stands, and the more so when they are relatively rich in species, the absolute percentage of covering of each life form tends to be low and the mutual relations and differences are not so manifest.
3. The result is to some extent dependent on the method of estimation and of personal views concerning the area covered by an individual or by a shoot (compare also 2.2.3): thus the results may vary appreciably if contour covering or actual covering is the basis of the estimation.

During the present investigation the weighted values were always based on the actual percentage of covering and the relative contribution of the various categories of every spectrum was indicated without the total coverage of the plant cover or of the vegetation layers being taken into account. The periodicity could be disregarded, the vegetation types having been studied at their optimal phase of seasonal development.

Additional data concerning the various species, and possibly relevant to the problem studied, could of course have been included in the lists of species to compile additional spectra, such as details relating to the pollination type or even to related organic compounds in the plants. The compilation of a pollination spectrum had to be abandoned for the lack of sufficient data, but this sort of information would possibly increase our knowledge of the ecology of biotic communities considerably.

2.3.1 Taxonomic spectrum

For each species belonging to the Bryophyta, Pteridophyta and Spermatophyta the o r d e r to which it belongs has been indicated. For the Bryophytes the system in MARGADANT (1959) was followed, for the Spermatophytes the system of PULLE (1952). For the classification of the Pteridophyta the views of various authorities were taken into account (e.g. PICHI-SERMOLLI, 1958; POELT, 1961). During data programmation the orders could be united into classes or phyla if this proved to be more convenient. In the list of species no species are distinguished within the group of *Rubus fruticosa* (aggr.) and in *Taraxacum*. If this had been done, some qualitative taxonomic spectra would have turned out differently, the representation of particularly the *Asterales* being much higher, and also other spectra would have to be altered to some extent.

2.3.2 Spectrum of degrees of ploidy

After cytogenetic examination had assumed considerable importance in the last decades, several attempts were made to relate cytology with phytogeography and, more recently, also with phytosociology. The well-known hypothesis that the number of polyploid species increases with the degree of latitude (HAGERUP, 1931) provides an example. The reliability of this assumption will not be queried here. PIGNATTI (1960) was the first to compile spectra of the degree of ploidy of a number of plant communities and to compare the spectra of relevés of the *Plantagini-Lolietum* (vegetation of trampled soils) of various plant ecologists from Ireland, the Netherlands, W. Germany and Italy. Several authors have attempted to relate the percentage numbers of polyploid forms to the degree of extremeness of the environment, e.g. STEBBINS (1942), CAIN (1944) and LÖVE & LÖVE (1949). Polyploidy is supposed to occur more frequently in "unfavourable" circumstances, and according to Stebbins, more particularly in "open" habitats and in sites where recently the local flora has been strongly modified. Pignatti distinguished three categories of species: diploids, polyploids and forms represented by both diploids and polyploids, and estimated their relative frequencies according to the respective presence of the species, from which he obtained an index for every series of relevés. This index (which can also be determined for a single relevé or other vegetation recording) relates the percentage of diploids to that of the polyploid forms. There is a relation between the type of formation and the index in question, which proves to be 0.2-0.4 in the case of the *Plantagini-Lolietum* (Pignatti), 1.0-1.9 for elder carr (the *Alnion glutinosae)* and 1.5-3.0 for the Mediterranean holm-oak forest (the *Quercion ilicis)*.

This might be taken as an indication of a lower incidence of polyploids in forests as compared to other formations and of a relatively frequent occurrence of polyploids in pioneer vegetation and in vegetation of much disturbed sites. This assumption is somewhat similar to the hypothesis proposed by BAKER & STEBBINS (1965) that the genetic plasticity of populations occurring in "open" environments is greater than that of populations living in more stable habitats. In "open" environments the number of species exhibiting both di- and polyploidy might be expected to be high, but such forms are not taken into account in Pignatti's index. It is especially this category which may prove to be of considerable importance for a possible correlation. It is, furthermore, highly probable that of a large number of species at present only known as polyploids (or considered to be so), also diploid forms exist and, to a lesser extent, the reverse may also be the case. Of necessity the question must be raised, therefore, if the estimation of spectra of the degree of ploidy is not premature at the present state of our knowledge. The compilation of recorded chromosome numbers of central and N.W. European taxa by Löve & Löve (1961) provides sufficient data to demonstrate that a re-estimation of the ploidy spectra would, at present, yield results which differ from Pignatti's figures.

As an example, I give here a table based on 31 relevés of the *Plantagini-Lolietum* made in the Netherlands by Sissingh (1950):

		Pignatti, 1960
Diploids	37.7%	18.9%
Di- and Polyploids	0.6%	30.1%
Polyploids	61.6%	50.9%
Index	0.6	0.370

According to the compilation made by Löve & Löve, Pignatti's index would be much higher than he himself had calculated. The number of forms with both diploid and polyploid races reported by Pignatti is strikingly large and this may be (partly) based on data from other sources such as that compiled by Darlington & Wylie (1955) who give more autopolyploid forms.

Some objections against the estimation of ploidy spectra are, accordingly:

1. Many species have not, or not thoroughly, been cytologically studied.
2. From certain areas (such as Scandinavia) a relatively greater number of records are available than from other regions.
3. The data reported in the literature on chromosome numbers may not be related to vegetation analysis without certain restrictions, because the chromosome counts were usually obtained from material from other populations and in many cases even from other intraspecific taxa, hailing from other ecological habitats and/or other geographical regions.
4. The data are, up to a point, dependent on the reliability of the sources consulted.
5. The interpretation of the so-called 'basic' chromosomal number of a given genus ('x') varies with different authorities, so that different degrees of ploidy may be attributed to the same species (compare, e.g., the basic numbers reported for ferns by Löve & Löve, 1961, with those of Manton, 1951).

Nevertheless I have made an attempt here to compile ploidy spectra with the purpose of starting a discussion, whilst realising that far-reaching conclusions are decidedly premature. The data are chiefly based on Manton (1951) for the Pteridophytes and on Ordnuff (1967) and Löve & Löve (1961) for the Spermatophytes. Additional data on ferns are taken from more recent publications (by Lovis and Vida). The lists in Löve & Löve are not always unequivocal and leave some doubt about the interpretation e.g. under *Cerastium* for which $x = 9$ is recorded, but $x = 18$ is equally probable.

The following categories are distinguished in the present paper:
diploid
di- and tetraploid
tetraploid
di-, (tetra-) and at least hexaploid
at least hexaploid
tetra- and at least hexaploid
hybrids
the remainder, including, e.g., unknown numbers, uncertain basic numbers (i.a., aneuploid series, apogamous or apomictic forms with 3n, 5n, etc.).

Data concerning bryophytes have been omitted, the cytological data being inadequate (in spite of some lists published in Trans. Brit. Bryol. Soc., Vol. 3).

2.3.3 Dissemination

The compilation of more or less reliable dissemination spectra meets with considerable difficulties emanating from the following points:

1. When giving due consideration to the dispersal of the diaspores one must establish the following: a. how do the diaspores get to a habitat where they may be able to settle, and, b. how is the species in question capable of maintaining itself in the new environment and, eventually, of increasing its area in the new site. One could thus distinguish between **migration** spectra and **population spectra**, but strangely enough this is never done.

For the population spectra such forms of dispersal as blasto-, auto-, and barochory, and also anthropochory, may be of considerable importance, whereas anemo- and hydrochory are usually more important for the migration spectra. Species of *Papaver,* for instance, may be dispersed through the action of man, but they usually spread in their natural habitat by autochory. Vegetative propagation is of course particularly important in connection with a population spectrum.

2. Many species exhibit polychory and the relative importance of the various modes of dissemination may be to a large extent dependent on prevailing local conditions, sometimes only incidentally or temporarily. Especially the importance of anthropochory in the establishment of a vegetation cover on walls can not even approximately be estimated without accurate observations.

3. The various categories of dissemination types can not always be sharply distinguished. The difference between baro- and anemochory is, for instance, not very clear in poppies, whose seeds are liberated by the rhythmic swinging of the dried fruit stalk caused by wind. Of frequent occurrence are seeds of relatively small size (usually under 2 mm long),

which are dispersed by wind over short distances (not exceeding a few metres or occasionally, in the case of storms, several tens of metres), but exhibit barochory when air currents do not exert any appreciable influence.

4. The mechanism of dispersal and the dissemination potential of a great number of species are insufficiently known.

5. A clear insight can hardly be obtained without numerous observations in many places where ecological studies are carried out.

All this infers that dissemination spectra are not very reliable unless a special study of the dispersal mechanism was made in the biotopes studied (see e.g. WOODELL & ROSSITER, 1959; RISBETH, 1948). Still, such spectra can give a general idea, because a number of relations are rather manifest:

Anemochory, for instance, is very likely to occur among species with diaspores as fine as dust and among forms with seed floss or with winged diaspores, endozoochory is common among taxa with berries, drupes or other fleshy fructifications, and anthropochory prevails among adventitious plants and agricultural weeds. Hydrochory is not rare among riparian taxa found on quay-side walls near the waterline, and myrmecochory can often easily be observed on walls inhabited by ants, and is always likely if the seeds are provided with an elaiosome or a caruncle.

Observations relating to dispersal were principally made on a number of walls in Holland and in the south of France. For the spectra, the following classification, mainly adopted from HEINTZE (1932), is used here:

1. **Vegetative propagation:** by means of rhizomes, stolones, bulbils, turions, etc.; examples: *Potentilla anserina, Allium* species, *Hydrocharis.*
2. **Autochory:** dispersal by means of an actively growing organ or tissue; examples: *Linaria cymbalaria* and species of *Geranium.*
3. **Barochory:** dispersal under the influence of gravitational force; example: *Aesculus.*
4. **Anemochory:** dispersal by air currents.
   a. forms with a high dissemination potential by having diaspores as fine as dust e.g. *Pteridophyta* and *Orobanche.*
   b. forms with a medium dissemination potential by having diaspores with floss e.g. many *Asteraceae* and *Epilobium.*
   c. forms with a low dissemination potential: diaspores winged or relatively large (usually more than 2 mm long), e.g. *Acer* and species of *Poa* - the "range" not extending a few tens of metres.
5. **Hydrochory:** dispersal by water; example: *Apium nodiflorum.*
6. **Zoochory:** dispersal by animals.
   a. **Endozoochory:** the diaspores pass through the digestive tract of

a frugivorous bird or mammal; examples: *Ribes* and *Fragaria,* but also many caryopsis-bearing grasses, etc.

b. **Epizoochory:** dispersal by adherence to the cover of hairs or feathers of the animal; examples: *Arctium* and *Cynoglossum.*

c. **Myrmecochory:** dispersal by ants; examples: *Chelidonium* and several species of *Veronica.*

7. **Anthropochory:** dispersal through the action of man (also indirectly by means of articles of trade and means of conveyance).
8. **Polychory:** dispersal not clearly restricted to only one of the other categories.

Other types of dispersal (such as diszoochory) are presumably of little importance among the species occurring in vegetation of walls. The arrangement of the species in the spectra is principally based on the migration spectra. Blastochory, autochory and barochory are only indicated in those rare instances in which these forms of dissemination occur frequently in the biotopes studied. For many species, whose mode of dispersal has not, or at least not unequivocally, been established, the entry is: "dispersal unknown". Many species, though probably more or less polychorous, are listed under one of the forms of dispersal if this is clearly their most frequent mode of dissemination.

If a species is referred to one of the categories, it does not necessarily imply that it always behaves in the same way in other habitats. In the dissemination spectra only the vascular plants are taken into account, the Pteridophytes almost without exception being anemochorous, although in certain cases **ombrochory**, i.e. dispersal through displacement by rain water, may be important for their dissemination. The same applies to the diaspores of higher plants in certain habitats such as the species occurring on the quay-sides of canals and moats in such towns as Amsterdam and Amersfoort, where paved roads run in the immediate vicinity of the quay-side walls.

2.3.4 Formation spectrum

The number of specifically wall-dwelling species being low but the total number of species recorded relatively large, it was deemed important to establish in which kind of formation the latter occur most frequently. In the case of the vascular plants the following categories of formations are distinguished:

1. Rock surfaces, walls and gravelly or stony slopes (particularly in connection with chasmophytes and epilithic forms).
2. Pioneer vegetation of dry sites including 'open' dune country.
3. Shoals and mud flats (halophytes).
4. Arable land including vineyards.

5. Ruderal places including fields lying fallow.
6. Banks, reed vegetation, broads and peat bogs.
   a. Riparian vegetation.
   b. Peat bogs and marshes.
7. Grassy sites including meadows, hay-fields, roadsides, dikes, etc. and *Hochstauden* vegetation.
   a. Grassy sites.
   b. *Hochstauden* vegetation (generally on peaty soils).
8. Dwarf shrub vegetation, scrubs and forests.
   a. Heaths, moors, garigues and macchia.
   b. Scrubs.
   c. Forests.
9. Vegetation of contact zones and various other cases.
   a. Contact zones of terrestrial formations and fresh water, or vegetation of habitats with a strongly fluctuating water-table, which periodically run dry.
   b. Contact zones of terrestrial formations and seawater (the foot of dunes along beaches, shoals and mud flats).
   c. Various, not necessarily restricted to one of the above-mentioned categories.
10. Other cases.
    a. Cultivated plants.
    b. Adventitious forms.
    c. Unknown or doubtful.

Only those species which are chiefly restricted to contact zones are included in this category here. However, several species also occur in situations where some form of natural or artificial disturbance prevails, such as the species of the *Agropyro-Rumicion crispi* e.g. species of *Juncus* and *Potentilla anserina.* The contact belts between seawater and dunes include the plant cover of beach ridges and old high water marks with such forms as *Atriplex littorale,* which can also be classified among the more or less nitrophilous, ruderal vegetation types.

Species predominantly occurring in both scrub vegetation and forest margins, such as *Clematis vitalba,* are classified among the elements of scrub vegetation; species of both the tall forest and the scrub stories are listed as forest elements.

To the category of the various unclassified species all forms are referred which do not clearly exhibit an optimum in a single formation type. Not infrequently these are species occurring in sites where sudden changes or even calamities have previously taken place (such as fires or changes in the agricultural or architectural management), or in ecotones (such as road sides and forest edges) - examples: *Tanacetum vulgare, Senecio jacobea, Chamaenerion angustifolium,* species of *Epilobium, Galeopsis tetrahit* and *Hypericum perforatum.* Several of these forms have the habit of a *Hochstaude* (a tall, often suffrutescent herb or a "soft shrub"). This growth form is characteristic of pioneer vegetation of secondary succession seres (SEGAL, 1969). However, this does not apply to such species as *Glechoma hederacea* and *Lamium album*. Among this category of species there are several forms which exhibit a certain preference for ruderal habitats. Forms characteristic of vegetation on felled patches in woods and of other disturbed areas within forest formations, such as *Digitalis purpurea* and *Luzula luzuloides*, are here referred to the category of forest-dwelling species. Also included among the cultivated plants are those species which are indigenous in other regions of Europe than the site of the relevé, such as *Buxus sempervirens* in the Netherlands. The rock-dwelling forms include coastal forms exclusively growing on rocks such as *Crithmum maritimum*.

This type of classification could not be applied to the bryophytes and the fruticose lichens: the ecological link with the substrate being much more pronounced, the relation to the formation types based on the higher plants is far less clear. Bryophytes and lichens are often elements of stenocoenoses which can be more or less independent of the total biotic community (see 4.1), so that a classification was chosen which is principally based on the nature of the substratum:

1. Terrestrial: soil-dwellers.
   a. General and of constant occurrence, not only in pioneer vegetation.
   b. In open, and generally dry, habitats.
   c. In moist sites, more particularly along the sides of ditches and streams and in marshy habitats.
2. Rocks, stones, boulders and walls (epilithic forms).
3. Tree boles (epiphytic forms).
4. Combinations, viz.
   a. terrestrial and epilithic.
   b. epilithic and epiphytic.
   c. epilithic and lignicole (on dead wood).
   d. indifferent or ubiquitous.

It was not necessary to distinguish different categories of lignicole forms or additional categories in the present study, because they are either completely absent or occur very infrequently. All species are referred to the category to which they most frequently seem to belong. No distinction was made between obligatory and facultative epiliths, etc. This would even be undesirable in any study other than a strictly local one, because a number of species show a higher preference to a single substrate only in certain areas and not necessarily elsewhere. As an example, in deciduous forest in Oeland (Sweden), according to SJÖGREN (1964), *Frullania dilatata* and *Orthotrichum diaphanum* are obligatory epiphytes and *Oxyrrhynchium swartzii* is predominantly terrestrial. However, these, and other, species exhibit an increasingly weaker preference to a certain substrate as they occur farther to the S.W. in atlantic Western Europe, where they are also found on rocks and walls. (Compare also the different behaviour of *Hymenophyllum tunbridgense,* which is usually epilithic in continental Europe, but not infrequently also epiphytic in the eu-atlantic regions of England and Ireland).

The distinction between the various categories, such as epiliths and terricoles, is not always very sharp. Saxicole mosses can be divided according to AMANN (1918) into:

1. *lithophytes,* which grow on the bare stone surface without an appreciable deposit of humus (examples: *Gyroweisa tenuis* and juvenile forms of other species);
2. *exochomophytes,* which grow on a more or less conspicuous layer of dust-, soil-, or humus particles;
3. *chasmophytes,* which live in rock crevices; and
4. tufficoles or tuff dwellers, such as *Eucladium verticillatum, Didymodon tophaceus* and *Cratoneuron commutatum.*

The exochomophytes constitute by far the largest group; however, they are often capable of establishing themselves as lithophytes, to grow out only after enough dust or humus has accumulated (such a transition is shown by e.g. *Barbula revoluta*).

A distinction between calciphile and calcifuge species is of no significance for the purpose of the present study, because nearly all species on mortar-jointed walls are calciphile or indifferent.

(N.B.: Some species qualified as calcifuge by Amann are not at all rare on walls e.g. *Grimmia pulvinata, Rhynchostegium confertum, Ceratodon purpureus* and *Trichostomum crispulum,* which have been recorded from habitats where the pH can be as high as 9 or, in the case of *G. pulvinata,* even reaches 11, probably because of the high calcium-content. *Tortula virescens* is sometimes found growing on basic substrates on stone roofs.)

2.3.5 Distributional area

It may serve a useful purpose to establish if the various phytogeographical elements of a given plant community are represented in different regions. We shall define an 'element' as the total number of species (or intraspecific taxa) which attain their limit of distribution in or near a certain area (such as a country) and whose main areas of distribution roughly coincide (HEIMANS, 1960). In the countries of western and central Europe an Atlantic, a Boreal, a continental, an alpine and an autochtonous element can be distinguished. For the recognition of the elements the geographic position of the total distributional area of these elements is also taken into account and in certain cases a partial area of distribution. A holarctic form, for instance, need not necessarily be distributed throughout Europe, but may be restricted to certain climatic regions, which do not necessarily correspond with, or are not always comparable to, climatic zones in North America or Asia. Generally speaking, the distribution of the species in these three continents agrees more or less clearly with the geographical latitude.

Schwickerath (e.g. 1931) compiled spectra of the type of distribution for a number of plant communities in the vicinity of Aachen and he found that in different vegetation types clearly other geographical elements prevail. In the present paper, both spectra of the distributional areas and spectra of the patterns of distribution are given. In the first group of spectra the data are related to the total area of distribution, in the second to the distribution in that part of temperate Europe which is situated W. of the meridian of 15° E. Long. and S. of 55° N. Lat., in which area most of the relevés were recorded. Apart from the adjoining Mediterranean and eu-atlantic regions, there are few regions where walls are encountered which carry interesting types of vegetation of vascular plants. Spectra of distributional areas can of course also be given for vegetation types occurring in a much more restricted area, such as Holland, by using the occurrence of the various species in the phytogeographical districts as a criterion [1]). For the compilation of the spectra the following phytogeographic types were distinguished:

Cosmopolitan forms: occurring all over the world.

Supraholarctic forms: i.e. forms also found outside Holarctis, but not cosmopolitan.

Holarctic: not only occurring in Europe (and sometimes N. Africa), but also in N. America and (mostly) also in temperate Asia i.e. in the Holarctic Floral Region.

Eurasiatic: found throughout Europe and a large part of Asia.

European: found in at least considerable areas in S., W., central and E. Europe.

1. W., S. and central European: principally occurring in an area which includes both Mediterranean and Atlantic to central European regions.
2. N. and central European: the main area of distribution includes both Boreal and central European regions.
3. W. and central European: the main area of distribution includes both Atlantic and central European regions (central European subregio *sensu* Meusel).
4. S. and central European: found mainly in Mediterranean and central European areas (in the latter often as "radiations" from the Mediterranean region).

---

[1]) This will form the subject of a forthcoming publication.

5. S. and W. European: the distributional area includes Mediterranean and Altantic areas (the latter often as extensions of the Mediterranean region).
6. Central European: (central European province *sensu* Meusel).
7. W. European: in the Atlantic belts of Europe.
8. S. European: in the Mediterranean region (Macaronesian-Mediterranean region *sensu* Meusel).
9. Alpine and montane: restricted in distribution to middle montane and alpine zones (alpine subregio *sensu* Meusel).

In addition, the following types are recognised here:

a. Neophytes: species that were not autochtonous, but became established in recent times and have come to stay in one or a few vegetation types.
b. Adventitious forms: introduced species which are not usually capable of maintaining themselves or of enlarging their area of distribution.
c. Species that have run wild: cultivated plants and garden escapes.

Cosmopolitan species, though spread all over the world, are not necessarily always common everywhere (such as *Anogramma leptophylla)*. To this category I have also referred several anthropochores with an originally much more restricted area of distribution such as *Plantago major.* Holarctic forms usually occur both in Eurasia and in N. America, but there are also some circumatlantic species e.g. *Calluna vulgaris.* Species occuring in Greenland as well as in Europe are included in the eurasiatic group. Eurasiatic species often extend into N. Africa; in this group are included many species whose main area of distribution lies in Europe (especially in the Mediterranean region) but has extensions reaching far into Asia. Many of the European species are lacking in N. and/or central Fenno-Scandinavia and a number of them (such as *Dactylis glomerata*) have become established elsewhere. A certain number of them are restricted to the coastal belts, such as *Halimione portulacoides.* Species extending to W. Siberia or to Asia Minor (or occasionally even to Persia) have been referred to the "European" group of species. The category of W., S. and central European species includes forms not found in the Boreal or continental parts of Europe. The N. and central European group includes those species which, outside these regions, only occur in montane zones. Mediterranean-Atlantic forms are often restricted in their distribution to the south-atlantic region (reaching N. only as far as Ireland and southern England), but not infrequently also occur in the subatlantic parts of Europe. Mediterranean-euatlantic taxa are mostly restricted to the coastal zones, such as *Elytrigia pungens* and *Phleum arenarium* (but not e.g. *Centranthus ruber*). Likewise, the Atlantic forms are but rarely restricted to the eu-atlantic zones (examples: *Centaurea nigra* and the littoral plant *Atriplex babingtonii*). If the radiation of distributional areas into the Atlantic region is only of rare occurrence (as is the case with *Matthiola incana, Asplenium onopteris, Adiantum capillus-veneris* and *Anogramma leptophylla*), forms with such a distribution pattern are included in the South-European group. The Mediterranean species include those taxa whose area of distribution extends to Pontic regions or as far as Asia Minor and N. Africa. No category was made for species which inhabit N.-, central and W. Europe, or N.- and W. Europe, because the number of such forms is too low (examples: *Achillea ptarmica* and *Matricaria maritima* ssp. *inodora, Empetrum nigrum* and *Lonicera periclymenum* and the littorals *Glaux maritima* and *Limonium vulgare;* of *Polygonum lapathifolium* conceivably other intraspecific taxa occur in N. Europe and W.- or central Europe).

Species included in the category of central European taxa sometimes have their principal area of distribution in more continental parts of Europe (examples: *Potentilla arenaria* and *Dianthus arenarius*).

Among the neophytes only those species are listed whose historical plant geography and recent spread are more or less known. These species have usually become established in the 19th or 20th century. Species whose area of distribution was initially restricted to a much smaller part of Europe and have recently spread to adjoining regions (such as the originally Mediterranean species of *Papaver* and *Geranium pyrenaicum*) have not been regarded as neophytes, and neither have the species which can nowadays be considered to be cosmopolitan or have at least a very wide distribution, such as *Erigeron canadensis* and *Matricaria matricarioides.*

The above-mentioned classification shows very little agreement with that proposed by MEUSEL, JAGER & WEINERT (1965), which was not very useful for the purpose of the present investigation for the simple reason that the rather singular aggregate of wall-dwelling species includes forms that are not restricted to a single phytogeographical region, subregion or province. Moreover, the distribution of many species is not sufficiently known in detail for a finer geographical subdivision. The Atlantic species predominantly exhibit an oceanity of 1, the central-European and Mediterranean ones an oceanity 2, the remainder a degree of oceanity of 1 or 2 (see under 3.3.1). For the mosses, separate spectra were drawn up, because:

1. the area of distribution of the individual species is sometimes considerably larger than the average area of species of vascular plants, and
2. the area is insufficiently known in many cases (this applies more particularly to regions outside Europe: the temperate zones of the southern hemisphere have been much less thoroughly studied bryologically than the holarctic regions and a number of forms recorded from Holarctis are elsewhere mainly known from the few regions, such as Tasmania and parts of New Zealand, that have been better explored bryologically).

2.3.6 Data processing of the spectra

The data recorded in the list of species were used to compile spectra of each of the 11 geographical areas distinguished, of the various plant communities, and of the material as a whole. The calculations are essentially simple but very time-consuming and that is why they were carried out on a computer with a great memorising capacity (viz. an "Electrologica EX 8" computer of the Mathematisch Centrum, Amsterdam). In the present publication only summaries of the results can be given, but the complete information is available upon request.

2.4 Ecological studies

The ecological studies have been restricted to a number of factors viz. those that could be observed or estimated by means of rather simple and not too time-consuming methods. More particularly, as many observations as possible were made concerning the nature of the substratum, the exposure and the inclination, the immediate surroundings and the age of the habitat. The exposure was determined by means of a compass with divisions of 1/16th of a circle. For the estimation of the angle of slant an improvised inclinometer was used consisting of a graduated arc and a water-level. Physical measurements and chemical analyses were only occasionally carried out. The methods employed will be briefly discussed.

Light measurements were performed by means of a Gossen Candela Lux-meter with direct recording provided with a flat radiation-sensitive selenium cell. The measurements were made perpendicularly to the direction of the wall whose vegetation cover was under examination, a similar measurement always being carried out in the open nearby and in the same direction so as to obtain a relative measure of the amount of incident light at the time of observation. The light meter, for which special diaphragms had been made for high intensities of radiation, was gauged at the Laboratory of Plant Physiology, Wageningen (Prof. E. C. Wassink).

Temperatures and relative moisture content were measured by means of thermocouples. Semiconductors proved to be less suitable for the purpose, because of their relatively great sensitivity to radiation which, on account of their fairly large surface area (diam. 1.1 mm as a minimum) and in spite of all possible precautions taken, appeared to be considerable. In order to preclude errors through radiation, following STOUTJESDIJK (1961), tiny thermocouples of copper/constantane or constantane/manganine were used with a wire gauge of 0.06 mm. For the estimation of the relative humidity a number of modifications was made to enable the continuous registration of the relative humidity or the temperature on a potentiometric recorder without any danger of desiccation of the "wet" welds as occurred rather frequently when Stoutjesdijk's system was tried out. The modified system was developed by D. J. van Weers who also experimented with the apparatus on a number of walls and recorded a number of data (VAN WEERS, 1965). Fig. 2 is a diagrammatic representation of the method.

For the estimation of wind velocities and convection currents in the air I used a laboratory-made anemometer based on the principle of thermocouples, as designed by the Institute of Sanitary Techniques, Indoor Climate Section, of the Organisation for Applied Scientific Research (T.N.O.) at Rijswijk, in which institute the instrument was also calibrated in the wind tunnel. The principle of this anemometer was described by DEN OUDEN (1958). The minimum sensitivity is about 2 cm/sec. and the readings are to a large extent independent of the direction of the wind.

Silt analyses were made of a number of humus samples in the Laboratory of Soil Science of the University of Amsterdam by Dr. T. W. M. Levelt through the kind co-operation of Professor J. P. Bakker. Of 53 stands of mural vegetation and of 15 stands from paved surfaces (streets, etc.) chemical analyses were carried out in the Hugo de Vries Laboratory, Amsterdam, for the estimation of pH, calcium and nitrogen content and in a number of cases also magnesium content. The chemical analyses of habitats and ash analyses of the plants have been described by SNEL (1964). Some analyses of plant material were carried out in the laboratory of the Royal Tropical Institute under the direction of J. Heesterman, the remainder by M. Snel.

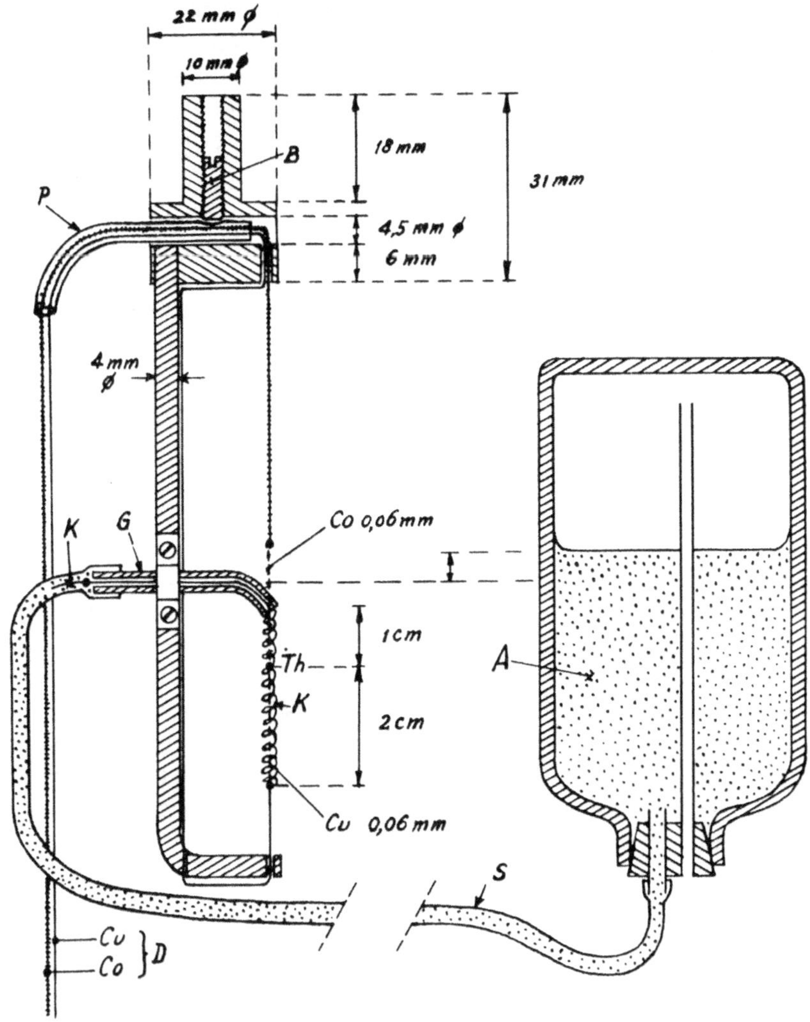

Fig. 2 Apparatus for the estimation of the relative humidity.
A: distilled water; B: securing screw; D: insulated thermocouple wire Cu-Co diam. 0.5 mm.; G: glass capillary tube; K: cotton thread no. 100/3; P: plastic insulating sneath diam. 4 mm.; S: plastic hose; Th: thermic welding point.

3 Ecology of wall vegetation

In the preceding chapter the influence of human activities on the development of wall vegetation and the resemblances and differences in comparison with vegetation of rock crevices was pointed out. In this chapter, the ecological factors will be briefly surveyed. The environmental factors are separately dealt with, but they are of course not all independent of one another. They constitute an interrelated complex of factors that is, especially in respect of the subject under discussion, very insufficiently known. More extensive studies seem to be indicated, the more so since the results might have some practical significance. Each ecological factor can be considered on at least three different levels viz. from a geographical point of view (compare: distributional area or macroclimate), from a local point of view (compare: landscape or local climate), and at the level of the habitat (compare: microclimate). One could even go further and consider the patterns within vegetation stands separately, or the ecology of the individual species present, or even various developmental stages. In the present discussion the emphasis is on the conditions prevailing at the level of the site of a vegetation stand, but the relations to the conditions found at other levels are not altogether disregarded. The various levels are not only recognisable in relation to space, but also in relation to time. The effect of time on the environment and the plant cover is very considerable, both being dynamic and the time lapsed being the basis of the alterations. The ecosystem, like an organism, has a physiognomic aspect that seemingly remains unchanged, because its average composition does not change to an appreciable extent. However, an exclusively static consideration of the complete system is rather meaningless in that all kinds of activities take place (in an organism, for instance, the ingestion of food, assimilation, dissimilation, protein synthesis, transpiration, etc.), and form and function are inseparable. In the ecosystem the relation between structure and environmental conditions must be indicated and explained, and inevitably all changes that take place must be included in the analysis. However, physiological time is less uniform than physical time, far less accurately predictable and far more difficult to study. The time factor will only be fleetingly mentioned in this chapter e.g. when historical factors (3.1.2), the accessibility and dissemination (3.1.8 and 3.1.9), decomposition (3.2.2), and biotic factors (3.4) are being discussed. In a forthcoming publication (SEGAL, 1969), special attention is paid to the factor 'time'.

Species living on rocks, stones or walls are called *petrophytes* or *saxicoles,* the more characteristic forms in rock- and wall crevices being distinguished as *chasmophytes,* and the species growing on small ledges on which detritus and humus have accumulated as *chomophytes.* There is no clear dividing line between the chasmo- and the chomophytes. Species dwelling on naked rock, such as crustaceous lichens, are called **lithophytes** (see GAMS, 1918; ELLENBERG, 1963).

3.1 Topography and history

Walls are built by man and are for that reason restricted to regions inhabited at one time or another. The oldest and most characteristically overgrown walls are, therefore, to be looked for in old settlements and include e.g. walls of, or near, old city limits, of cathedrals, churches, moats and canals. The possibility of combining biological studies with the cultural history of a region made the investigation doubly interesting.

3.1.1 Landscape and situation

In the area studied, the best examples of wall vegetation are situated in the eu-atlantic and submediterranean regions. In subatlantic and eu-mediterranean regions good examples are much rarer and where they occur the atmospheric humidity never remains low for long periods. This means that walls situated in the immediate vicinity of a body of water e.g. near rivers, streams, moats, etc., provide the best sites. Furthermore, the best specimens of wall vegetation are most frequently found on N.-, W.- or E.-facing walls because there humidity and temperature ranges are usually small. The same applies to well-shaded S.-facing walls for obvious reasons. In temperate western Europe the best specimens of mural vegetation are found in Ireland, southern England and western France (more particularly Brittany and Normandy and the neighbouring old culture regions such as the towns and the castles in the Loire valley). The conditions are not so favourable in the Low Countries on account of the slightly more continental climate. Nevertheless some well-developed examples of wall vegetation occur farther inland in W. Germany and Switzerland in the more subatlantic parts, but they rapidly become rarer towards the East. In more easterly situated regions well-developed mural vegetation is predominantly encountered in the drainage basins of the greater rivers such as the Elbe, Weser, Rhine, Meuse, Loire, Rhône and even the Danube, where the stands occur in more or less sheltered sites.

The distribution of a number of more or less characteristic species is likewise restricted to such localities, as was pointed out already for the species occurring in the Netherlands such as *Antirrhinum majus, Asplenium adiantum-nigrum, A. trichomanes, Ceterach officinarum, Cheiranthus cheiri, Corydalis lutea, Gymnocarpium robertianum, Parietaria judaica* and *Poa compressa* (SEGAL, 1962b).

The last-mentioned species has a more continental distribution and must have migrated in a downstream direction towards the Atlantic region. Other species must have followed the opposite (upstream) route, and for a third group the Atlantic coastal strip served as an approach route judging from their pattern of distribution. How this distribution pattern must be explained is not yet very clear, but undoubtedly anthropogenic and historical factors are of importance apart from topographic factors. Many of the oldest human settlements are situated near the coast and on river banks, although this does not explain why certain wall-dwelling species are not found further inland except rarely (which is the more surprising because they are partly anemochores, in the first place the ferns), so that one must assume that in addition ecotopographic factors play a role. It is noteworthy that species with heavier seeds which are intentionally introduced may occasionally develop but do not increase their population and do not behave as a "neophyte". This does not only apply to wall-dwellers, but also to hundreds of species (mostly more continental ones) of other habitats than walls, which exhibit the same distribution patterns and are so characteristic of the Dutch "Fluviatile District". In my opinion, the possible, though perhaps incomplete, explanation must be sought in the very varied geomorphological structure of the landscape along the rivers, their estuaries, and the coast. The variation is caused by the escarpments and inclines along the river valleys (of different exposure and inclination) which are responsible for many gradients of soil composition and soil texture, of moisture, and of periodic flooding. The climatological gradients presumably extend even in a direction perpendicular to the direction of the streambeds of the rivers. Moreover, such phenomena must be associated with the occurrence of special wind effects and/or aerial convection currents, which in turn are of importance not only in connection with the transport of diaspores towards and within the area, but especially for the inflow of soil- and humus particles which is a prerequisite for the establishment of the majority of the wall-dwelling species. The greater geomorphological variation is, accordingly, associated with the occurrence of a strickingly rich and varied flora in this region of Holland. Conceivably, it is the combination of a high relative humidity and other microclimatological factors which, apart from anthropogenic-historical causes, and perhaps natural accessibility factors, possibly from earlier climate periods, creates a favourable habitat for the more characteristic species of wall vegetation. The occurrence of steep inclines in any part of the flat· low-lying planes in western Europe probably creates ecological situations, for instance after a disturbance through human interference, that can serve as preserving sites or refugia for some species. It is noteworthy in the present context that the distribution patterns of ferns do not apply to their occurrence on walls in the lowlands, but also to natural chasmophytic vegetation. This is evident from the many distribution records in the literature, whose connection with geomorphological variation was hardly recognised, for that matter (see e.g. MEUSEL, JAGER & WEINERT, 1965; LAWALREE, 1950; SOPER, 1963). This distribution pattern is also exhibited by a number of ferns that are hardly ever (if at all) found on walls, such as *Asplenium septentrionale*. Most probably the higher relative humidity in the valleys of rivers and mountain streams is also a major factor. I found some indications for this conjecture when I recorded temperature and relative humidity in a number of places in the Sûre valley (Luxemburg) and in the immediate vicinity. Two additional factors must be reckoned with viz.

1. In river valleys the conditions are generally more favourable for the formation of rock crevices, because the valleys have been cut out and eroded by the rivers and smaller streams.
2. The possibility that the valleys are more thoroughly investigated than the usually much less accessible inclines and escarpments beyond them, must not altogether be precluded.

Walls situated at a greater distance from the coast or large river systems at first do not seem to differ much from similar walls standing closer to the sea or to an estuary in their ecological characteristics. The absence of the more faithful wall-dwelling species in more continental areas can only be explained by the consistently or temporarily greater dryness of the air and/or the lack of other favourable factors mentioned. Of a somewhat singular character are the walls of quay-sides, moats, river banks and canals which are situated below a road running overhead, such as the walls of the Amsterdam canals. The influx of soil particles, humus and various waste products (dung, urine, etc.) as a result of traffic and drainage of rain water, can be rather excessive and, provided other prerequisites are met, such walls (if not frequently cleaned or repaired) can become rapidly covered with vegetation which not rarely includes more or less nitrophilous species.

The adjacent landscape affects the ecology of all walls to a certain extent, if only by determining the accessibility of the site. The distance and the situation of the nearest locality of potential wall-dwellers decides if the diaspores will arrive at the site, but there are various other factors.

In the coastal dunes, for instance, sand is readily blown about and the sand grains deposited on walls behind the dune belt may favour the development of vegetation. Dependent on the shape of the sand particles and the presence of adhering compounds, diaspores of dune-inhabiting species may be carried along. On walls behind the more calcareous Dutch dunes *Streblotrichum convolutum,* and sometimes also *Tortula subulata* are commonly encountered. These two species do not occur in the more acid type of dunes and only occasionally on walls behind them. The influence of the sea can be important because the sea winds carry marine salts inland, the concentration of which can be appreciable in a belt extending to 10 to 15 km from the coastline. In regions where the "maritime" species *Crithmum maritimum* grows, it is sometimes also present on walls e.g. in ports in Brittany and the South of England. On the quay-sides near the sea front halophytes are sometimes found as wall-dwellers; *Plantago maritima,* for instance, was recorded from walls at Brixham and Topsham (Southern England) and at La Faou (Brittany). In the latter town I noted, apart from more stereotyped representatives of wall vegetation, such forms as *Triglochin maritima, Halimione portulacoides, Aster tripolium, Limonium vulgare, Puccinellia maritima, P. distans, Carex extensa* and *Atriplex glabriuscula.* Of particular interest was the occurrence of *Sonchus arvensis* var. *maritimus* (=*S. maritimus*) and *Agrostis stolonifera* var. (=*stolonifera* subvar.) *salina,* which morphologically agreed very well with the forms of these species in natural sites in the coastal area; this may be taken as an indication of the taxonomic significance of these forms which are clearly more than mere habitat modifications.

In the vicinity of the dunes in the area of The Hague, *Calamagrostis epigeios,* which is common in those dunes, is frequently met with on walls. In the freshwater tidal region of the Dutch estuaries, where *Angelica archangelica* is found in tall herbaceous vegetation, this species was encountered on quay-sides at Dordrecht and Kralingse Veer. Likewise, in the Dutch "Lagoon District" riparian forms, marsh plants and species of swampy hay-fields are found on canal sides, *Thelypteris palustris* having been recorded in about a dozen towns in the W. and the N. of the Netherlands and also in the Belgian towns of Ghent and Bruges. *Apium nodiflorum* was once recorded on a wall alongside a brook in W. France. On walls below the surface the more usual types of aquatics are not at all common, but various haptophytes (plants normally attached to solid substrates such as rocks and stones e.g. *Fontinalis antipyretica*) are not infrequently encountered. *Fontinalis* sometimes completely covers walls below the surface (recorded at e.g. Floursies, Dept. Nord, France). In towns escaped ornamentals and garden herbs are sometimes found growing on walls, especially *Apium graveolens,* but sometimes also *Petroselinum crispum.*

3.1.2 Historical factors

Vegetation can develop on a wall only after the conditions for settlement of plant species have become sufficiently favourable. Prerequisites are that the wall has been exposed to weathering long enough or that soil particles have accumulated (or both). It is to be expected that a wall constructed out of blocks of material that fairly quickly deteriorate (such as marl) soon carries some vegetation. However, nearly always the initial phases of vegetation development on brick walls, in the form of lichens or acrocarpic mosses but sometimes also of higher plants, takes place in the joints. Older types of mortar decompose especially fast. During the Middle Ages, insofar that walls were not built by stacking stones without cement, mixtures of lime, loam (clay) and hay or straw were used, but the composition varied considerably from place to place. For centuries a rather soft lime mortar on the basis of lime and sand, sometimes with an admixture of ground-up vulcanic ash, was used. Nearly all stone and brick buildings dating from the 15th till late in the 19th Century were jointed with lime mortar, but since the beginning of the present century a much harder kind of mortar came into use. These harder mortars, made with Portland cement and to a lesser extent with ground-up slags from blast furnaces, are not only less subject to weathering. They also do not crack so easily along the contact surface with the bricks which may happen as a result of differences in dilatation through differences in temperature or differences in water retention, especially when the wall becomes frosted over (see 3.2.2). In addition, the higher pH may inhibit the development of vegetation. After the introduction of Portland cement the building of very solid walls became possible. The originally hand-made, porous bricks with a wavy or grooved surface were also gradually replaced by the harder and smoother machine-moulded normalised bricks and still later by strip-pressed ones. Brickwork that is satisfactory from an architectural point of view is not so suitable to support plant growth. A good type of mortar prevents retention of water, whereas the older types of walls were thicker and more massive and could thus retain a good deal of moisture. Nowadays houses are often built with a ventilated hollow wall. It follows that the time lapsed after its erection before any proper vegetation has developed on a wall varies appreciably, the more so if topographical factors are involved and particularly if the inflow of extraneous soil- and humus particles is considerable. In general, an age or at least several scores of years is required before mosses can become established on a wall. However, lichens may have appeared after a few years. Occasionally, establishment of *Tortula muralis* occurs after a year or two and of *Bryum argenteum* after 4 to 5 years, but only on the horizontal upper surface of walls on which extraneous matter (sand, etc.) has accumulated e.g. near dunes and near building sites raised with pumped-up sand or sludge. The first higher plants to settle, usually *Asplenium ruta-muraria* or *Linaria cymbalaria,* do not often become established on walls jointed with lime mortar before the latter are between fifty and a hundred years old, except in the eu-atlantic regions where this may happen sooner. One might thus expect that the oldest walls or the oldest remains of walls in the area under discussion, which are usually of Roman origin, show the best development of wall vegetation. This is not the case, however; for after about two millenia walls often are so badly decomposed and have trapped so much soil and humus particles that — if the walls were not renovated — the vegetation cover has developed further and includes other taxa. These may include several shrubby forms that gradually oust out the original, more typical type of vegetation, in spite of the fact that the larger specimens of herbs and woody plants are often pulled out by man. Other factors, such as the leaning over of walls with age, or a slant resulting from other causes such as decomposition and the development of vegetation, also favour other vegetation elements which crowd out the more characteristic wall-dwellers (see under 3.1.5). The best examples of wall vegetation are encountered on walls between 100 and 500 years old.

Other historical factors are more extensively treated in connection with the influence of man on the development of vegetation (see under 3.4.3).

3.1.3 Dimensions

Although it is but seldom taken into account in phytosociological inquiry, the question of the minimum surface area of a vegetation fragment in order to obtain the greatest possible chance of recording a representative and well-developed stand is of paramount importance. This area is usually several times larger than the so-called minimum area (see 4.5), but there may be a correlation between this minimum area and "minimum size" among vegetation stands on a more or less horizontal substratum. If, however, the total area covered by vegetation of a certain kind in a certain site is small, there is not only a possibility of the presence of an empoverished fragment of a vegetation type much more luxuriantly developed elsewhere, but also the possibility of a strong influence of the immediate surroundings of the stand on account of the border line which is relatively long in relation to the surface area. This last factor does not play a major role on walls, in contradistinction to the otherwise rather similar paved roads, although occasionally the encroachment of vicinists can be noticed. This is to be expected because of the mere fact that a wall is placed perpendicularly to its surroundings: the wall habitat is nearly always very abruptly discontinuous with its surroundings, both in space and in time. Accordingly, a rather small area can correspond with the minimum size. On the other hand the total surface area of a wall, in contradistinction to other habitats, is rather limited, so that the minimal size is often not attained and fragmentarily developed stands are the rule rather than the exception. This holds more particularly for vegetation of vascular plants. A rule of thumb is that the minimal size of wall vegetation amounts to 20 m^2 to 100 m^2 for stands with a minimum area between 4 m^2 and 20 m^2. For vegetation of lichens and bryophytes the minimum size is about twice the size of the minimum area, of the latter usually 0.5 m^2 to 1 m^2 and of the former 1 m^2 to 2 m^2. Presumably the minimum size of stands containing larger life forms is larger in respect of the minimum area of those life forms and there must be some relation between the dimensions.

Fortunately in many sites the wall surface is large enough to show a well-developed stand; however, a large number of relevés of smaller vegetation fragments has substantially contributed towards a better understanding of the type of vegetation.

The above-mentioned values of the surface areas need some specification. The total surface area available for germination of diaspores is for most taxa very much smaller than the total surface of the wall, the sites of germination being restricted to the cracks in the wall (usually only in the joints) where the conditions for settlement are favourable. The values given above refer to a number of more or less ideal cases of walls with a well-developed and fairly dense plant cover. A sparsely developed stand on a sufficiently large wall may carry a more or less "complete" vegetation specimen but of necessity the minimum area, being primarily a function of species density, is larger (see under 4.5) and also the minimal size is proportionally larger in that case.

3.1.4 Kinds of walls

A concise survey of the various kinds of walls seems recommendable. One can roughly distinguish between walls which form an integral part of inhabited buildings, walls built up against earth, or forming parts of old fortifications, quay-sides or roads, etc., and isolated walls (usually fencing in court-yards, cemeteries, etc.). Walls of inhabited houses and castles usually do not support vegetation unless they are very thick. This is most probably to be ascribed to long periods of drought caused by the heating during the cold season; the temperature inside the house being more constant than it is outside, a cold period may cause desiccation of the outer walls. This phenomenon is even more pronounced in built-up areas of towns, all brick and stone buildings forming an artifical "desert" as it were (see under 3.3.2). In addition, atmospheric pollution caused by the production of poisonous gases and soot may be a strong deterrent to the settlement of lichens and possibly also of certain groups of bacteria (see under 3.1.7). Church walls on the other hand are often covered with some vegetation, not only because churches are not so often heated, but also because the walls are much thicker and, when the buildings are old, much more decomposed to form a more favourable substratum.

In the Netherlands there is a striking difference in the development of mural vegetation between Roman-Catholic and Protestant churches of approximately the same age. Mr. J. G. Hirs (in lit.) pointed out to me that damage by moisture and deterioriation of the masonry is worse in churches of Protestant parishes. This is indubitably associated with the lesser frequency of the services and the period of heating. The walls cool off more rapidly and take up water vapour from the congregations more easily. Roman-Catholic churches dating from the 19th and the 20th century situated in the predominantly Roman-Catholic parts of the Netherlands rarely support a well-developed growth of wall-dwellers, whereas coetaneous Protestant churches much more often carry vegetation of at least lichens or mosses. The moisture relations of isolated walls, if not erected in a region with a usually high atmospheric relative humidity, is not very favourable either, because the temperature can rise appreciably as a result of solar radiation, the evaporating surface is relatively large, and the supply of water almost restricted to the rain water which can easily run off. Thick walls have a greater heat-buffering capacity than thinner ones. On unattached walls in drier regions at best only lichens can establish themselves as a rule.

The most suitable habitats are accordingly provided by walls built up against a body of earth, for instance, along canals, moats, old fortifications and city walls, etc. From the earth there is usually a constant supply of moisture to the wall through cracks and by capillary action, and possibly also of nutrients and solid soil particles. For the development and maintenance of vegetation on walls emerging from a body of water (canals, moats, harbours), the capillary suction and the atmospheric water vapour are of course of great importance.

Roughly four zones can be distinguished on walls in a vertical direction:

1. the base;
2. a middle zone;
3. an upper zone, and
4. the apical zone.

The lowermost zone usually receives more moisture and certainly retains water longer than the more exposed upper parts e.g. on account of capillary suction of water from the soil, but frequently also on account of a longer period of overshadowing or a shorter daily exposure to direct solar radiation. In this zone the deposition of particles blown up by the wind is usually very important, sometimes so much that the nature of the substratum supporting plant growth is decided by the material of extraneous origin rather than by the masonry itself. For this reason, the composition of the plant cover on the basal zone of a wall is often atypical and tends to resemble the type of vegetation growing in the vicinity up to a point. As an example, the bottom parts of walls built on a poor sandy soil often support vegetation in which basiphilous and basitolerant taxa are lacking. Generally speaking the basal part of walls are relatively favourable sites for the development of bryophytes.

The uppermost zone of a wall is often covered by a sometimes protruding slab of a different material (such as a hard kind of natural stone). If the slab is of a different consistency this favours the occurrence of tension through unequal dilation of the different building stones under the influence of changeable climatic conditions. Frequently the first cracks and crevices develop at the dividing line between the wall and the layer of covering slabs so that the first plant growth also begins there. The plant cover may be instrumental in retaining rain water trickling down the wall so that the upper zone is often more humid than the middle zone. The partial overshadowing caused by a protruding slab may also promote the development of vegetation. In this zone the rates of evaporation and transpiration are thus also affected and usually reduced, which is associated with a different amplitude of temperature changes.

Walls bordering canals and moats are somewhat different as a habitat in that the water relations are more complicated on account of the considerable influence of the water below it and of the influx from above of humus, soil particles and waste matter usually rich in nitrogen. The submerged part of such a wall is often overgrown by haptophytic filamentous algae provided the water is not too strongly polluted and sufficiently clear. Immediately above the water-level, or in some cases in the zone of fluctuation of the water-level, often nitrophilous riparian species develop which are partly characteristic of habitats where fluctuation of environmental conditions play a conspicuous role, such as *Rorippa sylvestris, Rumex conglomeratus, R. obtusifolius* and *Lycopus europaeus*. In the upper zone nitrophilous vegetation is often encountered which shows some resemblance to vegetation between paving bricks and cobbles and includes such species as *Sagina procumbens, Poa annua, Bryum argenteum, Ceratodon purpureus, Dryopteris filix-mas* and several other species of ferns (see SEGAL, 1962b).

3.1.5 Inclination

The present study was chiefly restricted to vertical walls. The characteristic plant communities found on vertical walls are usually already "adulterated" or "disturbed" when the angle of inclination is only a little less than 90° by the appearance of elements of vegetation types of arable land and of ruderal places and to some extent of garden escapes. A vertical position of the wall is prohibitive to the development of a large number of species. It is nevertheless striking that so many species are encountered on vertical walls: in the Netherlands I recorded 386 species of vascular plants that were found growing on walls i.e. over 25% of the total number of indigenous species. In addition 37 escapes from cultivation were noted. If one takes into account that a not inconsiderable percentage of the species occurring in the Netherlands is made up by rarities, and that the occurrence of aquatics and of forms with bulbs and corms and other specialised groups such as parasites and saprophytes is of course impossible, the number of recorded species is indeed very high. However, the environment is selective to a considerable extent and only a small number of species is regularly found on walls, whereas a great many were recorded once or only a few times. Careful observation showed that in many such "incidental" records the germination had taken place in certain places of the wall where the steepness is much less such as cracks in which already some soil particles or humus had accumulated or decomposition products had been formed. Many plants do not develop properly if the radicle of the young seedling is prevented from growing downwards in a vertical direction and conceivably a development by means of a lateral growth of the radicle enables the characteristic wall-dwelling forms to develop on a vertical wall surface. The establishment of some species is noticeably favoured by a form of decay which results in a decrease of the angle of inclination (examples are *Cheiranthus cheiri* and *Poa compressa*). Other taxa, such as species of *Asplenium, Cystopteris fragilis* and *Corydalis lutea* disappear fast in such circumstances and are soon replaced by less particular forms.

Since walls usually form sudden discontinuities in the landscape in the vertical relief, the incidence of ascending and falling winds or air currents (thermic disturbances, etc.) is a common phenomenon contributing towards the extremity of the habitat.

There is, furthermore, a relation between the incident radiation and the angle of inclination. The radiation is highest on south-facing walls with an angle of inclination equal to the latitude (in the Netherlands it is about 52°).

The effect of the inclination on the composition and the structure of the plant cover has been studied in many places and nearly always appears to be very pronounced. Vegetation on vertical walls is almost invariably poor in species, their number usually not exceeding 15. This number increases (up to several dozen) as the slant of the wall is greater. There was a fortuitous case at Carteret (in Normandy) where I recorded vegetation of *Crithmum maritimum* and *Linaria cymbalaria* on two SSE-facing walls of which the one was situated on top of the other and the angle of inclination was 90° and 45°, respectively. The vertical wall supported only 11 species in all with coverages of about 25% of the herbaceous layer and 1% of the bryophyte layer, whereas on the slanting wall these values were 25, 35 and 3%, respectively. On a wall at Draguignan with an inclination of 86°, 46 species were counted (herb layer: 20% coverage, bryophytes: 35% coverage); on a wall near Sommières with approximately the same inclination 51 species (herb layer: 50%, moss layer: 30% coverage); approximately the same figures were recorded on a wall at Carcasonne with an inclination of 75%. The wall near Sommières was built up against a vineyard and the one at Draguignan against a hay-field, but the wall at Carcasonne stood free (on a free-standing wall the number of species is normally smaller than it is on a similar wall with about the same exposure and inclination which is built up against a mass of earth or forms part of an edifice).

3.1.6 Exposure

Exposure is of very considerable importance. A good few vegetation types are restricted in their occurrence to N.- and W.- or E.-facing walls, whereas on S.-facing wall surfaces the development of vegetation is usually very poor. It is noteworthy that this rule also applies to Mediterranean forms occurring in more temperate regions, which are sometimes mentioned in the literature as predominantly found on S.-facing walls, such as *Ceterach officinarum* and *Sedum dasyphyllum* (see e.g. OBERDORFER, 1962; HEGI). This contrast between the density of the stands on walls of different exposures is not so pronounced in the eu-atlantic regions as a rule, but usually also manifest. A great many wall-dwelling species are less luxuriant on S.-facing walls than when they are growing on N.-facing wall surfaces. In southern England I was able to compare the N.- and S.-facing sides of the same wall in a number of places e.g. at Yealmpton (Cornwall). The specimens of *Ceterach* on the N.-facing side were larger and their fronds were spread out during the daytime, those on the S.-facing side were smaller and the leaflets of their fronds rolled in, the scales on their ventral side forming a good protection against desiccation. In the same area, but farther inland, *Ceterach* is mostly restricted to the N.-facing wall surfaces. Presumably the fluctuations in climatological factors are of decisive importance: a S.-facing wall receives more solar radiation and also becomes warmer on a summer day, but it is, however, subject to more loss of heat through radiation during the night than a comparable but N.-facing wall surface for the very reason that a vegetation cover of sufficient density to level out extreme differences in temperature is lacking. A S.-facing wall is, moreover, more subject to desiccation. Differences in moisture content and in temperature, and especially their daily fluctuations, are considerably smaller in a N.-facing wall. The more constant environmental conditions result in the development of a more intricate vegetation structure, a phenomenon also known from the N.- and the S.-facing slopes of our coastal dunes (see 3.3.2).

The exposure is also of importance in connection with the prevailing direction of rain and sleet which is in turn dependent on the prevailing wind direction. In the Atlantic region, accordingly, the W.- and S.W.-facing walls are coloured green by a cover of microscopic algae. In particular representatives of the genus *Protococcus* (s.l.) decidedly prefer the sides that receive most rain. Several mosses are more often encountered on S.- and W.-facing walls than on N.- or E.-facing ones e.g. *Grimmia pulvinata* and *Tortula intermedia*. On N.-facing walls these forms are soon replaced by other species during the course of the succession seres (which mostly proceed faster than on S.-facing wall surfaces). On N.-facing walls and in sheltered places certain hypnid mosses and also ferns are fairly regularly present which are far less often seen on surfaces exposed to the south (see 6.6).

All these differences are primarily attributable to the different effects of the insolation, the intensity of the radiation being largely dependent on the angle of incidence of the sunlight which changes in the course of one year. According to Grafe (1956), who carried out measurements on vertical surfaces in Hamburg, a south-facing wall receives in winter (when the sun is low in the sky) more direct radiation than a horizontal surface of the same size, whereas between April and September the amount of incident radiation is lower (lowest on 21st June). The temperature may thus be relatively high in winter which prevents excessive damage by frost, the more so because frosts are more likely to occur when the skies are clear, so that in the spring the vegetation cover can develop fairly early and persist till late in the autumn, as is substantiated by my observations. A N.-facing wall receives the strongest radiation in early morning and during late evening, but the intensity and hence the evaporation is low.

An E.-facing wall is most intensively irradiated in the morning and a W.-facing one in the afternoon; they receive more direct radiation per unit of surface area than a horizontal surface during early morning and late afternoon, respectively. During the winter season, when the sun remains lower in the sky, the amount of radiation is more intensive and of longer duration than in summer. Early in the morning the air temperature is lower than in the afternoon, so that a wall exposed to the east shows a certain resemblance to a N.-facing wall in that the fluctuation in temperature during the day is not excessive and the absolute maxima are not very high. In addition, the spectral composition of the incident light may be important, as the evening light contains more infra-red (i.e. heat) radiation and morning light relatively more ultra-violet rays.

The effect of winds must also be taken into account, the prevailing direction of the winds in the greater part of the area under investigation being W. or S.W. The action of wind is partly a desiccating one and partly mechanical. The mechanical action is of course more detrimental to the composition and the structure of the plant cover on the S.- and W.-facing walls and this may be one of the reasons why there is in W. Europe a relatively strong resemblance between vegetation on N.- and E.-facing walls. The agreement is supported by data relating to dune vegetation (e.g. WESTHOFF, 1947).

The effect of the exposure will be elucidated by a few examples: - At Bricquebrec (Brittany) two walls situated alongside a river, facing S. and N., respectively, were compared. In either case the number of species recorded was 13 and the relative humidity is usually high. On the S.-facing wall the coverage of the herb layer amounted to 12% and that of the bryophyte layer to 7%, whereas on the N.-facing one these figures were 35% and 10%, respectively. The different density of the herb layer is presumably attributable to effects of differences in temperature rather than to differences in moisture content and perhaps indirectly through the weathering of the S.-facing wall which favours establishment of herbaceous plants much more than that of bryophyte vegetation. At Wapenveld (the Netherlands) the N.- and S. sides of a concrete bridge were compared:

Exposure	S	N
Percentage of coverage, lichens	80	30
,, ,, ,, , mosses	7	35
Number of acrocarpic species of mosses	5	8
,, ,, pleurocarpic ,, ,, ,,	nil	3

Such a situation is characteristic of many S. and N.-facing surfaces: a higher percentage of lichens on the side exposed to the S., provided the relative humidity is usually not very low, and a greater dominance of mosses (including some pleurocarpic forms) on the N.-facing side.

3.1.7 Atmospheric conditions

It is common knowledge that a large number of lichens avoid habitats where the atmosphere is polluted by smoke, tar and injurious gases produced by domestic coal fires, factories, refineries, mines, blast furnaces, and motorised vehicles. This was pointed out as early as 1892 by ARNOLD. Other organisms of course also suffer from the poisonous substances contained in domestic and industrial air-borne pollution. Although lichen vegetation is not included in the main subject of the present study, some attention is paid here to the effect of noxious atmospheric constituents on lichens, lichen vegetation often being the forerunner of vegetation of bryophytes and of higher plants and accumulated organic matter formed by the dying of lichen thalli assisting materially in the establishment of other plant forms.

The mesoclimate of a built-up area, especially of a large city, differs appreciably from that in the surrounding open country. According to RYDZAK (1958), the consequences of the climatic differences are much more important as regards the establishment, and the persistence, of lichen vegetation than atmospheric pollution (see also under 3.3.1).

Of the noxious gases in the atmosphere, which are also directly detrimental to living conditions for human beings, sulphur dioxide is the most important, but hydrogen sulphide, nitrogen dioxide and fluoric acid are sometimes present in considerable quantities, and unsaturated hydrocarbons and carbon monoxide of exhaust gases can accumulate to high concentrations in some places (especially when there is a dense fog). Even carbon dioxide is poisonous in higher concentrations. The concentration of poisonous gases in the atmosphere is usually expressed as the concentration of SO_2. According to BARKMAN (1961), the annual average in the centre of Rotterdam is 0.17 mg/m^3 (in the summer it is 0.07 mg/m^3 and in the cold season 0.24 mg/m^3), in Pernis (oil refineries!) 0.43 mg/m^3, in London 0.10 to 0.13 mg/m^3 and in Leeds 0.17 mg/m^3.

According to RYDZAK (1958), the concentration falls by half at the periphery of such towns as Leeds and London. Apart from the gaseous contaminants dust- and, more particularly, soot particles are often prohibitive to the establishment of certain species, not only by the formation of a deposit on walls, tree bark and any plant growth already established, but also by the presence of poisonous constituents such as SO_3. This is a very hygroscopic substance liberated during the combustion of heavy fuel oil that ultimately forms sulphuric acid. The acid in the air contributes to the corrosion of walls and this seems to be a serious problem in Paris.

According to J. G. Hirs (in litt.) the Rijksmuseum in Amsterdam produced about 250 kg concentrated sulphuric acid per day during the heating season, so that the masonry of the building is seriously corroded. Nearly all the acid produced is deposited on the building on a calm day with mist and drizzle. In many Dutch towns e.g. in Amsterdam, the relatively richest development of lichen vegetation is always present in the suburban zones to the S. and the W.

VARESCHI (1953) reached the conclusion that in an ideally laid-out town ventilation corridors with strips of greenery would have to be kept open. These corridors would have to be oriented lengthwise in the prevailing direction of the wind. The zones and sites where lichens are absent — Barkman speaks of "epiphyte deserts", but one could almost equally well say: "saxicole deserts" — are often elliptic in circumference (at least when lichens are concerned), the long axis of the ellipse running parallel to the prevailing wind direction. The effect of air pollution on the growth of higher plants has hardly been studied. The deleterious effect of some substances (such as fluoric acid and organic chlorine compounds) on certain susceptible crops (such as gladioli, grape vines, strawberry plants, fruit trees and barley), is manifest, and so is the damage caused by sulphur dioxide on alfalfa, oats, barley, legumes, and several kinds of fruit trees (THOMAS, 1964). Hardy (evergreen) plants may suffer a good deal more from atmospheric pollution, the more so because the concentration of poisonous gases may be considerable higher during the winter as we have seen. This is probably the main reason why many Conifers and *Ericaceae* do not thrive well in our towns as a rule. The yew, *Taxus baccata,* which is fairly regularly found on walls in western Europe, is usually entirely lacking in the centres of larger towns and cities. Lichens are also hardy and seem to suffer from a cumulative effect of SO_2 and HF pollution.

BLEASDALE (1959) also mentions the effect of atmospheric pollution on winter-growing species, including winter annuals. According to KENT (1960) this would explain why *Erophila verna* and *Saxifraga tridactylites* are of rare occurrence in London and Middlesex nowadays and must have been much more common in the past. According to GAMS (1938), species of *Asplenium, Ceterach officinarum* and *Cystopteris fragilis* are sentitive to aerial pollution and likewise most mosses (and lichens). This I could not substantiate as far as the ferns are concerned. It is true they are scarcer in built-up and in industrial areas than they are elsewhere, but in sites where atmospheric pollution is unmistakable they are not entirely lacking. *Ceterach officinarum* and *Asplenium adiantum-nigrum,* for instance, are found on the walls of the platforms of railway stations and can be covered with a deposit of soot, and *Cystopteris fragilis* and *Asplenium ruta-muraria* occur on walls over canals smelling of hydrogen sulphide.

A number of mosses are undoubtedly sensitive to pollution of the air. It is significant that more or less nitrophilous species such as *Ceratodon purpureus, Bryum argenteum, B. caespiticium* and *Funaria hygrometrica* occur relatively often in towns. These species, and to a lesser extent *Tortula muralis,* are most probably toxitolerant. GILBERT (1968) also mentions *Bryum capillare* and *Leptobryum pyriforme* in this connection. However, the former species is less common than *B. caespiticium* and the last-mentioned species is fairly rare in the towns of western Europe, but rather common in the more continental parts. *Streblotrichum convolutum* and *Barbula unguiculata* are probably also moderately toxitolerant, but *Ceratodon* usually seems to "replace" the species of *Barbula* and related forms (such as *Didymodon). Barbula vinealis* seems to be one of the most sensitive species. Also rather sensitive are *Grimmiales* and *Orthotrichales,* and probably *Polytrichales,* and in general the larger acrocarpic forms such as *Tortula ruralis, Mnium* div. spec. and *Encalypta* div. spec. Generally speaking, pleurocarpic species seem to be sensitive to air pollution. Even the common species *Homalothecium sericeum* is rare in the vicinity of towns. *Hypnum cupressiforme* is less sensitive, but its most frequently wall-dwelling f. *tectorum* is not so common in polluted regions as the var. *resupinatum.* Among the hypnid mosses *Brachythecium rutabulum* seems to be the most toxitolerant. These considerations are in good agreement with the findings of GILBERT (1968) who made a study of the influence of air pollution on bryophytes in Newcastle-upon-Tyne (England).

That atmospheric pollution in urban areas has an appreciable effect on the development of lichens, is evident from the presence of only a few species, usually with limited abundance and in sites where the relative humidity is constantly high (such as the canal sides in Amsterdam). On the walls bordering upon the Amsterdam canals, which together cover a length of 50 km, only four species were encountered viz. the nitrophilous *Candellariella vitellina, Physcia orbicularis, P. caesia* and *Xanthoria parietina,* of which the last can presumably be considered to be toxitolerant (Barkman calls *Xanthoria parietina* moderately toxiphobic). Conceivably, the production of hydrogen sulphide in the canals plays a role. This argument, which is strongly indicative of the effect of atmospheric pollution, has hitherto not been adduced by authors on this subject. It is the more convincing since the epilithic lichens are considered to be generally less sensitive to air pollution than the epiphytic ones on account of the neutralising effect on the often basic substrate of e.g. SO_3, H_2S or HF.

The bark of elm trees usually supports a rich growth of epiphytes, but on the boles of elms growing alongside the Amsterdam canals lichens and mosses are conspicuous by their absence, whereas *Protococcus viridis,* an indicator of a fair degree of humidity, is quite commonly found on these trees.

A last factor to be mentioned is the inhibition of light penetration in urban zones by the solid particles in the air. According to KRATZER (1956) in towns 10 to 20% more radiation is absorbed than in the extra-urban country around them, and according to FITTER (1945) even more (up to 40% more). The light absorption is highest at lower outside temperatures when more fuel is burnt for heating. On the other hand differences in temperature are somewhat levelled out, especially the minimum temperatures being higher on account of a reduced loss of heat through radiation. Soot and tar are deposited on leaves and may block the stomatal pores, especially plants with large or pubescent leaves being thus affected and forms with smooth and coriaceous leaves to a lesser extent (e.g. *Phyllitis,* which is decidedly somewhat toxitolerant).

3.1.8 Accessibility

The accessibility (HEIMANS, 1954) i.e. the attainibility of a site to viable diaspores, depends on an intricate complex of factors some of which have already been dealt with or will be discussed presently. Of the utmost importance are the historical and topographical aspects of the site, and of course the ease of diaspore transport of a number of species. The establishment of a given species is, apart from the potential favourable conditions for its development in a certain place, dependent on the occurrence of that species within a certain distance from that place. The bridging of this distance is primarily determined by the efficacy of the dessimination mechanism. The accessibility of the habitat, however, is determined by such factors as exposition, prevailing wind direction, presence (or absence) of water, the intermediary of certain animals, human interference, etc. In areas where the soil is rich in lime, for instance in parts of Dutch and Belgian Limburg and in northern France, the distance is of minor importance as far as calciphilous and lime-tolerant species are concerned. However, farther away from such regions one may expect that the calciphiles with anemochorous diaspores, such as many species of ferns and mosses, are the first to appear.

Well-developed wall vegetation is relatively often found in areas with soils rich in lime. The walls of quay-sides, canals and moats often carry vegetation of riparian forms, and especially such forms as *Lycopus europaeus, Ranunculus sceleratus* and *Rorippa sylvestris* are frequently found growing a little above the surface of the water. In regions rich in bogs and fens, such as the Dutch "Lagoon District", one encounters species such as *Thelypteris palustris* and *Alnus glutinosa,* characteristic of these habitats.

In a preceding paragraph (see 3.1.1) I have already given a tentative explanation of the frequent occurrence of more or less typical wall-dwelling species near the coast and in large river systems (see also SEGAL, 1962b). In the lowlands of western Europe *Antirrhinum majus, Centranthus ruber* and *Corydalis lutea* are fairly generally considered to be escapes from cultivation and this is sometimes also said of *Cheiranthus cheiri.* However, there are some phytogeographical arguments pleading for a different view. The principal area of distribution of *Antirrhinum majus* and *Centranthus ruber* is Mediterranean to submediterranean, that of *Corydalis lutea* submediterranean and that of *Cheiranthus cheiri* submediterranean to Atlantic. Here and there they appear in natural vegetation on rocks. The species under discussion are all rather common in southern England, but in Britain they are likewise regarded as "introduced" by CLAPHAM, TUTIN & WARBURG (1962). In France and in Germany their area of distribution becomes more and more restricted to the valleys of the great rivers, especially of the Meuse and the Rhine. Apart from the migration route along the rivers, *Antirrhinum majus, Cheiranthus cheiri* and *Centranthus ruber,* and also *Parietaria judaica* have followed a dispersal route along the Atlantic coast as is quite evident from a number of records in W. and N.W. France, Flanders and the S.W. of the Netherlands, and certainly also from their occurrence in southern England. A similar migration route must be considered probable for a number of other Mediterranean and submediterranean elements, as is clear from the distribution pattern of certain species of *Trifolium* and of *Medicago* in the same parts of Europe. In the past such species as *Phleum arenarium, Catapodium marinum, Euphorbia paralias, Calystegia soldanella* and numerous other characteristic species of the Atlantic coastal region most probably have followed such a migration route. It is not at all unlikely that a number of species which are rather characteristic of the Dutch "Fluviatile District", have followed the Atlantic migration route. They are also found in Kennemerland (the dune area to the north of Haarlem) and were formerly supposed to provide a phytogeographical argument in favour of the hypothesis of the presence of an old northern branch of the river Rhine (VAN EEDEN, 1867). The present distribution patterns, to my mind, support the assumption that the species mentioned can be regarded as autochthone elements in the Atlantic parts of western Europe, even if they have most probably only reached the Low Countries in historical times, because natural habitats are rather scarce and must also have been rare in the past. However, man, by building walls, has created a 'secondary' form of habitat which provides even more favourable conditions than their original, more restricted natural habitat on or in rocks, and the species concerned subsequently must have followed human culture. It is even conceivable that the use of some of the more attractive species as ornamentals only began after their natural establishment (and not the other way around!). All these forms are more or less nitrophilous and this characteristic feature has presumably furthered their dispersal into anthropogenic habitats. An additional argument in favour of my point of view is provided by the morphological characters and probable intraspecific taxonomic status of some of the plants in question. Of *Cheiranthus cheiri* almost always the typical "wild" form is found on walls, which is in several respects quite distinct from its modern cultivars. The same holds for *Antirrhinum majus*: the majority of the wall-inhabiting specimens agrees in every respect with the typical subspecies (ssp. *majus*). Its cultivars can of course run wild, as was observed on few occasions e.g. on the marl walls of Neerkanne Castle S. of Maestricht, where all sorts of floral colours of the snapdragon were present. It is theoretically possible that cultivars after having run wild are subjected to such a strong selective pressure that ultimately a phenotype prevails which resembles the original wild form very closely, but experience has taught us that many cultivars can maintain themselves for a short while only. In central Europe where it occurs scattered, *Hieracium amplexicaule* is often considered to be an escape from culture (e.g. by HEGI). VAN SOEST (1934) has pointed out that this composite with its pappus-bearing achenes could easily have bridged the distance from the original area of distribution to the scattered central-European localities. *Hieracium amplexicaule* is a Mediterranean species found in rock crevices, which extends northwards to the Alps, the Jura Mts. and the Black Forest. It is not at all unusual that the dispersion of a species increases towards the far borders of its area of distribution. The ease of migration of a given species is chiefly dependent on its capacity of dissemination and of the accessibility of the site. Efficacious dispersal mechanisms are of major importance when it comes to establishment in a new habitat as I shall discuss presently (see under 3.1.9). The attainibility of a habitat is associated with the degree of "open-ness" of the environment. A steep wall is inaccessible in several respects, even if only by the unfavourable water supply, the considerable fluctuations in temperature as a result of insolation and heat radiation, the insignificant amounts of substratum and the often extremely high pH. Still, many forms manage to get established for a short time (but only a few maintain themselves much longer) and apparently the attainibility is no great deterrent. Diaspores present at a site usually do not get a proper chance to germinate, but as several types of diaspores, such as fern spores, retain their capacity to germinate for a long time (10 to 20 years according to GAMS, 1938), they may remain dormant only to start germinating as soon as the habitat has become suitable e.g. when the light intensity becomes lower or fluctuations in moisture content are reduced.

The so-called Law of Beyerinck can not be applied to higher forms of life without certain restrictions. The habitat is selective, it is true, but "everything" is not "everywhere". There are too many barriers of both a topographical and an ecological nature. Every species indeed tries to extend its area of distribution, time being an important factor. One had better start from the more generally valid principle: "Everything tends to be everywhere". In this simple rule both accessibility and habitat (environment) selection are taken into account (SEGAL, 1969).

3.1.9 Dissemination

Anthropochory, dispersal through the action of man, is of primary importance as we have seen. Many spermatophytes found on walls are in fact anthropochorous, anemochory being of importance among bryophytes, pteridophytes, composites and grasses. *Cyperaceae* and *Juncaceae* are of rare occurrence on walls because most representatives of these two groups are geophytes and hence their establishment on wall surfaces is very difficult. However, a not negligible number of geophytes are found on walls, though always in small numbers and with a low degree of presence. It is, therefore, noteworthy that the annual species of the *Cyperaceae* and *Juncaceae* also occur only sporadically on walls. These two families are even rare on slanting walls or stone slopes with the exception of *Carex otrubae* which is fairly common on walls near water. Of the grasses, a surprisingly small number of species are encountered on steep walls; the number increasing appreciably as the steepness declines. Relatively common are some species of *Poa*. *Asteraceae* are upon the whole not so discriminating and a large number of species were recorded. The considerable representation of species of the genus *Epilobium* on walls is striking, all European species having been recorded on walls at one time or another. *Zoochory* is often important, especially myrmecochory. The seeds of *Corydalis lutea, Chelidonium majus* and the Mediterranean *Parietaria lusitanica,* and of several species of the genus *Veronica* e.g. of *V. hederifolia* and *V. cymbalaria,* are provided with an elaiosome which appears to exert a great attraction to ants. *Parietaria judaica, Mercurialis annua* and *Glechoma hederacea* contain essential oils in the seed skin, which are presumably the cause of their myrmecochory. The seeds of *Lamium amplexicaule* (which in the Mediterranean area is often found growing on walls) have (like those of *L. purpureum* and *L. album*) a pseudostrophiole: a part of the floral axis attached to each mericarp functions as an elaiosome. I have repeatedly observed the displacement of the seeds of all these species (and of *Cheiranthus cheiri)* by ants. The largish seeds of *Parietaria* can, incidentally, also be dispersed by the action of wind across small distances. *Hydrochory* is only of importance for the dispersal of plants usually found growing at a short distance above the surface of water. The transport is aided by small fluctuations of the water-level and by some wavedash causing splashing of water against the wall e.g. as a result of passing water craft. *Autochory* is found among e.g. *Geraniaceae,* represented on walls fairly regularly by a few species. In many species the autochory may have a secondary dispersing function, for instance in ferns: once a specimen has become established and managed to produce viable spores, the ballistic effect of the annular cells when the sporangium bursts open flings the spores actively about. The common wall-dwelling *Linaria cymbalaria* has a form of autochory brought about by the negatively heliotropic curvature of the fruiting pedicel, which pushes the capsules into cracks and cavities of the wall. The seeds can also be dispersed through the action of wind, however. What dispersal mechanisms can still be classified under autochory is a matter of opinion. Many plants have some system of effective dispersal across small distances. When the capsules of *Viola* species burst open, the inrolling fruit valves actively push the seeds out. The dead and desiccated fruit stalks of *Papaver* and several other plants move to and fro with every gust of wind because they are somewhat flexible, and thus shake the seeds out of the capsules and deposit them sideways at a certain distance from the stalk. This example is not generally accepted as a case of autochory, because no active ballistic mechanism is present, but it will be clear that this is a border-line case and that the distinction between autochory and anemochory has more of an academic than a practical nature. The same holds for other forms of dispersal mechanisms for that matter. A good few species exhibit more than one type of diaspore dispersal: *polychory*.

3.2 Substratum

The question of what must be considered to represent the substratum of mural vegetation cannot so easily be answered. It is not only a matter of the masonry, of brick and mortar, because naked brick or stone supports very little plant growth if any. Before mosses can settle — and this holds a fortiori for higher vascular plants – soil-, dust- or humus particles must have been deposited, or the wall must have been subjected to decomposition. This process is partly brought about by the action of micro-organisms, in the first place bacteria. Moreover, the water relations are of paramount importance for the maintenance of life on a vertical wall.

The initial development of plant growth takes place in small holes or crevices, in cracks or pores, or on rough surfaces where decomposition products or deposits of extraneous matter have accumulated.

3.2.1 Wall and mortar

The hardness of the substratum and the roughness of the surface, the porosity in connection with water retention, and the chemical composition (in connection with resistance to weathering and as a potential source of mineral nutrition) all have a considerable, direct or indirect, bearing on the establishment of vegetation on a wall. The chemical composition is particularly mentioned in connection with the often basic reaction of the mortar — the pH of the currently used Portland cement is usually between 11 and 12. At such high pH values not many forms of life are possible. A soft porous stone (or brick with an irregular surface) is favourable for the establishment of vegetation, as it offers the best possibilities for decomposition and for the retention of dust particles whilst it can also retain water. Rainwater being rich in carbon dioxide, it can, upon penetration, neutralise the lime and cause a lowering of the pH. Both the kinds of building stones (and bricks) and the various mortar mixtures vary considerably in their properties, but an extensive survey would be out of place here. (Curiously enough, relevant literature is scarce.) An enumeration of various types of cement and of their composition and properties was given by Lea & Desch (1935). Only a few remarks will be made here. The walls involved in the present study mostly belong to one of the following types:

a. brick walls, mortar-jointed;
b. walls of lumps of natural stone, usually mortar-jointed;
c. concrete walls.

Red brick is most frequently used. Freshly baked red brick has a fairly strong alkaline reaction. The reddish colour is caused by iron compounds. Yellow brick is usually not very alkaline (when it is made of strongly leached acid clay), but some modern types of yellow brick, however, are decidedly alkaline ($pH > 9$). During processing the iron is replaced by lime (as in the commonly used Dutch IJssel brick). The hardness usually lies between 2 (gypsum) and 6 (feldspar; scale of Mohrs); inferior i.e. excessively soft brick is scratched by gypsum crystals.

The mortar must, if possible, have such a composition that, after hardening, the physical properties approach those of the brick or stone used, so as to avoid as much as possible cracking along the planes of contact. The strength of a wall is much more dependent on the cohesion of the horizontal joints than on that of the vertical ones.

The composition of cement has altered considerably in the course of time. In the Middle Ages and also before that time, insofar no stacked walls smeared with loam were built, one used mixtures of lime with sand, loam and straw in various proportions, dependent on the quantities of these materials available in the neighbourhood. Masonry from that time, as far as it is preserved, was almost always repaired at a later date. By about 1600 one changed to lime mortar in western Europe, which type of mortar is still used but has largely been replaced by Portland cement after the properties of this kind of cement had been discovered about 1870. Originally various mixtures of Portland cement, lime and sand were used until about 40 years ago when the use of lime gradually decreased. Still later other substances were used for mortar such as the cheaper blast furnace cement, which is even more weather-resistant, because it has a lower free-lime content. This cement contains 70 to 85% ground slag from blast furnaces and 30 to 15% Portland cement; it contains less CaO (viz. about 48%) than Portland cement (which contains about 64%) and relatively more MgO (viz. 3% against 1% in ordinary cement). Blast furnace cement can be recognised by the smell of H_2S liberated when hydrochloric acid is placed on it, because it has a fairly high sulphur content. In addition, it contains relatively large amounts of magnesium, vanadium, phosphates and (other) trace elements. Portland cement usually contains 0.13-0.81% Na_2O and 0.31-0.61% K_2O (NICOL, 1947). It has a stronger alkaline reaction than lime mortar and may contain various admixtures such as $CaCl_2$, $MgCO_3$, Na_2SO_4 and other sodium and potassium salts, iron and aluminium compounds, diatom earth and other mineral soil types. Trass is added in certain cases as a mortar-hardener when water repellency is required. It contains MgO, K_2O and Fe_2O_3, among other things.

The composition of lime mortar varies between wide limits. An analysis, carried out in 1957, of the mortar joints of a stone dam at Kamerik near Woerden dating from the middle of the last century yielded the following results (values, except pH, in per cents). These data have kindly been put at my disposal by J. G. Hirs of the Institute for Building Research T.N.O., Rijswijk.

	E.-facing side	W.-facing side
pH	7	7
Moisture content	26.2	24.1
Loss in weight after combustion	14.9	16.2
CaO	12.8	16.4
$Al_2O_3 + Fe_2O_3$ (sesquioxides)	4.3	3.5
SiO_2	5.7	8.2
Insoluble (sand)	61.8	55.7
CO_2	8.1	10.4
SO_3	0.4	0.3
Organic compounds loss in weight after combustion minus CO_2	6.8	5.8
$CaCO_3$ calculated from CO_2	18.4	23.7
CaO „ „ $CaCO_3$	10.3	13.3
CaO bound to other substances than CO_2	2.5	3.2

The interpretation of these data is not easy, because there is no basis of comparison and next to nothing is known of the availability of ions or their possible interchangeability, of the possibility of establishment of an equilibrium between the various constituents of the substratum and the water, etc. The weight ratio lime/sand must have been about 1:2. The extraneous detritus and the action of the plants must have contributed to the complete neutralisation of the lime in the mortar. The eastern face of the dam carried a fairly dense stand of e.g. *Asplenium trichomanes* and *A. adiantum-nigrum,* the western face a sparser stand of *A. adiantum-nigrum.* The plants retain extraneous matter blown in by air currents, they form litter and humus, and improve the water relations (the E.-side has a higher water- and sesquioxide content and contains more organic matter than the W.-side), but on the other hand the plants extract minerals they require for their nutrition from the substratum (a lower calcium- and silicic acid and a higher insoluble fraction in the E.-facing wall). Samples of the humus-containing decomposed substrate from between the roots of plants growing on a wall at Libeek (S. Limburg, Netherlands) were subjected to a soil analysis. The wall in question supported on its nothern surface vegetation with a coverage of 60 % herbs and 15% bryophytes and lichens, including e.g. *Asplenium trichomanes, A. ruta-muraria, Tortula muralis, Ceratodon purpureus, Rhynchostegium murale* and *Lepraria* cf. *aeruginosa* (compare Table 30, no. 10), and on its southern surface vegetation of *Asplenium ruta-muraria* with 60% coverage and of mosses and lichens with 5% coverage including the same species, but with only a sparse representation of *Rhynchostegium*. As a comparison the results of analyses of two samples in the city of Luxemburg are shown. The wall at Libeek was built of brick jointed with lime mortar, the walls in Luxemburg of local sand stone likewise jointed with lime mortar.

	Illite	Kaolin	Quartz
1. Libeek, N. side	80	10	10
2. Libeek, S. side	80	nil	20
3. Same wall, material scratched off from the surface	75	10	15
4. Same wall, fraction passed through meshes $< 2\ m\mu$	75	10	15
5. Luxemburg, S.-facing wall	75	15	10
6. Luxemburg, top of wall	95	trace	5

The figures indicate percentages (error about 5%). The absence of kaolin in sample no. 2 and the small quantity in sample no. 6 might indicate that this mineral is liberated or formed during the decomposition of the stone (or at least of the mortar). Kaolin has the composition $Al_2O_3.2SiO_2.2H_2O$. Illite is a more complicated compound with the provisional composition: $2K_2O.3MO.8R_2O_3.24SiO_2.12H_2O$ (Twenhofel, 1950). The wall from which sample no. 5 was obtained supported a herb layer (coverage 7%) with e.g. *Asplenium ruta-muraria, Linaria cymbalaria,* some *Parietaria judaica, Poa compressa* and *Campanula rotundifolia,* and a moss layer of 15% with e.g. *Schistidium apocarpum, Tortula muralis, T. intermedia* and *Bryum argenteum;* the surface of the stones was covered with lichen vegetation (50% coverage). Such vegetation constitutes the initial phase of a denser stand of *Parietaria judaica.*

The wall from which sample no. 6 was obtained supported a herb layer (coverage 10%) with *Sedum forsteranum, Arenaria serpyllifolia* and *Poa compressa* and a layer of mosses and foliaceous lichens with *Homalothecium sericeum, Tortula muralis, Schistidium apocarpum* and *Grimmia pulvinata.* The chemical composition of walls built of natural stone is usually decided by the available material, often locally quarried stone being used. In Brittany and in Cornwall, for instance, granite is extensively used, in Devon the Devon limestone, in the Ardennes and Eifel schists, in Luxemburg the Liassic 'Luxemburg sand stone', in the French Jura a Jurassic limestone, and in Belgium and Dutch Limburg it is often marl (Maestrichtien, Upper Cretaceous). The composition and the physical properties of these kinds of rock vary a great deal and this is reflected in the developing plant cover, especially of the epilithic bryophytes and lichens. Marl, for instance, is soft, rich in calcium and magnesium and susceptible to weathering. It often supports vegetation of basiphilous mosses, such as *Barbula revoluta, Gyroweisia tenuis* and *Gymnostomum aeruginosum*. These species are also encountered on the not so basic Luxemburg sandstone. On these two kinds of stone, in moist situations and especially near ground-level where water is sucked up by capillary action, characteristic liverworts, such as *Conocephalum conicum* and *Preissia quadrata* are found. Other organisms prefer weathered granite surfaces e.g. certain species of *Grimmia* and *Orthotrichum*. Despite the effect of the basic mortar, *Erigeron mucronatus* seems to be restricted to walls built of granite. Many characteristic wall-dwellers, such as *Corydalis lutea* and *Cheiranthus cheiri* show a preference for walls built of limestone or a calcareous sandstone without being restricted to this kind of substratum. The alkalinity of such walls can roughly be tested by means of phenolphthalein (colour change at pH 8.3); the reaction is often especially distinct on a fresh fracture. Basalt is often employed as campshot of canals and harbour quays, because it is usually a very hard stone with a relatively high resistance against the combined action of wind and water. On this hard rock, in cracks and corners where dust and soil particles accumulate, vegetation of e.g. *Grimmia pulvinata, Orthotrichum anomalum* and *O. diaphanum* is often encountered. Concrete, a mixture of cement, sand and fine gravel, is also alkaline, but although basalt is also basic, it is most probably its hardness that accounts for the same composition of the bryophyte flora as occurs on basalt blocks.

On walls, or partly embedded in them ,various objects occur which may hamper the development of vegetation. In the immediate vicinity of iron bars (cramp irons, braces, etc.) the bryophytes are far less opulent, presumably on account of a poisoning effect. Masonry produces a decaying product rich in calcium and magnesium and the most characteristic wall-dwellers are clearly calcicole to lime-tolerant. Only in well-advanced successional stages, and in special circumstances (such as after the accumulation of blown-up sand poor in mineral substances), can calcifugous species appear. A number of species sometimes supposed to be calcifugous are commonly found on walls. According to McVean & Ratcliffe (1962), for instance, *Dryopteris filix-mas* and *D. dilatata* are calcifugous in Scotland, but this is certainly not the case in a large part of their distributional area. On the other hand, *Cystopteris fragilis, Asplenium adiantum-nigrum* and *Sagina nodosa* are not calcicole or even 'manifestly' calcicole, as some workers seem to think, and one should bear in mind that of such species physiological races or other intraspecific taxa may exist with a preference to either more acid or more alkaline habitats.

3.2.2 Decomposition

Decomposition of a wall is primarily caused by climatic factors: rain (snow, sleet), wind, differences in temperature, frost. In addition, certain biotic factors play a considerable role. The atmospheric precipiation has an eroding effect on account of the presence of small quantities of anions, the most important being the presence of carbonic acid formed by the carbon dioxide dissolved in the water. The soluble lime in the wall is slowly degraded: at room temperature, 1 l of water saturated with carbon dioxide can dissolve 1 g of lime. Such high concentrations of anions do not usually occur, but the time factor is of course also important. The detrimental effect of precipitation containing sulphuric acid on walls has already been mentioned (see under 3.1.7); the more soluble calcium sulphate is derived from carbonate by the acid. Apart from the action of acids, the oxidising action of hydrogen peroxide and of ozone contribute towards the decomposition. The decomposing action is less noticeable on vertical surfaces (to which snow does not adhere and along which rain water quickly runs down) than on horizontal ones.

Solid particles carried along by air currents act as abrasive miniature projectiles and, conversely, the wind carries detached particles away. The combined action of wind and precipation is doubly aggressive: the rain drops are flung against the wall and can penetrate rather far into the masonry if this is sufficiently porous. The effect of fluctuations in temperature on the decomposition is very great. Stone (and brick) has a high heat conductivity, but a small heat capacity. A wall irradiated by the sun, therefore, becomes easily heated up, but it also cools down rapidly when it irradiates heat itself. Masonry being constructed out of different materials with different physical (and chemical) properties viz. the bricks (or blocks of stones) and the mortar, different tensions may be present, which act most strongly in the places of contact between these materials where, accordingly, the first crevices and cracks appear. Frequently such physically caused deterioration occurs between a covering slab and the wall below it. This is why the first development of vegetation often takes place below a covering slab where also the locally provided shelter and the possibility of some water retention are important. Especially when a covering slab has oblique sides, the insolation can be rather excessive if one takes into account the insolation is strongest on a S.-facing surface with an inclination of 52° in the Netherlands.

The effect of the temperature and temperature fluctuations will be discussed under 3.3.2. The weathering reaches its maximum when the walls contain water and frost sets in, because the water in the pores and cracks expands when it turns into ice and the tension forces cleave the stone or brick. Decomposition is furthermore accomplished by the action of plants, animals and man. Every kind of building stone has its individual decomposition pattern: in limestone crevices and cavities are formed, sandstone crumbles, schist cracks along cleavage planes, and granite turns into a fine gravel. The decomposition of lime mortar resembles that of limestone to some extent, but Portland cement mortar and concrete behave more like granite in this respect. During decomposition the volume and the weight of masonry usually increases. The structure becomes less compact and the particles may become bulkier and heavier by the incorporation of, principally, oxygen, carbon dioxide, organic carbon compounds, water and SO_3 (TWENHOFEL, 1950). In limestone, however, the loss of dissolved calcium carbonate may more than compensate for the increase in volume.

The effect of plants is rather complicated, because they can accelerate the decomposition in more than one way. The plants are capable of catching small particles carried along by air currents, they produce litter and humus, they penetrate with their roots or with other attachment organs into the wall *(Hedera helix)* and further chemical decomposition e.g. by producing CO_2. Accumulated extraneous materials favours the rate of growth of the available plants and the chance of settlement of new ones. Humic acids can bind cations of the wall material and the roots of the higher plants crack open crevices and widen them. The forces exerted by e.g. rhizomes of *Cheiranthus cheiri, Hedera helix* and *Taxus baccata,* but also of more herbaceous forms such as *Parietaria judaica* and *Dryopteris filix-mas,* on a wall must not be under-estimated. Moreover, certain forms ultimately build up a rhizosphere, which accelerates the chemical decomposition of the wall. On the other hand, a plant cover may level out extreme fluctuations in temperature and moisture content, and may thus exert a preserving action.

Before a plant cover of some significance can get a foothold on a wall, many things must have happened, however. Weathering and decomposition must have somewhat progressed and there must be enough granular material to enable germination. The first living beings are bacteria and blue-green algae. On this pioneer vegetation other algae and lichens usually develop and only then the conditions become favourable for the establishment of bryophyte vegetation to be followed, much later again, by vegetation of higher forms.

Little is known of the decomposition activities of micro-organisms.

According to Paine (1936), three groups of autotrophic organisms are involved: nitrifying, sulphur-oxidising, and sulphate-reducing bacteria. These groups correspond with the microflora of normal types of soil. Parker (1947) has surveyed the activities of chemo-autotrophic sulphur-oxidising bacteria on walls. *Thiobacillus thioparus* can, among other things, oxidise sulphur compounds to H_2SO_4, even at pH 10. The metabolic products of this bacterium lower the pH of the culture medium from pH 8.5 to 3.3. At lower pH values (under 6.5), *Thiobacillus concretivorus* takes over, especially when the pH falls below 5. This species oxidises H_2S to H_2SO_4. *Thiobacillus concretivorus* attains optimum development at 31° C and its occurrence in W. Europe has not been established.

It is not at all improbable that the decomposition of the walls of Dutch canals is associated with the liberation of H_2S. After bacterial oxidation calcium sulphate is formed, which reacts with the calcium aluminate ($3CaO.2Al_2O_3$) to form the mineral ettringite, which crystallises out with 31 mol. crystal water and naturally increases very much in volume, bursting the mortar asunder (Bing, 1947).

The pH of fresh cement is often as high as 11 or 12. It is known that these extreme conditions are not altogether prohibitive to the activity of bacteria, but a penetrating study of the, presumably highly specific, responsible bacterial forms is badly needed. Düggeli (1930), who investigated the lactic- and butyric acid bacteria in fourteen different kinds of carbonate-containing rocks in Switzerland, is of the opinion that *Streptococcus lactis* and *Bacillus amylobacter* are wide-spread and that they contribute substantially to the decomposition. Under laboratory conditions *Streptococcus lactis* is capable of dissolving 18.2% of the weight of marble (127 g in 0.5 l of liquid medium) in 36 days and theoretically all the marble would be completely disintegrated after 199 days. I do not know how far these acid bacteria play a role in the decomposition of masonry, but presumably they are of minor importance, because the presence of assimilable carbohydrates in the substratum is required.

The influence exerted by animals on decomposition is usually negligible, the burrowing of ants nesting in walls and sometimes of bees and wasps probably being the most important disturbing activities. The inflow of substances through the action of these and other animals such as insects, spiders and birds, their excrements and other remains including the slime tracks of snails and slugs, indirectly contribute to the decomposition. The human influence is varied, but often aimed at counteracting decomposition and deterioration. However, some measures have the opposite effect, for instance when walls are cleaned with hydrochloric acid or hypochlorite. This may cause a strong attack of certain compounds (such as carbonates) and sometimes also promotes the establishment of plant growth by the liberation, or the addition, of mineral nutrients.

Walls originally jointed with lime mortar are later partly restored, renovated or plastered over with Portland cement. Owing to their different properties of the two types of mortar such a treatment is not always efficient, because the new patches do not tightly stick on and may become detached, the newly formed cracks offering opportunities for the germination of diaspores. It is for this reason that sometimes lime mortar is used for the restoration of old, historical buildings. If it is desirable from the point of view of "nature conservancy" that a previously vegetation-covered wall that has become barren after unavoidable cleaning- or restoration activities becomes decomposed again in a relatively short time, a treatment with suitable chemicals is to be recommended. The alkalinity can be lowered by treating the walls a few times with a 1% solution of phosphoric acid (a suggestion made by J. G. Hirs of the Institute for Building Research T.N.O., Rijswijk).

3.2.3 Sediment and humus

The composition of the fine-particles fraction of the substratum depends on:

1. the nature and the amount of extraneous particles from the immediate vicinity (and thus, indirectly, on the adjacent soil types), on the nature of the landscape and on the action of wind;
2. the composition of the masonry and the stage of decomposition; and
3. the development of vegetation and the accumulation of humus.

In the neighbourhood of dunes or other places with moving sand an inflow of sand particles takes place, and elsewhere other kinds of soil particles, if movable, are deposited on the walls. In urban and industrial areas the deposition of dust, soot, tar and coal ash is very important. The relative position of such areas in respect of the walls and the prevailing direction of the wind often decides the rate of inflow. Humus is often introduced from the top of the wall if it is built up against an earth wall or a street, detritus particles being washed down by rain. This is, for instance, of considerable importance in Dutch towns where the walls of canals receive the downwash of the streets above them. The "dunging" must have been even more important in the past when horses and other domestic animals were much more common in the street scene than they are to-day. In such places the inflowing material must be rich in nitrogen judging by the frequent occurrence of nitrophilous species such as *Sagina procumbens, Poa annua, Bryum argenteum, B. caespiticium, Ceratodon purpureus* and *Parietaria judaica.* A number of ash analyses of plants from such sites and of their substrata point in the same direction. A practical difficulty, when samples are to be taken for chemical analyses, is that the composition of the substratum may vary across small distances, particularly in the direction perpendicular to the brick or the mortar and also perpendicular to the root systems of the plants. It is, moreover, not always such an easy matter to indicate the border-line between the wall proper and the accumulated mass of litter and humus, the more so when there is a layer of decomposed wall material between them.

Detritus particles can also be introduced by water.

Chemical analyses were carried out on 53 samples of substrata of mural sites and on 15 samples from between paving stones (or bricks). That local variation may be considerable is demonstrated by the analyses of samples obtained from the same wall, partly around the root system of higher plants and partly from below cushions of mosses, taken at a distance of only 0.5 m apart:

Sample no. 1, from a wall facing the church at Rinteln, W. Germany, with *Tortula muralis, Homalothecium sericeum, Rhynchostegium murale* and other species.

Sample no. 2, from a wall facing the railway track at Fortezza, Italy, with *Asplenium ruta-muraria, A. trichomanes, Cystopteris fragilis, Poa compressa,* and other species. Concentrations given in per cents of the dry material.

	Sample 1		Sample 2	
pH	7.1	7.3	7.7	8.2
Ca	2.40	3.18	2.70	2.73
N	0.27	0.77	0.44	0.07

The differences in nitrogen content are especially very great. The interpretation of such analyses is a rather risky undertaking. The method of taking the samples alone is already a problem by itself, considering that the proportions of decomposition products, solid soil particles and humus vary from place to place. Samples can only be taken in places where a sufficient quantity of loose substratum has accumulated, unless one decides to grind up the hard stone or the mortar. Differences in composition are also brought about by the plant cover. Calcium is hardly ever a limiting factor, but conceivably nitrogen compounds frequently are. A low concentration of assimilable nitrogen *may* indicate that nitrate has been absorbed by the roots of plants.

Table 1 gives a survey of estimations of various factors, all species that (barring a few exceptions) have been recorded at least twice on a sampled wall (and not necessarily in the immediate vicinity of the

Table 1 Estimations of pH and total nitrogen in substrate of mural vegetation and of vegetation of trampled sites

NSa = number of samples; UW = ultimate weights; MW = mean weights

		pH			N-total		
	NSa	UW		MW	UW		MW
<u>VASCULAR PLANTS</u>							
Achillea millefolium	2s	7.5	7.5	7.5	0.07	0.25	0.18
Agrostis gigantea	2s	6.3	7.0	6.5	0.21	0.78	0.50
Agrostis stolonifera	3	6.0	7.0	6.3	0.21	1.08	0.69
	1s	-	-	7.2	-	-	0.38
Arenaria serpyllifolia	4	7.1	8.3	7.3	0.35	0.63	0.48
Asplenium adiantum-nigrum	2	7.0	7.0	7.0	0.16	0.60	0.39
Asplenium ruta-muraria	22	5.7	8.7	7.4	0.16	1.33	0.51
Asplenium trichomanes	14	7.0	8.3	7.3	0.16	1.33	0.52
Asplenium viride	1	-	-	8.4	-	-	0.24
Athyrium filix-femina	8	5.7	7.3	6.4	0.46	1.08	0.70
Betula verrucosa	2s	5.2	7.1	5.6	0.31	0.32	0.32
Bidens tripartitus	2	6.3	7.0	6.5	0.21	0.78	0.50
Bromus madritensis	2	6.7	7.3	6.9	0.55	0.63	0.59
Bromus sterilis	3	6.3	7.0	6.5	0.22	1.47	0.72
Calamintha nepeta	2	7.3	8.3	7.6	0.46	0.63	0.55
Campanula rotundifolia	2	5.7	8.1	6.2	0.51	1.33	0.92
Capsella bursa-pastoris	7s	7.1	7.5	7.3	0.07	0.38	0.23
Catapodium rigidum	2	6.4	8.3	6.8	0.46	1.47	0.97
Centranthus ruber	1	-	-	6.4	-	-	1.47
Ceterach officinarum	3	7.0	8.3	7.4	0.46	0.63	0.52
Chamaenerion angustifolium	3	6.9	7.1	7.0	0.24	0.99	0.62
	2s	5.2	7.4	5.5	0.10	0.31	0.21
Cheiranthus cheiri	1	-	-	6.4	-	-	1.47
Chelidonium majus	2	5.7	7.0	6.0	0.51	0.63	0.57
Chrysanthemum parthenium	2	6.4	7.0	6.6	0.23	1.47	0.85
Corydalis lutea	1	-	-	8.4	-	-	0.44
Cystopteris fragilis	8	5.7	8.2	6.7	0.24	1.33	0.60
Dactylis glomerata	3	6.3	7.3	6.7	0.22	0.63	0.48
	1s	-	-	7.5	-	-	0.28
Dryopteris carthusiana	6	6.0	7.7	6.6	0.28	1.08	0.67
Dryopteris filix-mas	19	5.7	8.2	6.7	0.16	1.08	0.49
Epilobium montanum	7	5.7	8.4	6.5	0.16	0.99	0.53
Epilobium roseum	4	6.0	7.3	6.5	0.07	1.08	0.49
	2s	5.2	7.4	5.5	0.21	0.31	0.26
Eragrostis poaeoides	1s	-	-	7.5	-	-	0.28
Erigeron canadensis	4	5.7	7.4	6.2	0.37	0.88	0.59
	6s	5.2	7.5	6.0	0.07	0.31	0.19
Erigeron mucronatus	1	-	-	6.7	-	-	0.55
Geranium robertianum	3	7.0	7.9	7.3	0.20	0.60	0.39
Geranium rotundifolium	2	7.3	8.3	7.6	0.46	0.63	0.55
Gymnocarpium robertianum	6	5.7	8.1	6.4	0.24	1.33	0.52
Helichrysum italicum	2	6.7	7.0	6.8	0.46	0.55	0.51
Herniagia glabra	1	-	-	7.1	-	-	0.32

		pH			N-total		
	NSa	UW		MW	UW		MW
Hieracium lachenalii	2	7.0	7.2	7.1	0.49	0.63	0.56
Holcus lanatus	3	6.0	7.0	6.4	0.16	1.08	0.61
Hypericum perforatum	2	6.7	7.0	6.8	0.46	0.55	0.51
Linaria cymbalaria	15	6.0	8.1	6.8	0.07	1.47	0.55
Lolium perenne	3s	7.3	7.5	7.4	0.10	0.28	0.20
Lycopus europaeus	6	6.0	7.0	6.5	0.21	1.08	0.49
Matricaria matricarioides	4s	7.1	7.4	7.2	0.21	0.38	0.30
Mycelis muralis	3	6.0	8.0	6.4	0.34	1.08	0.68
Parietaria judaica	9	6.3	8.3	6.8	0.27	1.47	0.53
Phyllitis scolopendrium	9	6.0	7.1	6.5	0.16	1.08	0.66
Plantago coronopus	2s	6.3	6.5	6.4	0.17	0.18	0.18
Plantago lanceolata	2s	7.2	7.2	7.2	0.11	0.38	0.25
Plantago major	5	6.9	7.4	7.1	0.16	0.88	0.39
	9s	5.2	7.4	6.0	0.14	0.32	0.24
Poa angustifolia	2	5.7	7.4	6.1	0.51	0.88	0.70
Poa annua	17	5.7	7.4	6.5	0.16	1.08	0.52
	15s	5.2	7.5	6.3	0.07	1.38	0.21
Poa compressa	5	5.7	8.2	6.4	0.27	0.77	0.51
	2s	5.2	7.3	5.5	0.21	0.31	0.26
Poa pratensis	7	6.0	8.1	6.7	0.21	1.08	0.45
	3s	7.2	7.5	7.3	0.21	0.38	0.29
Polygonum aequale	3s	7.3	7.5	7.4	0.07	0.14	0.10
Polygonum heterophyllum	9s	6.3	7.5	6.9	0.07	0.29	0.17
Polypodium vulgare	2	6.9	6.9	6.9	0.35	0.64	0.50
Polystichum lonchitis	2	8.0	8.1	8.1	0.24	0.34	0.29
Ranunculus sceleratus	2	6.0	7.0	6.3	0.22	1.08	0.65
Rorippa sylvestris	2	6.0	6.9	6.3	0.35	1.08	0.72
Rumex crispus	3	6.3	7.0	6.6	0.21	0.78	0.45
Rumex obtusifolius	2	6.0	6.9	6.3	0.35	1.08	0.72
Sagina procumbens	18	5.7	8.1	6.5	0.16	1.08	0.46
	12s	5.2	7.4	6.2	0.10	0.38	0.22
Scutellaria galericulata	2	6.3	6.9	6.5	0.35	0.78	0.57
Sedum acre	3s	6.5	7.3	6.8	0.18	0.32	0.24
Sedum album	2	7.0	7.1	7.0	0.46	0.47	0.47
Sedum dasyphyllum	3	6.9	7.2	7.0	0.24	0.46	0.35
Senecio viscosus	2	7.3	8.3	7.6	0.46	0.63	0.55
Sonchus oleraceus	10	5.7	7.3	6.4	0.16	0.78	0.39
	3s	7.2	7.5	7.4	0.07	0.28	0.15
Stellaria media	3s	7.2	7.2	7.2	0.11	0.38	0.26
Tussilago farfara	2	6.3	6.6	6.4	0.31	0.78	0.55
Umbilicus rupestris	3	6.3	7.3	6.6	0.22	0.63	0.47
Urtica dioica	2	5.7	6.0	5.8	0.51	1.08	0.80
Veronica hederifolia	2	6.7	7.3	6.9	0.55	0.63	0.59

		pH			N-total		
	NSa	UW		MW	UW		MW
<u>Bryophytes</u>							
Aloina rigida var. ambigua	2	7.3	7.5	7.4	0.17	0.63	0.40
Amblystegium serpens	9	6.0	7.3	6.7	0.07	1.47	0.68
Amblystegium varium	2	6.9	7.5	7.1	0.17	0.28	0.23
Barbula acuta	1	-	-	7.3	-	-	0.63
Barbula unguiculata	5	5.7	7.5	6.3	0.12	0.63	0.40
Barbula vinealis	8	6.4	8.3	6.9	0.35	1.47	0.72
Brachythecium rutabulum	6	6.0	7.3	6.6	0.07	1.47	0.64
Brachythecium velutinum	4	6.4	7.1	6.8	0.64	1.47	0.98
Bryoerythrophyllum recurvirostre	4	5.7	8.1	6.4	0.28	1.33	0.65
Bryum argenteum	19	5.7	8.2	6.7	0.21	1.35	0.59
	10s	6.3	7.5	6.9	0.07	0.38	0.18
Bryum caespiticium	18	5.7	7.7	6.6	0.07	1.47	0.66
	4s	7.2	7.5	7.3	0.21	0.28	0.26
Bryum capillare	16	5.7	8.2	6.5	0.07	1.47	0.60
	3s	5.2	7.5	5.8	0.07	0.31	0.22
Ceratodon purpureus	22	6.0	7.7	6.7	0.07	1.47	0.60
	10s	5.2	7.5	6.2	0.07	0.32	0.24
Conocephalum conicum	2	7.0	8.1	7.3	0.63	1.33	0.98
Didymodon trifarius	2	6.9	7.0	6.9	0.28	0.66	0.44
Distichium capillaceum	2	6.9	8.1	7.2	0.24	0.24	0.24
Encalypta streptocarpa	3	7.4	8.1	7.7	0.24	1.33	0.82
Funaria hygrometrica	5	5.7	8.4	6.4	0.35	0.88	0.51
Homalothecium sericeum	7	6.3	7.3	6.7	0.07	1.47	0.50
Hypnum cupressiforme	6	6.4	7.3	6.9	0.07	1.47	0.55
Leptobryum pyriforme	6	5.7	8.5	6.5	0.28	1.33	0.68
Marchantia polymorpha	2	6.0	6.9	6.3	0.35	1.08	0.72
	1s	-	-	7.1	-	-	0.28
Orthotrichum diaphanum	4	6.0	8.2	7.1	0.27	1.35	0.74
Oxyrrhynchium swartzii	2	7.4	8.1	7.6	0.24	0.88	0.56
Rhynchostegium murale	3	7.1	7.4	7.3	0.07	0.88	0.46
Streblotrichum convolutum	12	5.7	8.4	6.7	0.16	0.88	0.49
	2s	7.1	7.5	7.3	0.28	0.32	0.30
Tortella tortuosa	2	7.1	8.1	7.4	0.24	0.47	0.36
Tortula marginata	4	5.7	8.5	6.4	0.17	0.51	0.33
Tortula muralis	42	5.7	8.7	6.5	0.07	1.47	0.55
Tortula ruralis	2	6.9	8.1	7.2	0.24	0.24	0.24

sampling spot). It was always tried to accumulate loose material from as large as possible a surface area by means of a putty-knife to make up a sample. All data relating to Ca-, Mg- and K-content are omitted because they are too erratic to be of importance. At concentrations of Ca exceeding 0.7% there was no correlation with the pH values, which mostly lie between 7 and 8, but at concentrations below 0.7% Ca a correlation was found. The highest concentration of Ca was 16.20%, of Mg 0.40%, and of K 0.23%, but one should not attribute too great a significance to these values, as it is not at all unlikely that they were rather different at the time when the seeds or spores of the plants found growing on the wall germinated. In suspensions of pulverized stone on which mosses were sometimes growing a very high pH value was recorded viz. up to 9 in the case of *Ceratodon purpureus,* up to 10 for *Tortula ruralis,* and for plantules of *Grimmia pulvinata* on a single occasion even 11. Presumably the pH values are rapidly lowered by the action of plants and by the accumulation of humus. The lowest pH recorded was 5.7 in substratum from a wall and 5.2 in the joints of a paved road. It is possible that the substratum in joints of paved surfaces is of a more constant composition than that of jointed masonry. Generally speaking, the nitrogen content is lower in the substratum of trampled sites than it is in the joints of masonry, conceivably as the result of a higher production of biomass by such forms as *Poa annua, Sagina procumbens* and *Bryum argenteum* in the former habitat. The pH of walls supporting stands of pioneer vegetation with species appearing early in the sucession seres (such as *Asplenium ruta-muraria)* is relatively high as a rule. The pH of substrata supporting stands of *Centranthus ruber, Cheiranthus cheiri, Chrysanthemum parthenium,* and other species that usually appear rather late during succession was not often estimated, but it was usually rather low, which could have some connection with the "late" appearance of these species.

3.2.4 Water

Almost needless to say, the presence of an adequate quantity of water in the substratum is one of the primary conditions for the maintenance of life, and hence for the establishment and maintenance of vegetation. The moisture relations are dependent on the nature of the substratum, the exposure of the wall, the (micro-)climate and the absence or the presence of vegetation. The composition of the substratum, especially the porosity of the building stones or bricks, determines the water-retaining capacities and the rate of evaporation. Fine pores further capillary suction. If the wall stands against water, the capillary suction carries the water upwards and now and again water is splashed up against the wall, but at any rate the relative humidity always tends to be high which is of course of importance for the vegetation cover. Water running down a wall e.g. from leaky of overflowing rain pipes and gutters, can cause local differences, not infrequently vegetation being better developed in the immediate vicinity of such a leaky pipe or drain. Water is, furthermore, sucked up through capillary action from the soil at the foot of the wall and it is not at all surprising that a strip of vegetation is developed along the foot end of a wall, the more so because humus particles are readily blown up against that part of the wall. In western Europe, walls facing to the W. or S.W. receive most of the atmospheric precipitation and this is often substantiated by the frequent occurrence of the green alga *Protococcus viridis* (sensu lato). A south-facing wall is subjected to the most excessive desiccation as a result of direct insolation. A north-facing wall, on the other hand, can retain water much more easily because the temperature does not reach such high values, the effect of wind is usually unimportant and the vegetation cover is frequently better developed. This holds true for an east-facing wall too, but to a lesser extent. Everything else being equal, a horizontal surface retains more water than a vertical wall. For that reason, the plant growth on a wall top is often more lush as compared to vegetation on a vertical wall surface. The relative dampness is also dependent on the temperature, the humidity of the air and the action of the wind, because these factors largely determine the loss of water through evaporation. In towns the evaporation from walls (those facing canal water excepted) is high on account of the relatively high temperature and the concomitant low relative humidity. Notwithstanding the reduced wind velocity, the greater turbulence of the air currents increases the evaporation. In towns precipitation in the form of dew is usually less than in the surrounding countryside.

The effect of vegetation on evaporation from a wall surface is considerable: in the immediate surroundings of a plant the insolation and the evaporation from the wall surface is lowered, the quiet mass of air between the plants acts as an insulator and accumulates water vapour. In the majority of the cases the moisture content of a wall is the limiting factor for the establishment of plant growth and no visible development takes place. Plants can only maintain themselves if they have enough water at their disposal, in other words, if they are sufficiently adapted to an extreme environment. Such an adaptation is often manifest among the small moss plantlets, which are provided with vitreous hairs and thus can surround themselves with an insulating layer of air, such as *Tortula muralis* and other species of *Tortula, Orthotrichum diaphanum, Bryum caespiticium, B. capillare* and species of *Grimmia.* Such vitreous hairs can also catch and accumulate raindrops. *Tortula muralis* has, in addition, an inrolled leaf edge and *Orthotrichum diaphanum* (like e.g. *Homalothecium sericeum*) has folded leaf blades which reduce the evaporation and further the insulating capacity of the leaves. The leaves of *Ceratodon purpureus* and of species of *Orthotrichum* are strongly recurved and this provides, like the inrolling of the edges e.g. in *Bryum capillare,* a protection of the chloroplasts against excessive incident light. The mode of growing in dense tufts of many acrocarpic mosses is another protection against too much transpiration and irradiation (also it provides shelter against the action of wind). The cells of many species are relatively small and the cell walls thick and papillose *(Tortula, Orthotrichum)* or the leaf edges are thickened. Other adaptations are the presence of a rhizoid felt e.g. in some species of *Bryum,* the presence of hyaline cells near the leaf base in which water can be stored (and thus serve to accomplish the hygroscopic movements of the leaf e.g. in *Encalypta, Tortula* and *Tortella*), and the crowded insertion of the leaves (e.g. in *Bryum argenteum).* Many of the species mentioned can loose their water almost entirely without dying and can be revived after imbibition of water. GREBE (1918) has published a survey of adaptations of mosses to periods of drought. In *Tortula muralis,* the vitreous hairs are the longest both absolutely and relatively. When the leaf continues growing the mesophyll encroaches upon the vitreous hair. This is a beautiful example of the adaptive regulatory action of a plant. I shall return to this subject later on (see 9.10.20).

Generally speaking, walls that stay dry for long periods do not support vegetation of any importance, walls that are not subjected to desiccation for very long periods but are periodically very dry carry vegetation of lichens and drought-resistant acrocarpic mosses; however, only if the periods of extreme desiccation are short or non-existent do walls get covered with vegetation of pleurocarpic mosses (or sometimes foliaceous liverworts) and higher plants. A fairly large number of wall-dwelling higher plants have xeromorphic adaptations. Succulent leaves are found in species of *Sedum,* thick and fleshy leaves in e.g. *Umbilicus rupestris* and *Parietaria lusitanica,* which can, in addition, be pubescent in e.g. *Hieracium amplexicaule, Mentha rotundifolia* and species of *Verbascum.* Glandular hairs have e.g. several species of *Geranium, Cheilanthes fragrans* and *Gymnocarpium robertianum*; other secretory organs or secretion ducts occur in species of *Euphorbia* and in *Chelidonium majus,* calcium oxalate crystals in the leaves of *Parietaria judaica,* well-developed sclerenchyma in *Catapodium rigidum* and *Poa compressa,* and also in *Festuca ovina* which, in addition, has inrolled leaf margins like the lepidote fronds of *Ceterach officinarum.* A thick waxy cuticle is found in e.g. species of *Fumaria, Sonchus oleraceus* and *Matthiola incana*; and lignification occurs in many species. The species mentioned are, at least locally, of common occurrence on walls. A considerable number of winter annuals also show such adaptations, which is not at all surprising, because during a part of their life cycle they are subjected to extreme conditions; examples are e.g. *Erophila verna, Saxifraga tridactylites, Senecio vulgaris, Hornungia petraea* and species of *Veronica.*

3.3 Physical factors

The system environment-biotic community is a very complex one in which the various factors are interdependent and do not change independently. This already holds true for abiotic factors alone e.g. in places where organisms are absent. It was unavoidable to touch upon certain aspects before giving a more systematic discussion of the physical factors, and in this chapter more particularly the effect of the climate and the light will be dealt with, the balance of radiation being mentioned in passing. A thorough study of these aspects has not been made, but I feel sure that it is exactly the study of microclimate and radiation balance which is of essential significance for an understanding of the special environment. However, the situation is too complicated to enable a fairly rapid acquisition of satisfactory results. It is hoped that such studies will be carried out in the not too distant future; the results may be useful for the building trade.

3.3.1 Macroclimate

For data concerning the climate in the various areas the reader is referred to Walter & Lieth (1960). Favourable conditions for the development of mural vegetation are encountered in areas where changes in temperature are not excessive, a relatively high winter temperature prevails (with few frost days) and prolonged periods of drought are almost completely non-existent. This is the principal reason for the optimum development of mural vegetation, and of the richest characteristic wall-dwelling flora, being found in the Atlantic and submediterranean regions of Europe. Meusel, Jager & Weinert (1965) give a classification of the 'oceanity' and the 'continentality', each in three classes, which is used as an aid in the diagnosis of the area of each individual species. Such a first approximation, provided it is applied with due caution, may serve as a starting point for the comparative qualification of areas. In the present context, the following classes are important:

Degree of oceanity 1:
S. Iceland, Great Britain (the greater part of it), Ireland, Norway, S.W. Sweden, Denmark, W. Germany, the Low Countries, Luxemburg, W., central and N.E. France, Portugal, N.W. and S.W. Spain, W. Italy, W Yougoslavia and the central alpine region.

Degree of oceanity 2 (continentality 3):
Sweden, Finland (the greater part), E. Germany, the Baltic, Danube and Balkan countries, Italy (the greater part), S.E. France, S.E. Spain, S.E. England and N. Iceland.

Degree of oceanity 3 (continentality 2):
N. and E. Finland, Russia, S.E. Europe, N. Italy, Spain (most of it) and the coastal areas of Asia Minor.

Degree of continentality 1:
Siberia and Asia Minor (except the coastal zone).

In both the continental and the orographically elevated areas specific mural vegetation is presumably entirely lacking, whereas the regions with a degree of oceanity 1 (provided they are not more northernly situated than Holland) offer the most favourable conditions, the most oceanic climatic regions even more so. From a phytogeographical point of view the best conditions within the 'oceanic' regions are provided by the British and probably also the Hibernian subregions *("Bezirksgruppen")* of the Midatlantic subprovince, whereas in the Flemish-Jutlandic subregion the conditions are already less suitable. The conditions in the Subatlantic province and in the Submediterranean subregio of the Macaronesian-Mediterranean region are, furthermore, somewhat less favourable. Towards the more continental countries the areas of distribution of the more specific mural communities taper out into the valleys of the great rivers, where the relative humidity is higher and the conditions better, especially if the sites are favourably situated with respect to the prevailing direction of the wind. One may assume that the degree of 'oceanity' is determined by a combination of periodic changes in the (average) temperature (both annually and daily) and the amount of atmospheric precipitation. When the fluctuations in temperature are small and the annual rain fall is high, the climate is oceanic. When temperature changes are considerable and the precipitation low, the climate is continental. On this basis various authors have proposed an "index of hygrothermy"; that of AMANN (1929) having the advantage of being a fairly simple one. This index (H) is expressed for a given place as follows:

$$H = \frac{P \cdot T}{t_H - t_C}$$

in which relation P represents the average annual rain fall in cm, T the average annual centigrade temperature, t_H and t_C the average centigrade temperature of the warmest and coldest month (July and January), respectively. According to PROCTOR (1960), the values of this index correspond satisfactorily with the distribution pattern of many mosses in Britain: *"Many common Atlantic species begin to appear in quantity where H exceeds 60-70"*. For the development of mural vegetation this index must also be fairly high and at least exceed 40. The index alone does not give a sufficient indication of the probability of the development of mural vegetation, however, because the microclimate is so important in this connection. REY (1956) has pointed out that in France the influence of the temperature and the precipitation can per site be divided into an Atlantic, a Mediterranean, a continental, and a montane influence, for each of which the following combinations of environmental conditions are supposed to be characteristic: warm and moist, warm and dry, cold and dry and cold and damp, respectively. The four effects, placed in a tetrangular form in a system of co-ordinates, are said to enable a characterisation of every locality. This graphic representation is far too simplified, however. The special effect of excessive summer temperatures in continental areas, for instance, is not properly accounted for. Nevertheless, as a first approximation, this idea can be employed in combination with the proposals made by Meusel et al. It appears, namely, that a classification of mural vegetation can be made on the basis of phytogeographical criteria, which is quite satisfactorily compatible with the distribution of the various vegetation types in different climatic regions. This classification is shown in Fig. 1 (Appendix I). In this connection the following areas can be classified according to a decreasing degree of oceanity (only as far as they occur in the regions under investigation; the ciphers refer to the symbols in Fig. 1). Atlantic and subatlantic areas with a degree of oceanity correspond with class 1 of Meusel et al.:

110: S.W. Britain, Normandy and Brittany — For phytogeographical reasons the English and the French parts of this eu-atlantic region are divided into 111 and 112 (North and South). 111 also includes the Edinburgh area (Scotland).

120: The N.W. European lowland plain: N.W. France, Flanders, western Holland.

130: The most continental parts of the region with oceanity class 1 according to Meusel et al.
In 131, the northern portion, the continental influence prevails. 132 shows a relatively large Mediterranean influence and includes the southern portion of the area with W. and central France.

400: Parts of Germany, Poland and Czechoslovakia included by Meusel et al. in oceanity class 2 (continentality class 3).

200: The alpine zone, with oceanity class 1 of Meusel et al.

300: Predominantly submediterranean areas.

310: Submediterranean areas with a relatively strong montane or Atlantic influence: large parts of S.W. France and N. Italy.

320-330: Regions with a relatively strong Mediterranean character, of which 332 includes mainly the departments Hérault and Gard in France, 331 the French department of Var, and 320 Tuscany in Italy.

In this survey the effect of precipitation is emphasised, but it is not only the total annual rainfall but also its distribution over the seasons of the year that is important. In Mediterranean France, for instance, the annual rainfall is about the same as in the region 130, but the precipitation falls mainly in the autumn and the winter and is much lower in the late spring and summer.

3.3.2 Microclimate

The balance of radiation energy of a wall is determined by the exposure, the inclination, the degree of overshadowing (if any), the reflected incident radiation and the heat radiated by the wall. In addition, the temperature of the wall is dependent on the physical characteristics of the wall, the effect of wind (direction and velocity), the air temperature near the wall, and the rate of evaporation (through the latter, indirectly, the relative humidity). On this complex of abiotic factors the degree of development of vegetation has a very marked effect. The temperature near the surface of a wall is mainly determined by the temperature of the air and of the substratum and by the strength of the locally prevailing air currents. According to KETTENACKER (1930), the influence of the temperature and the moisture content of the wall is restricted to a layer of air near the wall of about 8 cm in thickness in calm air. A number of records of temperatures near walls, made by the present author in the Netherlands and in southern France, confirm a rapidly declining influence of the wall, at a distance of 10 cm on average from the wall; however, that distance proved to be rather variable and to be dependent not only on the intensity of the reflected and re-radiated energy and on the physical characteristics of the wall, but also on the effect of the vegetation cover which sometimes makes the radiated energy felt farther away from the wall. So far the number of observations have been too small to permit more than preliminary conclusions. The relative humidity in close proximity of the wall depends on the general air humidity (which is largely determined by the structure of the landscape) and of the rate of evaporation, the latter in turn being decided by the physical properties and the water content of the wall, and of course by the effect of wind and turbulent air currents. The low heat retention capacity and the high heat conductivity of the substratum are the cause of considerable daily fluctuations in the temperature of, especially, south-facing walls, which as a rule are much greater than the daily. fluctuations of the air. A wall becomes rapidly warmed up by irradiation, but does not retain the heat very well. As soon as the insolation decreases or ceases altogether, the temperature at the wall surface reacts almost immediately. During the day a wall is fairly intensively heated by radiation and the heat can penetrate rather deeply into the masonry, but during the night the exchange of heat progresses rapidly and the temperature may fall rather sharply owing to radiation of heat from the wall. Unattached walls are subjected to the greatest fluctuations in temperature.

Dark rocks heat up sooner than light coloured types, and that is why the differences in colour between stone or brick and mortar promote the physical decomposition. Of the incident solar radiation a portion is reflected and another part absorbed and transferred into heat. According to PLATT & GRIFFITHS (1964), only 55% of the sunlight between the wavelengths of 0.3 and 2.5 μ is absorbed by red brick, as against 92% of the long-wave emittance exceeding 2.5 μ. The heat produced within the masonry is partly used to heat up the wall and the adjacent air, partly to evaporate water, partly re-radiated (as heat radiation), and small quantities are transferred into other forms of energy. The increase of the amount of caloric energy in the plant cover is not only proportional to the intensity of the irradiation, but to a large extent dependent on the absorption capacity of the stand. Only long-wave emission reaches the substratum below the leaves. On a S.-facing wall not only the temperatures are extreme, but also the loss of water (both of parts of the plants and of the substratum) is usually high because of evaporation. Species that can live in such extreme circumstances often possess morphological adaptations to prevent an excessive loss of water (see under 3.2.4). A north-facing wall is more suitable for supporting vegetation in several respects, but mainly in the first place because the temperatures are less extreme and the differences in temperature not so large as to induce great loss of water through evaporation. On the other hand, decomposition does not proceed so fast on N.-facing walls as it does on S.-facing ones (compare 3.1.6). The strong emission of radiation by walls may produce low temperatures, so that the effect of night frosts is often sooner evident on walls than in other (neighbouring) sites. Night frosts simulate a dry environment (physiological drought) and thus produce drought damage. The aerial parts of such species as *Linaria cymbalaria* and *Parietaria judaica* soon die off after the first night frosts of the autumn.

Walls are often situated in urban areas, and the larger and more densely built-up a town is, the more the local climate resembles a desert climate. All buildings, paved roads, squares, etc. form large bodies of brick, stone and masonry comparable to bare rocks which have replaced the soil, causing a different heat- and water balance. The humidity is usually lower than in adjacent rural zones because most of the atmospheric precipitation is rapidly drained off and but little is retained. The resulting relative decrease in evaporation also reduces the loss of heat. The rough surface of stones or bricks promotes the absorption of solar radiation and, especially in winter, there is an extra supply of heat in the form of coal fires and industrial furnaces. The latter also produce soot, tar and noxious compounds which get mixed with exhaust gases and the dust blown about by passing traffic. According to GEIGER (1961) the temperature in a town is upon the average 1° C higher than in its suburban areas, but during the cold season the difference can be much larger. RYDZAK (1958) mentions the following differences in the annual mean temperature: in London 1°.4 C, in Berlin 1°.0 C, in Moscow 0°.7 C, in Karlsruhe 7° C, and in Grunow 10° C. The differences in temperature are partly caused by the differences in topography and they can, therefore, vary considerably over short distances. The heat produced and accumulated by a town can be dispersed by reflection between houses and streets and by turbulent air currents. The higher mean temperatures in the towns cause the vegetation period to last longer in autumn than in comparable rural areas, but for the rest the microclimate in cracks and slits of walls is always milder than on the surface of the masonry.

The effect of a low relative humidity upon vegetation, more particularly upon lichens was discussed in great detail by several workers e.g. BARKMAN (1958) and RYDZAK (1958). The lichen-free areas ('lichen deserts') can be caused either by drought or by atmospheric pollution (or by both), but MROSE (1951) and RYDZAK (1958) believe that the effect of pollution is decidedly overrated and the effect of a low relative humidity is more critical in this connection. However, their arguments are not convincing. Rydzak reported, for instance, that in the suburban zone of Manchester, where the mean SO_2-content of the air is 2 mg per m^3, lichens occur, whereas the centre of Saarbrücken, with only 1.5 mg SO_2 per m^3, is lichen-free. He overlooked, however, that the average values may well be less important than the extreme values and do not give any information of the duration of the incidence of such higher degrees of aerial pollution. Conceivably it is the extreme values that are decisive, and the prevailing direction of the wind is likewise important. The microclimate of walls outside towns is somewhat comparable to the 'urban' climate, but the changes in temperature are relatively great, the relative humidity is low, and the effect of frost appreciable.

The influence of wind and thermic air currents on the microclimate is manifold: the displacement of air lowering the temperature but also the relative humidity; preventing the condensation of dew, and furthering the transpiration of plants and the evaporation of the substratum i.e. the desiccation. Thermic air currents may be of importance in connection with the transportation of diaspores e.g. along the surface of canal walls where during insolation considerable difference in temperatures over small vertical distances may occur. This was studied by attaching pieces of cellophane thinly smeared with vaseline on to a wall and by examining them after 24 hours under a microscope. The particles caught by the vaseline consisted chiefly of pollen grains, spores, and minute seeds. A few experiments with species of yarn made sticky with glue (which can catch larger seeds) yielded similar results. Convection currents could be measured by means of a heated-wire anemometer, but the interpretation of the experimental data requires further study and in any event a good deal more readings. During calm weather along the surface of canal walls displacements of air of several centimetres per second were always demonstrable. Near corners of walls, the effect of wind turbulent air currents is of course very much stronger and, as far as could be ascertained, vegetation is always not so well developed here as on more protected parts of the same wall (or may even be completely lacking).

Some species are found in places where the wind blows in sharp soil particles. In such situations it is especially *Streblotrichum convolutum,* and to a lesser extent *Tortella tortuosa,* which can survive. Not infrequently their plantlets are strongly eroded. *Sedum acre* is another taxon appearing on a shifting substratum, but it is usually found in vegetation of paved surfaces and not so often on walls.

For the time being I only wish to point out that one may anticipate that the microclimate in certain sites where a species occurs, either near the extreme boundary of its distributional area or in an area where its localities are far apart, resembles the macroclimate in the main area of distribution of that species. This simple principle does not only improve the insight into the ecology of a site, but also give some idea of the factors which determine the environment, or are the limiting factors, elsewhere. In continental regions, 'atlantic' species and vegetation types are, for instance, almost exclusively found in sites where the relative humidity is particularly high (e.g. near large bodies of water, in streambed regions, etc.), and conversely, "continental" species or vegetation types occur in Atlantic regions in places where the relative humidity tends to be low for long periods.

Some species reputed to be calcifugal are, presumably, rather hygrophilous (e.g. *Polypodium vulgare,* compare MEINDERS-GROENEVELD & SEGAL, 1967, and *Pteridium aquilinum*), and other forms said to be calciphiles may rather be thermophilous (e.g. *Cynanchum vincetoxicum* which occurs in western Europe on limestone slopes but grows on sandy soils poor in calcium in eastern Europe: compare e.g. HEGI).

3.3.3 Light

For mural vegetation light is but rarely a limiting factor, excepted at the base of walls if these are strongly overshadowed. For well shafts made of masonry, which are very suitable for the development of vegetation as far as the high relative humidity and the constancy of the temperature are concerned, the lower light intensities may become the limiting factor. Generally speaking, and everything else being equal, a decreasing light intensity leads to the disappearance (or absence) of forms in the following order: Spermatophytes, Pteridophytes and Bryophytes. If the light intensities are very low, only a few Algae, Fungi and Lichens (of the latter usually the poorly developed 'imperfect' forms, often indicated as *"Lepraria aeruginosa"*), and of course Bacteria, can persist. This sequence is also evident in various publications dealing with plant growth in deep holes and caves (see e.g. MORTON, 1955). A light gradient, such as that prevailing in wells, offers a good opportunity to study the optimal or suboptimal conditions for the development of certain species. The number of light measurements, carried out by means of a flat selenium cell with about the same spectral sensitivity as the human eye (greatest sensitivity between 500 and 600 $m\mu$), is still too small to justify more than preliminary conclusions (see Table 2).

Table 2 Light measurements in brick wells

species	number of readings	mean maximum of decrease in light intensity in per cent	minimum light intensity (measured in Lux)
Adiantum capillus-veneris	2	64	208
Asplenium ruta-muraria	17	29	1050
Asplenium trichomanes	21	70	445
Athyrium filix-femina	15	88	86
Chelidonium majus	3	21	1780
Cystopteris fragilis	9	81	89
Dryopteris filix-mas	24	79	120
Epilobium montanum	11	32	1520
Geranium robertianum	11	71	120
Gymnocarpium robertianum	4	38	560
Hedera helix	7	32	1510
Linaria cymbalaria	16	29	980
Parietaria judaica	8	61	310
Poa compressa	5	19	3060
Phyllitis scolopendrium	12	88	51
Rumex scutatus	3	16	5100
Sedum acre	6	20	2850
Amblystegium serpens	14	94	52
Streblotrichum convolutum	4	24	1980
Conocephalum conicum	3	35	1220
Homalothecium sericeum	8	38	470
Tortula muralis	17	18	1640

All measurements were carried out during periods of fairly clear or diffuse light between 11 and 15 hrs, the majority in France and in Holland. The minimum intensities recorded correspond with the situations in which vegetative development was barely noticeable and they all refer to plants with a much reduced vitality. Usually a marked fall in vitality e.g. in the height of the specimens, can be observed when the light intensity drops below a certain value. On the other hand, an optimum development does not necessarily coincide with a maximum of illumination; *Phyllitis scolopendrium,* for instance, seems to grow well when the light intensity is reduced by 40% and thus reaches the same intensities as those recorded by DEPASSE (1957) in localities of this species in forests, and *Amblystegium serpens* does not seem to suffer from a reduction of the intensity by 30 to 50% (the highest value pertaining to its var. *saxicola*). The latter species is, in addition, not very resistent to desiccation.

Most of the wall-dwelling species have diaspores that germinate in the light. Although, upon the hole, fern prothalli prefer dimmed light (GAMS, 1938), those of wall-dwelling forms, such as *Asplenium* spec. div. and *Cheilanthes fragrans,* are not so sensitive to light.

Species only seldom found on much overshadowed walls include e.g. species of *Sedum* (*S. dasyphyllum,* which prefers shady habitats, excepted), *Cheiranthus cheiri, Erigeron mucronatus, Centranthus ruber, Antirrhinum majus, Capparis spinosa, Grimmia pulvinata* and *Tortula intermedia* (and other species of *Tortula* subgenus *Syntrichia*). *Linaria cymbalaria, Tortula muralis, Bryum caespiticium, B. capillare* and *Ceratodon purpureus* occur both in exposed and in shady sites.

3.4 Biotic factors

Not only does the environment determine the nature of the plant growth its supports, but the presence of vegetation also involves changes in the habitat. In addition, various animal organisms gradually become established in this changing microcosmos. The effect of human intervention has repeatedly been mentioned already.

3.4.1 Phytogenic factors

The effect of vegetation on the environmental conditions is quite considerable: the presence of a plant cover diminishes the intensity of the insolation of the substratum, levels out differences in temperature and promotes the prevalence of a more constant and high relative humidity. Vegetation catches and retains external matter including many diaspores, forms litter and humus and by its decomposing activities furthers the settlement of new individuals or even of previously unrepresented species. These factors will be discussed in more detail under the heading of development of vegetation (see 7).

3.4.2 Zoogenic factors

The presence of ants is of considerable importance for the development of mural vegetation, many wall-dwelling species being more or less clearly myrmecochorous: *Linaria cymbalaria, Corydalis lutea, Cheiranthus cheiri, Parietaria judaica, P. lusitanica, Antirrhinum majus, Geranium robertianum, Bromus tectorum, B. madritensis, Campanula rotundifolia* and *Mercurialis annua* (see also 3.1.9). *Lasius flavus* and *L. niger* are presumably the most important vectors, partly because they often choose a wall to build their nests in and fill in cracks with humus and other materials. Ants are very commonly found on walls, especially in the Mediterranean region. Birds are usually only of importance as depositors of viable seeds, but quite commonly they enrich the environment with nitrogen compounds to some extent by their excrements, especially in towns where flights of pigeons, sea gulls and starlings roam about. Predominantly ornithochorous are e.g. *Taxus baccata, Rubus* spec., *Fragaria vesca, Sambucus racemosa* and *Solanum nigrum* and sometimes also, as garden escapes, species of *Cotoneaster* and *Berberis* and many 'casuals' such as *Symphoricarpus albus* and *Ilex aquifolia* in Amsterdam. These forms, and especially the larger ones, usually only appear after a fairly large quantity of litter, humus and introduced extraneous matter has accumulated in cracks and voids of the wall or when the vegetation cover has developed to a certain successional phase, but they show, generally speaking, a marked preference for more slanting parts of a wall (where sometimes also species of *Crataegus, Sorbus aucuparia* and *Rosa canina* are encountered).

The importance of birds as producers of dung is usually not so great on vertical walls, but on canal walls the washed-down excrements of gulls, pigeons, sparrows and starlings may contribute largely to the nitrogen dunging. Nitrophilous forms, such as *Parietaria judaica* and *Umbilicus rupestris* are sometimes found in rock crevices where some dunging by birds takes place, but more often human influence is demonstrable when these species occur (especially the first).

More particularly on limestone walls in protected sites, with a well-developed stand of vegetation, molluscs are not infrequently present, in spite of the predominantly low humidity. Representatives of the families *Clausiliidae* and *Vertiginidae* are common in the first place, some of which are more frequently found on walls than in other biotopes. *Clausilia biplicata* is one of the most common snails and *Lauria cylindracea* is frequently encountered among stands of *Parietaria judaica* or *Cheiranthus cheiri* on walls in the Mediterranean area, in western France and in England. In Holland, where only two localities were known in the southwestern part of the country, I discovered this species on a wall with vegetation of *Parietaria* at Buren (central Netherlands). On walls covered with ivy, *Vertigo pusilla* and *Ena obscura* were repeatedly found. The more commonly encountered molluscs include *Clausilia bidentata* and *C. parvula, Pyramidula rupestris, Helicigona lapicida, Cepaea nemoralis* and *C. hortensis, Balea perversa* (especially on walls with *Hedera helix*), *Pupilla muscorum, Vallonia costata, Discus rotundatus, Papillifera solida, Retinella nitidula* and *R. pura.* Other species have been recorded by BOYCOTT (1934) from walls in England, by GASCHOTT (1925) from walls in southern Germany, by JAECKEL (1943) from walls in Belgium, Luxemburg and France, by KUIPER (1953) from walls in the French Pyrenees, and by KRAUSP (1962) from walls in Thuringia (Germany). According to Gaschott (see also Kuiper) ruins can serve as *"Brennpunkte des Molluskenlebens"*. In areas brought under cultivation ruins often constitute refugia. Kuiper mentions several species for which ruins are the type localities. Holes and cracks in walls are often inhabited by various insects other than ants, such as caterpillars of whites, turtoise shell and peacock butterflies, hibernating queens of wasps and bumble bees, ground beetles and ladybirds (SCHMITT, 1950). In the cracks spiders and harvest men may dwell permanently and in moist protected places also woodlice. These are also the favourite haunts of the centipede *Lithobius forficatus,* which lives on worms, woodlice and insect larvae. On drier south-facing walls other animals are sometimes found, such as the Argus-butterfly *Pararge megaera* and, particularly in S. Europe, the wall lizard *Lacerta muralis* (and in more northern regions the common lizard *Lacerta agilis*). The solitary bee *Chalcicodoma muraria* prefers south-facing walls where it, like the wasp *Eumenes coarctatus,* builds clay nests. Other wall-dwelling wasps are e.g. *Pseudagenia carbonaria* and *Polistes gallicus* (SCHMITT). In walls jointed with loam a considerable number of bees and wasps can make their holes, such as digger wasps *(Sphegidae)* and *Andrena agilissima,* and various other insects (RODE, 1962).

3.4.3 Anthropogenic factors

Walls are always built by man and hence human influence is a primary one. In several other respect human agency, directly or indirectly, remains strong, so that it involves almost every ecological factor as is evident from several paragraphs of this chapter. For the development of mural vegetation as a landscape element human interference can have a positive as well as a negative side. Positive aspects, apart from the building of the walls, are the dispersal of diaspores of autochtonous species and of neophytes, the translocation of detritus, humus and soil and building activities, roadmaking etc. in the immediate surroundings (important e.g. for walls of canals). Negative ones are restorations, renovations and other disturbances of the habitat, such as atmospheric pollution and the human influence on the climate. Deposits of soot and tar on walls may be prohibitive to the settlement of e.g. lichens, or, together with the adhering dust particles, hamper the transpiration of the plants. The establishment of cultivated plants and ornamentals is often the result of human activities, some species being introduced with garbage and discarded articles of food, in Amsterdam e.g. many species including tomatoes *(Lycopersicon esculentum)*, potatoes *(Solanum tuberosum)*, strawberries *(Fragaria* spec.*)* and even the orange *(Citrus aurantium)*. Several anthropochorous plants with spiny seeds or with seeds covered with hooked protuberances, such as *Galium aparine, Bromus sterilis, Hordeum murinum,* species of *Arctium* and *Circaea lutetiana,* can be directly disseminated by man.

The increasing tendency for restoration and renovation in Holland and elsewhere is a death-blow to mural vegetation. The use of modern mortar mixtures (also in renovations of old masonry) does not promote decomposition, and hence the development or re-settlement of vegetation is retarded. Periodic wall cleaning is often detrimental to mural vegetation. The situation in the Netherlands has almost advanced to the total eradication of all good stands of mural vegetation, much having been destroyed in the last 10 to 15 years. This stands to reason, because it is always the most decayed, and thus most dangerous, walls usually supporting the best developed stands. In Amsterdam, for instance, 40 localities of *Phyllitis scolopendium,* 8 of *Asplenium trichomanes* and of species of *Gymnocarpium* have disappeared since 1954. In two places in Holland attempts have been made to execute restorations in such a way that the stands of mural vegetation could be preserved. Nevertheless one must anticipate the total disappearance of these picturesque landscape elements as human "civilisation" progresses and thus levels out the interesting changes it had originally generated.

3.5 Mural vegetation as an ecosystem

Vegetation is interrelated with the environment in several ways and thus the two constitute an ecosystem. A much simplified diagram of such a complex of relations is shown in Fig. 3. In this diagram neither the climate nor the habitat has been subdivided into a number of partly interrelated components, and the interactions between the plants and the microflora of the substratum are also disregarded, but even so the diagrammatic representation appears to be rather intricate. To multivariate interrelations anything but simplify the investigation of such a system and do not render the interpretation of the results an easy matter. Vegetation has also been characterised as a system of "interacting niche-differentiating species".
It is evident from a study of the preceding paragraphs that the wall habitat is an extreme environment in many respects such as the available room for settlement, the hardness and alkalinity of the substratum, the scarcity of soil and humus, the inclination, the temperature and the humid-

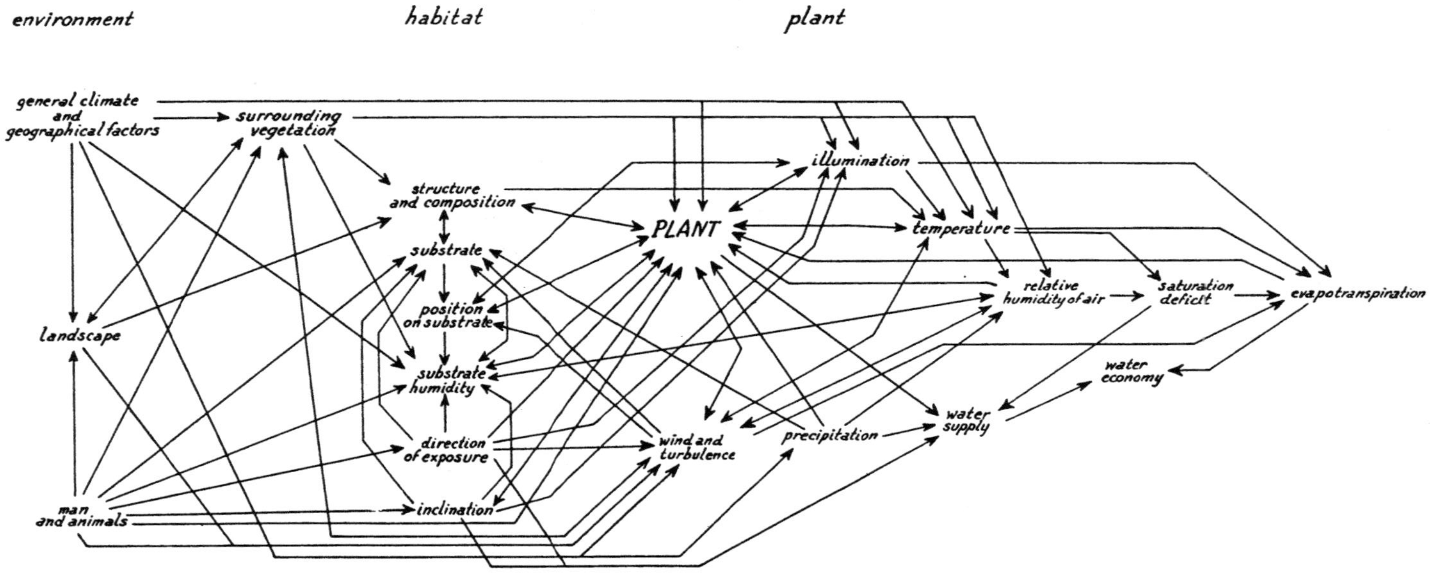

Fig. 3 Concise scheme of the ecosystem of mural communities.

ity, and generally the effect of climate and wind. This renders the habitat suitable for only a limited number of taxa adapted to the extreme conditions. Apart from a number of mosses and higher plants adapted to a life in dry sites (see under 3.2.4), these include the highly specialised chasmophytes. Only these forms can persist for long periods, provided there are no appreciable changes in the ecological environment (i.e. there is no increased stabilisation of the environmental conditions). Other species only appear as dwarfed specimens, such as *Dryopteris filix-mas* and a number of other non-chasmophytic ferns. An extreme environment is, generally speaking, synonymous with a rather unstable und structurally not very complicated habitat, that supports only a few species and shows a marked dominance or co-dominance of one species or of two (SEGAL, 1969). It is an open system with a not very efficient utilisation of the inflowing energy. There is only an incomplete cycle of organic and inorganic compounds, because a considerable part of the dead leaves and other litter is not retained in the system and thus the substratum does not receive any benefit from it.

4 Structure

4.1 The concept of structure

In so far as its relations to plant ecology are concerned, the concept of structure is being related to both distributional (i.e. spatial) and physiognomic aspects of the species constituting a community. The structural elements involved are life- and growth forms, stratification, horizontal distribution and dimensions. Occassionally the term 'structure' is employed in the sense of: 'the combination of species', but in this meaning the term is superfluous and even misleading. If the time factor is being taken into account, one must anticipate certain variations and fluctuations (which may show periodicity, as is the case with seasonal phenomena), or changes showing a distinct trend in a certain direction, which changes are often irreversible as in succession seres.

According to BRAUN-BLANQUET (1964), life forms and growth forms are almost identical notions. He remarks in this connection (p. 141):
"Während aber die Lebensformen die Individuen zu leichtfasslichen Typen vereinigen, welchen sich die gesammte Pflanzenwelt einordnen lässt, können die Wuchsformen nach den verschiedensten Merkmalen unterschieden und gruppiert werden."

His line of thought suggests that the distinction between life forms and growth forms is anything but clear-cut. However, a simple and convenient approach is to define a growth form semantically as a morphological type, and a life form as an ecological type in accordance with its adaptations to its environment. These definitions can be extended to include pollination and dissemination types (ecological flower types and diaspore types) if one does not prefer to restrict the criteria to the vegetative parts of the plants, which chiefly decide their habit form as a rule.

Horizontal distributions are often manifest in, among other things, social patterns such as sociability, aggregation, the formation of polycorms, etc. Patterns in a horizontal plane can be mathematically expressed as deviations from a random distribution of the individuals of a species (GREIG-SMITH, 1964).

The shape and the nature of the limits (and demarcation lines) can also be important, and increasingly so as the vegetation fragment, or the community, is relatively smaller in relation to its structural elements, or more clearly confined within certain limits, or if its surroundings conspicuously differ from that fragment (or community). In gradient situations the concept of a border-line assumes a different meaning.

The above-mentioned structural features are more or less clearly represented in the *minimum area,* which can, accordingly, be considered to be a secondary structural characteristic.

4.2 Stratification

Vertical structural patterns are usually not very well developed in wall-dwelling vegetation. As a rule only one or two layers of vegetation can be distinguished. If two layers are present, the layers are often clearly separated e.g. in a moss layer and a herb layer. If a vertical stratification can be recognised at all in the structural elements of the herb layer, there is mostly a carpet layer, containing such species as *Sagina procumbens* and *Soleirolia soleirolii,* shrubs and trees being of rare occurrence on vertical wall surfaces. A species such as *Cheiranthus cheiri,* which sometimes produces strongly lignified stems, prefers the more slanting parts of a wall or grows on top of walls. Trees presumably represent — at least in a number of cases — elements of a succession of which the typical mural vegetation types form the initial phases and which proceeds more rapidly when decomposition has progressed rather far.

A vertical stratification hardly ever occurs in the vegetation cover of lichens and bryophytes, larger and taller forms, if they develop, usually dominating rapidly over the smaller ones. This can also happen when seed plants are dominant: a high stand density of e.g. *Parietaria judaica* suppresses the bryophyte layer more or less completely.

4.3 Horizontal patterns

The individual specimens of any given species but rarely show a random spatial distribution. Within vegetation types a certain degree of overdispersion is the rule, which is partly expressed in the sociability of the species. The sociability is in many cases a specific characteristic of the taxon. Overdispersion is often promoted by vegetative reproduction e.g. in the form of rhizomes or stolons. Therophytes mostly grow solitary, even if the individual plants may be distributed in the form of recognisable patterns. In other words, sociability can be assured at more than one level. Tussocks can also be arranged in distinct patterns (BARKMAN, DOING & SEGAL, 1964). On the surface of a wall, a spurious overdispersion is frequently observed, which is caused by the regular pattern of mortar joints and bricks (or other rectangular building stones), the joints often providing the only germination sites and the germination, accordingly, taking place in a reticulate pattern. The bricks, or the stones, constitute an entirely different environment, and the vegetation cover they may carry (usually a community of crustaceous lichens) need not, and in most cases must not, phytosociologically be combined with the vegetation fragments of the joints between the bricks. In these joints the environmental conditions differ, sometimes substantially so, from the brick (or stone) surfaces of the wall, among other things, in terms of moisture relations and irradiation. In mural vegetation the proportion of rhizomatous geophytes, which, generally speaking, have a marked effect on the development of horizontal patterns, is usually small to negligible. More often hemicryptophytes are encountered, particularly those forms whose vegetative shoots or corms are aggregated into smaller or larger tufts or (especially mosses) form cushions.

The phanerophyte *Hedera helix* can cover a wall almost completely, but almost invariably roots in the soil. *Hedera* can attach itself to the wall by means of its special adhesive roots. The distribution of the individual specimens, and of the various species, is naturally to a large extent dependent on the local differences in microclimate and in other environmental conditions. Small areas where the decomposition is relatively more advanced or the relative humidity fairly high, are favourable sites for the germination and subsequent development of many species. However, a floristic homogeneity in communities poor in species is not necessarily indicative of an ecological homogeneity. In such cases there is often some master-factor, which contributes more towards the moulding of the ultimate structure than the variation of the other environmental conditions, provided the latter are not limiting factors, or else it is precisely the presence of such a limiting factor that thus becomes the master-factor. On walls the prevailing drought is often the limiting factor.

4.3.1 Sociability

Barkman, Doing & Segal (1964) proposed an objective scale, based on the number of vegetative shoots or corms per group, to express the sociability in 5 or in 10 steps. They pointed out that the degree of sociability is in many cases a characteristic of every botanical species. This relation renders an overall evaluation possible in a number of cases by simply assessing the sociability value of every species as the degree of sociability most frequently encountered in mural vegetation. This can of course not be done in the case of species in which this character varies appreciably; however, the number of such species is limited. For technical reasons, viz. for data processing, the sociability has been assessed as belonging to one of five classes as follows:

Symbol:	Number of shoots, or corm units per group:
1	1—3
2	4—30
3	31—3×10^2
4	3×10^2—3×10^3
5	>3×10^3

The degree of sociability is often quite high in mosses and, the individual corm units being small, this has a direct bearing on the abundance (see under 5.3). The sociability of acrocarpic mosses is frequently 3, that of ferns usually about 2 (assessed on the number of fronds), that of wall-dwelling herbs only in exceptional cases 4 or 5 (and in the latter cases it is usually a 'composite sociability' e.g. when a wall is more or less completely covered by *Parietaria judaica* vegetation but the individual plants grow in tufts). The sociability of therophytes is 1 in most cases and occasionally 2, that of geophytes usually 1, of hemicryptophytes 1-3, of chamaephytes 1-2, and of phanerophytes normally 1.

In the discussion of the various growth forms I shall return to this point (see under 4.6).

4.3.2 Boundaries and limits

Up to now little attention has been paid to the course and the place of demarcation lines or zones between vegetation types, which features can also be considered to constitute a structural characteristic. In this connection the work of PFEIFFER (1943) is of major importance. This author pointed out, among other things, that vegetation covering large surface areas is often surrounded by special vegetation types forming linear "belt" patterns that have but rarely been recognised as autonomous units. Such forms of contact vegetation occur more particularly near ecotones, i.e. near sharp limits between habitats caused by a more or less sudden change in one environmental factor or in several such ecological factors. This type of situation differs appreciably from the ecoclines, i.e. from situations in which gradual transitions (gradients) play an important role. There are intermediate cases between ecotones and ecoclines; moreover, the limits (and with them the zonal patterns) may shift periodically or aperiodically.

Which types of border lines are of importance in connection with mural vegetation? In order to be able to answer this question, it is necessary to dwell a little longer on the question of the level of observation. We can consider the boundary between a given individual plant specimen and its immediate surroundings (such as its rhizosphere or phyllosphere), between stenocoenoses (see 5.2) of various kinds and their environment, between different vegetation types and, also, from the point of view of the special habitat under discussion, for instance, between brick and mortar, between moister and drier parts of the wall (exposure), or between wall surface and soil. A considerable number of these demarcation lines are quite distinct and exhibit a *"Konturenkomplexität"* (an intricate contour structure, compare RAMENSKY, 1925, PFEIFFER, 1943), which may be regarded as typical of e.g. pioneer vegetation. Zonations *("Verschwommene Komplexität")* are of frequent occurrence on walls bordering upon a mass of water, the capillary water and the splashes maintaining a moisture gradient.

The question whether border lines are of any importance at lower structural levels of organisation can probably be ascertained by studying the distributional ecology of organisms of small dimensions such as bacteria and microscopic algae. It could be worth while to study this point more thoroughly.

The most suitable for an examination are the border zones between walls and the adjacent substrata: soil or water. Dependent on the moisture content of the substratum, including its fluctuations in time associated with differences in exposition and in overshadowing, a narrower or a broader transitional zone occurs at the foot of the wall. Apart from moisture relations, other factors of course also contribute towards the ecological environment. In particular the attachment of the substratum to the wall (by being blown against the wall, by splashing, or by soil cultivation), and the inherent decrease in the angle of inclination at the base, may be mentioned.

In this zone not infrequently bryophyte vegetation develops, in which in relatively dry sites such species as *Streblotrichum convolutum* occur and in moister situations *Brachythecium rutabulum* and *Rhytidiadelphus squarrosus.* The border zone between a wall and an adjacent mass of water is usually characterised by the presence of more or less hygrophilous species such as *Leptodictyum riparium.* On not too steep quay-sides and canal walls, especially in sites rich in nitrogen compounds, species of the *Bidention* may settle (see 6.9.1.4). The limit between the places which are constantly waterlogged and the 'dry' parts higher up is a shifting zone and all the *Bidention* species in question are often encountered in unstable habitats elsewhere, which is not at all surprising. Such typical zones of contact have not been included in the present investigation, and neither has the vegetation cover of the soil adjacent to the foot of walls which frequently also shows a typical contact vegetation type in a narrow belt parallel to the wall surface and often also exhibits a ruderal character by the presence of such species as *Chenopodium murale* and *Hordeum murinum.* Only in exceptional cases do the representatives of mural vegetation penetrate into this zone; this happens in particular when the surface of the substratum near the foot of the wall is paved and the ecological conditions in the joints between the bricks of the wall and the joints between the flag stones or bricks of the pavement are thus rather similar. In the cracks of the pavement such species as *Linaria cymbalaria* and *Parietaria judaica* sometimes develop, albeit with a reduced vitality and fertility (compare 8.2.6).

4.4 Dimensions

Presumably there is a causal relation between the dimensions of dominant organisms and the stability of the habitat (compare SEGAL, 1969) and, besides, there is probably an approximate relation between the dimensions of an organism and its life expectancy or the duration of its generation cycle (BONNER, 1965). These assumptions are based on observational data, but the time is not ripe as yet to furnish more than indications and the number of as yet unexplained exceptions is probably rather high. The tenability of the last-mentioned causal relation can conceivably be put to the test by the processing of a great number of factual data, the adequate information being easily quantifiable. This is by no means the case with the concept of "stability". Conceivably, one could attempt to establish relations between biomass, production, and energy flow, but hardly any research has been carried out in this field. The various types of mural vegetation usually have the character of pioneer vegetation and the habitat is upon the whole highly unstable, the instability showing itself in e.g. the extreme differences in moisture content and in temperature at various times of the day or during different seasons.

Species growing on walls are usually of a small stature, but their dimensions increase when a succession series develops. In the most extreme cases, apart from lichens, only some acrocarpic mosses and small winter annuals are present; in less extreme situations one also finds pleurocarpic mosses and hemicryptophytes. Chasmophytes and phanerophytes are mostly lacking or only appear later in the succession seres, when the walls have become more strongly decomposed. The heigt of the more specific wall-dwelling plants therefore seldom exceeds half a metre.

On walls, smallish specimens of species are frequently encountered which become much taller in some other environments. Such small forms as a rule also exhibit a reduced fertility. Good examples are *Phyllitis scolopendrium, Dryopteris filix-mas, Athyrium filix-femina* and a number of other ferns, which usually show optimal development when growing in forest soil. Such species rarely grow taller than 20-30 cm on walls in the Atlantic regions of western Europe, but are nevertheless good differential species of certain types of mural vegetation. Other species, such as *Linaria cymbalaria* and *Asplenium ruta-muraria,* are not underdeveloped in respect of specimens occurring in their 'natural' (original) habitats such as rock crevices, nor are they less fertile; and *Parietaria judaica* may even grow taller on walls than it does elsewhere. A reduced vitality and fertility can be observed in a large number of more or less incidentally wall-dwelling species for which the wall habitat is of course anything but an optimum one.

4.5 Minimum area

RAMENSKY (1925) drew attention to the fact that the smallest representative area of any given type of vegetation is smaller if one takes only the specific composition into account, than if one also wishes to record the relations between individuals or mass relations among the constituting species. MEIJER DREES (1954) distinguished a *qualitative minimum area* in which (to all intents and purposes) all species are represented, and a *quantitative minimum area* in which all the timber-producing trees (in a tropical rain forest) are represented by individuals which have reached such dimensions that the area can be used for prospective timber estimates. Unfortunately, he did not use the terms 'qualitative' and 'quantitative' minimum area in quite the same sense as Ramensky did. References in the phytosociological literature pertaining to a "minimum area" usually intend to convey approximately the same concept as what is here called the *qualitative* minimum area. This means, in practice, that the minimum area is normally related to the (co)dominant species with the lowest density, i.e. the species with the greatest size. The smallest area estimated by also taking account certain numerical and quantitative biomass relations, referred to by BEEFTINK (1965) as the quantitative minimum area, could be called the *representative minimum area* if, moreover, the representative occurrence of structural patterns within the limits of this area is made a prerequisite. (If desired, one could of course distinguish as many different kinds of minimum areas as there are criteria or prerequisites).

One should preferably start from the smallest idiotaxonomic units (and, if necessary, from their varieties of forms) and determine their relation to the surface area. As a rule this is impracticable, however, since the intraspecific entities are insufficiently known or the genetic-taxonomic significance of the morphological variation have not, or not adequately, been investigated. This concept of a 'representative minimum area' almost coincides with the definition of minimum area by CAIN & CASTRO (1959, p. 167): "The smallest area that provides sufficient space or combination of habitat factors for a particular stand of a community type to develop its essential combination of species or its characteristic composition and structure". If we also take into account a number of quantitative characteristics apart from qualitative ones, we arrive at the following definition: *The qualitative minimum area is the area which, even after a progressive increase, at most yields a relatively small increase in the number of species.* Within that area approximately all species of the community are represented.

The representative minimum area is the smallest area that provides sufficient space or combination of habit factors to develop its characteristic composition and structure both in a qualitative and in a quantitative sense. The practical disadvantage of this definition of a minimum area is that its quantification is more difficult and its estimation much more subjective than is the case with the qualitative minimum area.

For practical purposes the representative minimum area is most probably at least twice as large as the quantitative minimum area. The minimum area must always be related to a homogeneous stand of vegetation. However, homogeneity in the absolute sense of the word can hardly be expected in a vegetation cover. One can at least distinguish various degrees of homogeneity, dependent on the types of structural elements in the system and their topological relations, for example different levels of sociability or different horizontal and vertical distributional patterns. Furthermore, homogeneity is — at least at the level of total structure and overall composition of a (complex) vegetation unit — non-existent in gradient situations, and the latter are most probably of more common occurrence, both on a large and a small scale of observation, than one is wont to believe. The probability of incidence of homogeneity is also smaller as the total surface area of a system is smaller in respect of its constituting structural elements and the influence of adjacent systems thus relatively more important. On walls the total surface area of a system, especially its vertical extension, is often small in relation to the size of the wall-dwelling plants. Several workers, such as HOPKINS (1957), doubt whether the minimum area is a definite and invariable characteristic of a plant community; for, they find no clear indications of the existence of any surface area that indeed comprises practically all constituting elements of that community. Some of these authors, and others as well, came to the conclusion that it may be better to relate the number of species to the logarithm of the surface area, which procedure is also in agreement with the actual method of analysis by counting the number of species occurring in surface areas which increase exponentially; see e.g. GLEASON (1925), HOPKINS (1955), KILBURN (1963). Their conclusions are not always conformable, however. According to PRESTON (1962), for instance, the relation between number of taxa and surface area is exponential and can be expressed in the form of a $y^n = ax$ relation, but GLEASON (1922) and HOPKINS (1955) arrive at a logarithmic relation for surface areas which are relatively speaking not too small (in the majority of the herbaceous vegetation fragments they must be at least 1 m^2).

The minimum area is primarily a function of the density of the species in the stand and secondarily of their structure. These two factors often being correlated, it is possible to establish the minimum area once and for all for those vegetation types in which one or only a few life- and growth forms dominate, and provided their dimensions are already known (Segal, 1965). The relation between minimum area and density of the stand can be studied from models, but also in natural vegetation. It is, for instance, quite obvious that on shoals the minimum area of the pioneer phase of *Salicornia europaea* vegetation with a very sparse distribution of the individuals is considerably larger than in the zone of optimum development of *Salicornia* in which the number of plants per unit of surface area is sometimes hundreds of times higher.

That the minimum area of a stand of forest is appreciably larger than that of herbaceous vegetation types, and these in turn are much larger than that of bryophyte vegetation, depends on the size of the largest of the dominant species, but the stand density is directly related to this size. If one prefers to do so, one could ultimately relate the concept of a minimum area with a single species within a vegetation type, or relate it to the various spatial patterns of distribution exhibited by one of the constituting elements.

It is, therefore, at least of importance to study the effect of the constituting structural elements upon minimum area size, which relation can be better demonstrated by a logarithmic plotting of the surface area against the number of species (see e.g. Gleason, 1922).

It thus becomes possible to recognise certain structural elements in the diagrams; for, in a not too simply constructed system, more minimum areas can be recognised which show themselves in the stepwise course of the saturation curve (Barkman, 1968). In this way more information is obtained than from the corresponding normal species/area curve. This could be substantiated by studying various vegetation types of different structural complexity in the halo-, hygro- and xeroseres. The plotting of the logarithm of the number of species against the surface area or the logarithm of this area (e.g. Preston, 1962; Van der Maarel, 1962) is methodologically pointless and only results in loss of information. Estimations of minimum areas must necessarily start from a surface area which is proportional to the smallest structural elements to be analysed. This means that in bryophyte vegetation an initial surface area of 1 cm^2 is sometimes already on the large side. It is also necessary, if the surface areas are small, to do several repeats of the analysis using different experimental areas instead of only a single estimate (Goodall, 1952), the effect of chance being far too great in the latter case. In an extremely sparse stand, the number of individual determinations must of necessity sometimes be very high.

Calleja (1962) related the proportional (average) increase in the number of species in respect of the increase of the corresponding surface area: $\left(\frac{\Delta y}{\Delta x}\right)$ to that surface area. Mathematically speaking, this is presumably a very useful method, which is similar to the currently followed method of counting the number of additional species after every increase of the experimental surface area. The ideal diagrams obtained in this way are hyperbolic and the parameters of each hyperbole are supposed to be characteristic of the corresponding type of vegetation (and actually of an structural pattern). The threshold value of the increment (i.e. $\Delta x = \Delta y$), would be the characteristic point. However, even in this case one would first have to come to some agreement concerning the choice of the system of co-ordinates to be used. Another possibility is to plot the proportional increment quotient $\left(\frac{\Delta y}{\Delta x}\right)$ against the logarithm of the area. The results of a comparative study of this method will be discussed in a forthcoming paper.

What are now the consequences of these considerations in connection with mural vegetation? I have already pointed out that inhomogeneity not infrequently impedes the determination of the minimum area and that the total area covered by the stand plays an important role as well.

As a matter of fact on walls only vegetation fragments can usually be studied, so that an attempted estimation of the minimum area does not yield a saturation at the level of the largest structural constituents.

In vegetation types exclusively containing bryophytes and lichens this saturation point is reached much sooner than in a plant cover containing herbaceous or even woody higher plants. Another factor to be reckoned with is the relation between the potential area of germination, usually confined to the joints in the wall, and the total surface of the wall (which is of course far larger). It is impossible nevertheless to relate consistently the minimum area to the joints only, because prostrate and creeping forms overgrow the bricks or stones and may even cover them completely. Generally speaking, the joints of brick walls contribute from about 15% to 30% (the average would be about one fifth) towards the total surface area of the wall. (However, one should bear in mind that also in other vegetation types the potential area of germination is usually rather limited in size and associated with certain structural patterns in the habitat such as the dependence on shrinkage cracks in the soil surface and other relief structures, the absence or the presence of other species i.e. interference phenomena, and the local enrichment of the soil by decaying deposits of vegetable litter, animal excrements, etc.)

During the present study, the total surface area of a wall was taken as the starting-point and the relative area of the joints was determined afterwards (which can be fairly accurately achieved by measuring the width of the joints and the length and breadth of the average exposed surface of a brick). The representative minimum area for the following types of vegetation was fixed as follows:

Minute acrocarpic mosses	1 dm^2
Larger acrocarpic mosses	0.25 m^2
Pleurocarpic mosses and foliose hepatics	0.5 m^2
Thallose hepatics	1 m^2
Small vascular plants (such as *Anogramma leptophylla, Asplenium* spec. div., *Soleirolia soleirolii, Poa annua, Sagina procumbens, Campanula erinus* and *Sedum dasyphyllum)*	1 m^2
Larger vascular plants (such as *Campanula rotundifolia, Epilobium* spec. div., species of *Fumaria* and *Gymnocarpium, Cystopteris fragilis, Linaria cymbalaria* and *Lycopus europaeus)*	2 m^2
Large herbs, suffrutices and shrubby forms (such as *Capparis spinosa, Centranthus ruber, Cheiranthus cheiri, Corydalis lutea, Parietaria judaica, Ficus carica* and *Taxus baccata*	4 m^2

For the estimation of the specific abundance, these values should not be exceeded since this might have considerable consequences for species of a minute to small posture and a high degree of sociability. In small tussocks of *Bryum argenteum,* for instance, up to 400 individual specimens may occur per cm^2. Bearing this in mind, the specific abundance has been recorded in proportion to the minimum area of each minimum taxon and not necessarily in proportion to the minimum area of the largest constituents of the stand.

4.6 Life forms and growth forms

Life forms can be defined as ecological (adaptive) types, and growth forms as morphological types. For both of them several classifications have been proposed which are based on as many different criteria. Among the better known classifications of life forms are those of Raunkiaer (1918) and of Iversen (1936), and that of Gams (1918) who attempted to classify both plants and animals in one system.

Growth form classifications have been proposed by Dansereau (1951) for vascular plants, by Frey (1924) for thallophytes, by Ochsner (1928) and by Meusel (1935) for bryophytes, and by Hilitzer (1925) for lichens.

If the definition of a life form is not restricted to ecological adaptation involving only the vegetative structural elements and early generative stages, different pollination and dissemination types can also be regarded as life forms.

Growth forms include categories with different types of tissues (IVERSEN, 1936) and also fruit, seed, or pollen types, but even such criteria as flower colour may be useful.

For the assessment of a life- or growth form spectrum one can also follow different principles:

1. The spectrum can be estimated on the basis of the number of species belonging to every category represented in the system i.e. on the basis of the qualitative representation of the species. In this way RAUNKIAER (1918) calculated his 'biological spectra' of the floras of certain floristic regions. This method can of course be applied to the composition of a plant community, an ecosystem or any other landscape unit.
2. A clearer picture of the physiognomy, or the physiological and ecological functions within a biotic community, can be obtained by taking into account the quantitative representation of each species in that community, for which the adapted estimates of the assessment scale used for abundance and coverage can be used (see under 2.2.3).
3. The life form, and sometimes the growth form, is not always a constant feature of the species, but to a certain extent dependent on the prevailing environmental conditions or on the developmental stage. In a number of cases this variation is caused by the occurrence of infraspecific taxa with more or less clearly different phenotypes, but upon the whole very little is known about this relationship. It is, in any event, necessary to discern various life forms and growth forms in the field, and to project them as it were on their specific environment, rather than basing our conclusions on purely floristic evidence. It is, however, difficult to refer any given species to more than one category of a system for technical reasons, particularly when for some "overall" (or general) analysis a great many data are to be processed at some later stage. In such cases, during the present investigation, the species is always referred to the category to which its most frequently occurring wall-dwelling form belongs. The various dissemination and pollination types will not be dealt with in this chapter, because they have hardly any bearing on the structure of the vegetation cover as a whole.

According to SAUVAGE (1966), a system of life forms would be meaningful if the hierarchy of adaptive features is clearly stipulated according to the following criteria:

1. General environment (soil, water, air).
2. Mode of carbon compound metabolism (autotrophic, heterotrophic, parasitic, etc.).
3. Life span (therophytes, perennials, etc.).
4. Unfavourable period in the vegetation cycle (Raunkiaer).

The idea of such a hierarchy seems to have certain merits if one wishes to arrive at a uniform system. It is, perhaps, preferable to make the last category a variable one, so that different criteria for the definition of life forms can in turn be substituted here. One could also consider adding a fifth category representing a different classification, based, for instance, on the growth forms.

4.6.1 Life forms according to Gams

The system of life forms designed by GAMS (1918) is founded on the mode of attachment of the taxon to its substratum; *errant, adnate,* and *radicant* types being distinguished. The fungi, algae and lichens mostly being left outside the scope of the present investigation, this classification is not of great importance here, but it is mentioned because (1) adnate lichens and algae are sometimes very important pioneer species during the initial phases of the succession seres and form vegetation types (communities) which require closer study, and (2) Gams gained the insight that the environment requirements of adnate types are much more discriminating than those of radicant ones, so that we can raise the question whether mosses and small prostrate vascular plants make higher demands on the habitat than e.g. erect vascular plants. It is, at any rate, to be expected that the physical environmental factors tend to vary between more extreme values in adnate vegetation than among a growth of radicant, erect and taller forms.

4.6.2 Life forms according to Raunkiaer

RAUNKIAER's system (1918) is based on the diverse adaptations of the vegetative plant body to the most unfavourable period prevailing during its life cycle. One of the principal criteria for distinguishing categories among the terrestrial vascular plants is the localisation of the organs producing new shoots after the unfavourable climatic phase, or phases, for which the distance to the surface of the soil was chosen as the discriminating feature (which is, incidentally, also partly a morphological characteristic). This system was repeatedly amended in the course of time, e.g. by BRAUN-BLANQUET (1928 et seq.). ELLENBERG (1956) published a very elaborate version in which the main subdivision is based on life forms in the sense of Gams and the seasonal cycles received more attention than previously. This system thus combines various principles (of which the last is a very useful addition), including some suggestions made by Braun-Blanquet. The consecutive refinements are mostly based on morphological features and consequently the classifications tend to assume the character of a combination of life- and growth forms, the same growth form repeatedly occurring in different categories of life forms (e.g. 'chamaephyta reptantia' next to 'hemicryptophyta reptantia' and 'therophyta reptantia', etc.). Such a complex classification can of course also be dissociated from a life-form system proper and this attitude is preferred here, the life form categories thus being reduced to the original version of Raunkiaer:

1. **Therophytes:** Plants that pass the unfavourable season only as seeds.
2. **Geophytes:** Plants hibernating by buds formed below the surface of the soil.
3. **Hemicryptophytes:** As 2, but buds formed at soil level.
4. **Chamaephytes:** All buds are formed above the level of the soil, but almost invariably at a distance of under 25 cm from the soil surface.
5. **Phanerophytes:** Woody forms with at least some of their buds situated more than 25 cm above the soil surface, separable into:
   a. **Nanophanerophytes:** which form their buds mostly between 25 cm and 2 m above the soil surface, and
   b. **Macrophanerophytes:** which form most of their buds from 2 m above the soil surface upwards.

For practical reasons those biannuals are included in the therophytes which are annual or biannual depending on the prevailing conditions such as *Poa annua, Cynoglossum officinale, Capsella bursa-pastoris* and *Ranunculus sceleratus.* Some frequently annual species occasionally hibernate as a hemicryptophyte and behave as biannuals or perennials or vice versa (such as *Daucus carota, Dipsacus pilosus* and *Beta vulgaris*). Others tend to become hemicryptophytes (instead of annuals) in a more temperate climate (*Corynephorus canescens,* usually annual in the Mediterranean region). There are also species which occur facultatively as an annual, a biannual or a hemicryptophyte, such as *Picris echioides, Crepis* spec. div., and *Lactuca serriola.* In the Mediterranean area, where these latter forms are more frequently encountered on walls than farther to the north, they behave relatively more often as (bi)annuals. Almost always it could be ascertained which life form occurred most frequently on walls and the species as a whole was accordingly referred to this category. Likewise, *Ficus carica* is included among the nanophanerophytes because the species when found on walls (in the Mediterranean countries) does not often grow taller than 2 m, although it usually bears fruit. Other trees growing on walls usually do not flower and remain very small as a rule, which is indicative of a much reduced vitality and fertility, and they are classified as macrophanerophytes.

4.6.3 Life forms according to Iversen

IVERSEN's system (1936) is based on the different adaptations to the water factor. According to whether the plants are more or less clearly adapted to drier or to wetter conditions, differences of a morphological nature (involving the root system, the degree of pubescence, the leaf shape, etc.), differences in the anatomy (scleromorphic structures, aerenchyma, etc.) and physiological differences (water relations, turgor, stomatal regulation of evaporation, oxygen supply, etc.) can be discerned. Iversen based his classification mainly on the various degrees of scleromorphy, on the absence or the presence of aerenchyma, on turgor variations and on the development of the root system.

Terrestrial, riparian and marsh plants, whose assimilatory shoots are adapted to a life in atmospheric air, are classified as follows:

A. **Terriphytes:** A distinct aerenchyma is lacking.

1. **Seasonal xerophytes:** Plants with a superficial root system capable of surviving a prolonged interruption of the uptake of water unscathed e.g. species of *Sedum* and *Sempervivum*. The duration of the interruption is understood to be of the order of at least several weeks.
2. **Euxerophytes:** Plants with a very well developed root system in respect of the assimilating surface, that can withstand an interruption of the supply of water for a comparatively short time only. This group includes more or less succulent forms as well as scleromorphic, felty-pubescent and mesomorphic species e.g. *Plantago maritima, Elymus arenarius, Artemisia campestris* and *Lotus corniculatus*.
3. Plants with a moderately to weakly developed root system that can withstand an interruption of the water supply for a short while only.
   a. **Hemixerophytes:** Plants with pronounced xeromorphic features. This group includes succulents, scleromorphic species and forms with a felty indument e.g. *Atriplex hastata, Festuca ovina, Calluna vulgaris* and species of *Gnaphalium*.
   b. **Mesophytes.** Mesomorphic forms which can not stand any appreciable loss of water and soon wilt completely e.g. *Ranunculus ficaria* and *Alliaria petiolata*.

B. **Telmatophytes:** Plants with a strongly developed aerenchyma distinguished as:

Scleromorphic	telmatophytes	e.g.	*Phragmites communis*
Succulent	„	„	*Aster tripolium*
Mesomorphic	„	„	*Scirpus maritimus*
Hygromorphic	„	„	*Lycopus europaeus*

Amphiphytes and limnophytes are not treated here because they do not occur on walls.

This classification is not always a very clear-cut one. For a proper application more research would be needed than was possible in this study, concerning, among other things, the behaviour of the species after an appreciable drop of turgor caused by an excessive loss of water. For data processing, the morphological features were considered to be representative, under the silent assumption that in this (always rather coarse) classification no very great errors were made.

The application of the life form classification to mosses and to lichens meets with some difficulties, the most important being the incomplete information concerning the reaction of mosses to water. Quite a few species can endure a not too prolonged desiccation and would, according to Iversen's criteria, have to be included in the group of the hemixerophytes, but other ones are seasonal xerophytes or hygrophytes (and a few are hydrophytes). The rigidity of mosses is not so much determined by turgescence or by the presence of peripheral strengthening tissues as it is invascular plants, but by a central core of strengthening cells (cf. *Polytrichum*).

The *fluctuations* in the moisture content of the substratum and the air are even more important for the development of mosses than for that of higher plants. In many habitats the uppermost layer of the substratum is subjected to regular or irregular periods of desiccation.

In order to account for the water relations one would have to apply the 'hygrobia' classification of Iversen, but this would require a large number of observations in many different habitats. It is to be expected that the ecological behaviour in relation to the moisture factor is in itself ecologically determined. *Barbula revoluta,* for instance, in W. and central Europe often occurs on dry limestone in an environment with a constant, rather high atmospheric relative humidity, but in S. Europe it is much more frequently encountered on a damp substratum in habitats where the atmospheric humidity can periodically be comparatively low. Barkman (in Landwehr, 1966) has proposed a classification of mosses based on the water factor with the following categories: rheophytic, littoral, aquatic, hydrophytic, aerohydrophytic, mesophytic, xerophytic, and tropohydrophytic taxa. Several species fall into more than one category owing to the fact that so many species can withstand appreciable fluctuations in the available amount of moisture. In the present study no attempt is made to propose a classification which is more strictly based on the hygrobial types than Barkman's. On walls the xero-, meso- and hygrophytic forms are mainly represented, a number of species being both xero- and mesophytic, such as species of *Encalypta* and *Orthotrichum, Streblotrichum convolutum,* and *Zygodon viridissimus,* and other ones both meso- and hygrophytic, such as species of *Amblystegium* and *Oxyrrhynchium.*

Such 'dual' types could be referred to special categories e.g. 'mesoxerophytes' and 'mesohygrophytes'. However, there are species occurring in both dry and wet environments or in sites with a strong fluctuating degree of moisture. These will be referred to as poikilohygrous.

Examples are: *Bryum caespiticium, B. capillare, B. argenteum, Mnium rostratum, Ceratodon purpureus, Funaria hygrometrica, Brachythecium rutabulum* and *Calliergonella cuspidata*. (The latter species is, however, hardly ever found in mesohygrous habitats.) Nearly all *Pottiaceae* are also more or less poikilohygrous, particularly *Bryoerythrophyllum recurvirostre* and, to a lesser extent, species of *Didymodon* and *Barbula*. These species prefer a habitat where at least the relative humidity is not often very low, as in the eu-atlantic regions, but also elsewhere near water. Bryophytes and lichens are referred here to four different categories on the basis of the environment in which they are usually encountered. My observations do not always agree with Barkman's in every respect, most probably because he and I studied different kinds of habitats (Barkman studied epiphytic vegetation in particular) and because we worked in different geographical areas. A species is always referred to those categories in which it occurs most frequently. Strictly speaking the following classification is not quite compatible with Iversen's system of life forms. However, in the case of both vascular plants and bryophytes the criteria are not always judged by exact observations, and the xerophytic mosses are comparable to seasonal xerophytes, the mesophytic species to hemicryptophytes and the hygrophytic ones to the mesophytic vascular plants:

Xerophytic forms: Species occurring in dry sites e.g. species of *Aloina* and *Grimmia*.

Mesophytic forms: Species found growing in moderately damp places, such as *Brachythecium velutinum* and *Rhynchostegium confertum*.

Hygrophytic forms: Species found on a moist substratum e.g. the majority of the liverworts and the species of *Drepanocladus*.

Poikilohygrophytic forms: Species occurring in both drier and moister habitats, often in sites with a fluctuating moisture content of the substratum and/or of the air.

4.6.4 Growth forms

As early as the beginning of the 19th century a classification of plants into a number of growth forms was proposed by Von Humboldt & Bonpland (1807). The classification of life forms according to Raunkiaer and its emended forms are to a large extent based on growth forms as far as the definition of the subtypes is concerned.

Van der Maarel (1966) proposed a classification of growth forms based on the direction of growth, on the position of the leaves in respect of the substratum, and on the degree of sociability. This system seems inconsistent, to me, because these factors are probably much less clearly correlated than is usually assumed. Corresponding growth forms of different sizes are considered to be equivalent (e.g. *'pulvinatae'* next to

'Bryo-pulvinatae') and on the other hand the various criteria are not subordinate (as in Dansereau's classification), so that the units within the same order of magnitude are heterogeneous. The difference between *'caespitosae'* and *'pulvinatae'*, for instance, is in several respects smaller than the difference between *'pulvinatae'* and *'erectae'*. The units are mostly derived from the subtypes of the amended Raunkiaer system of life forms. The idea of a distinction of habit forms within growth types, such as trees, shrubs, herbs, bryophytes and various thallophytic morphotypes, has its merits. The question remains if one should strive after a classification which is equally well applicable to all existing types, or at least to a number of them.

Generally speaking, vegetation consisting of representatives of special taxonomic groups, in the first place of thallophytes but often also of bryophytes, is studied by specialists who meet their own requirements and make their own systems. I have also followed this principle of using different classifications, viz. one for Spermatophytes and Pteridophytes combined, one for bryophytes and one for lichens. For vascular plants the sociability is the starting-point, this criterion having the considerable advantages of objectivity and easiness of quantification. The scale of estimation used is that of BARKMAN, DOING & SEGAL, (1964; see also under 4.3.1). It goes without saying that many species exhibit a different degree of sociability in different circumstances. However, it is a specific character of a number of taxa. Whenever a terminology is required other than a scale of numbers one would have to draft different systems for trees and shrubs and for herbaceous forms, using existing classifications as a starting-point. It is recommendable to employ, as is already done for lichens, the principal groups of growth forms distinguished by Gams for such a classification, more particularly the distinction between adnate and radicant types. This would, in the case of higher plants, mainly have certain consequences for the classification of aquatics (see e.g. SEGAL, 1965). For the classification of higher plants (as an alternative to Dansereau's system) the criteria used for the distinction of Raunkiaer's subtypes would be useful. Arborescent forms, shrubs, lianas and epiphytes have been almost completely left out of consideration in the present discussion, as they are hardly represented among the wall-dwellers. Only the following principal groups are distinguished:

Pinids: Coniferous trees;
Fagids: Hardwood trees (arborescent Dicots);
Ericids: Shrubs (or shrublets) with scale-like or needle-like leaves.
Vacciniids: Shrubs (or shrublets) with flat leaves.

Herbaceous taxa are distinguished by their mode of aggregation, the subordinate criterion being the general direction of growth:

A. **Degree of sociability 1**

Lilliids: Shoots erect, without a basal rosette of leaves e.g. *Solanum nigrum,* the majority of the *Liliaceae* s.s., and most species of *Epipactis.*

Echiids: Shoot erect, with a basal rosette of leaves or at least many leaves crowded in lowermost parts of stems e.g. many *Brassicaceae* and several species of *Echium, Silene, Verbascum* and *Hieracium.*

B. **Degree of sociability 1 or 2**

Poids: Shoots erect e.g. many grasses and species of *Dryopteris.*

Aspleniids (a subtype of the Poids): Chasmophytes such as many species of *Asplenium.*

Fumariids: Shoots partly decumbent and ascending, forming aggregates e.g. *Fumariaceae, Oxalis* species, *Scleranthus* and many suffrutices such as *Thymus.*

Fragariids: Shoots predominantly prostrate or plants with runners, such as *Fragaria* and many species of *Potentilla, Agrostis stolonifera* and *Sagina procumbens.*

C. **Degree of sociability 3**

Corynephoids: Shoot erect; examples: many grasses, sedges and *Juncus* species e.g. *Deschampsia caespitosa, Carex paniculata* and *Juncus effusus.*

Androsaceids: Cushion-forming plants, such as the typical alpine cushion plants belonging to the genera *Silene, Minuartia,* etc. but also many densely developed suffrutices.

D. **Degree of sociability 4 or 5**

Soleiroliids: Carpet-forming plants such as *Soleirolia soleirolii.*

It remains to be seen if there is a need to subdivide category D further. A high degree of sociability is usually deducible from a relatively high density of the shoots (or the tussocks or tufts) as in e.g. *Phragmites* vegetation or in a dense stand of *Parietaria judaica* on walls.

Obligatory chasmophily occurs especially in group B. This classification is fundamentally independent of the classification of plants as life forms, but inescapably there are some correlations here and there. The group of the therophytes, for instance, includes many liliids and echiids, but therophytes also occur among group B *(Scleranthus annuus).* Geophytes mostly belong in group A.

The system proposed here is by no means intended as the ultimate solution of a growth form system of herbs (and suffrutices), but rather as a possible contribution towards a discussion on such a system for which there is presumably a need.

I am well aware of the fact that the system in the present form is open to several alternative interpretations. A certain latitude in the preliminary proposals is preferable to rigid definitions which would not make the number of doubtful cases (in this system at least) any smaller for that matter.

In the system proposed here, such terms as erectae, scaposae, rosulatae, aggregatae, decumbentiae, repentiae, velantiae, scandentiae, caespitosae, pulvinatae and tapetae have been avoided for two reasons:

1. The degrees of sociability of forms whose sizes fall in dimensions of a different order of magnitude often differ appreciably. The number of shoots in a turf of *Carex paniculata,* for instance, is many times smaller than the number of plantlets in an average turf of *Grimmia pulvinata.*
2. The use of superfluous latin names for categories can be avoided by adding the suffix '-ids' to the stem of a generic name if the genus in question includes a number of species referable to the type of growth form concerned. This nomenclature, already previously employed in a classification of aquatics (DEN HARTOG & SEGAL, 1964), has also been applied to the growth form system of bryophytes and lichens.

The system of the bryophytes is primarily based on that of MEUSEL (1935), particularly the subdivision into orthotropous and plagiotropous forms, and for the subordinate groups I have heavily borrowed from GIESENHAGEN (1910), FREY (1924), HILITZER (1925), OCHSNER (1928), GIMINGHAM & BIRSE (1957), and BARKMAN (1958). The latter has given an elaborate comparative survey of the various growth form classifications of epiphytic bryophytes. His own system, which is more than a mere compilation, is rather detailed, as was of course necessary for his special subject of study. For the mosses (predominantly epilithic) encountered during my studies of wall vegetation, I propose the following system (the definitions being partly adopted from Barkman and from Gimingham & Birse):

A. **Orthotropes:** Thallus axes ('stems') more or less perpendicular to the surface of the substratum, growth by lateral innovation, gametangia terminal.
   1. Turf-forming bryophytes: Erect, acrocarpic mosses with parallel, simple or sympodially branched thallus shoots, on walls represented by:

      Buxbaumiids (*Protonemamoose* of Meusel): soots short, distinctly separated from one another, of ten produced by the persistent protonema (hence: 'rhizoid strand mosses'); example: *Buxbaumia.*

      Bryids (*Kurzrasen* of Giesenhagen, "*Bryum*-forum" of Frey: forming low turfs; examples: many *Pottiaceae, Bryaceae* and *Dicranaceae.*

      Polytrichids (*Hochrasen* of Giesenhagen, "*Polytrichum*-form" of Frey): forming not very extensive turfs, with thallus shoots normally over 2 cm tall; examples: the majority of the species of *Polytrichum* and several species of *Mnium.*

2. Cushion-forming bryophytes: Erect, acrocarpic bryophytes, with primary thallus shoots radiating from a central point of origin and their branches similar, adopting the same directional growth as main shoot and branching especially from near apex of shoot, thus adding to the size of the cushion.

   Grimmids ("*Grimmia*-form" of Frey): growing in small cushions; examples: *Grimmiaceae* and numerous species of *Orthotrichum* and *Tortula.*

   Leucobryids ("*Leucobryum*-form" of Frey): producing large cushions, normally attaining a diameter of at least 5 cm; examples: *Leucobryum* and *Eucladium.*

B. **Plagiotropes:** main thallus shoots creeping over the substratum and showing a differentiation in principal axis and lateral branches.

   Neckerids ("*Neckera*-form" of Ochsner): thallus shoots and their lateral branches complanate, arranged parallelwise in a horizontal plane, thus forming platforms; examples: species of *Neckera* and *Plagiothecium.*

   Leucodonids ("*Leucodon* type" of Barkman): branches patent at right angles, erect or ascendent, straight or curved upwards, mostly simple and forming loose mats; examples: *Leucodon, Anomodon viticulosus, Leptodon smithii, Homalothecium, Orthotrichum lyellii.*

   Isotheciids ("*Isothecium*-form" of Ochsner): thallus shoots sympodial, at first "stoloniferous", becoming erect with the erect portion unbranched below, bearing "scale leaves" and as a rule abundantly ramified above to form a kind of canopy; examples: *Isothecium, Climacium, Scorpiurium.*

   Hypnids ("type des *Hypnum*" of Hilitzer): thallus shoots prostrate or partly ascending, forming compact mats; examples: *Brachythecium, Rhynchostegium, Hypnum.*

   Radulids ("*Radula*-form" of Ochsner): prostrate forms with their thallus shoots and branches strongly appressed to the substrate; examples: the majority of the foliose hepatics, species of *Amblystegium, Oxyrrhynchium* and *Rhynchostegiella.*

   Metzgeriids ("*Metzgeria*-form" of Ochsner): prostrate, thallose hepatics; examples: *Marchantia, Pellia.*

Meusel's subdivision into orthotropous and plagiotropous forms was originally intended for mosses, but has here been applied to liverworts also (teste Barkman). As far as the mosses are concerned this classification almost coincides with the main subdivision in acrocarpic and pleurocarpic forms. According to Meusel, the *Orthotrichaceae* had perhaps better be referred to the leucodonid taxa. Since the prostrate shoots of most species are very short, however, and these forms physiognomically link up in every respect with the grimmiids, they have been placed with the latter (t. Barkman). The subdivision of A1 and of A2 is, strictly speaking, illogical. It might be better to distinguish bryids *sensu lato* and grimmiids *sensu lato* and to disregard the dimensions altogether. However, since the classification proposed here has been followed by several bryologists such as Giesenhagen, Frey, Gimingham & Birse, and Barkman, it seems to meet a certain demand. Gimingham & Birse and earlier GIMINGHAM & ROBERTSON (1950) already distinguished two principal categories viz. turfs and cushions, for that matter.

The classification of lichens as proposed below, which may be applicable to other groups of thallophytes, is likewise an opportunistic system, mainly intented as a broad survey, as the lichens are not so well accounted for in the present study. BARKMAN (1958) has compiled various classifications into a comprehensive survey. The classification given here is based on ideas of FREY (1924), HILITZER (1925), SCHULZ (1931), OMURA (1950), MATTICK (1951) and BARKMAN (1958), the definitions being partly also borrowed from the latter.

A. **Gelatinous types**

Collemids ("Collemataceenform" of Schulz): gelatinous (casu quo, Algae), if lichenous with preponderant algal component, capable of losing much water upon drying without apparent damage whilst shrinking appreciably, firmly stuck to the substratum but without rhizinae; examples: *Collema, Gloeocystis.*

B. **Crustaceous types**

Leprariids (Straubflechten of Matticke "Leprose type", including pulverulent protococcoid algae, of Barkman): lichenous or algal forms for the greater part consisting of loosely aggregated unicellular algae, in the lichenous forms associated with Fungi; thallus absent or indistinct and never distinctly margined, upper surface leprose; examples: *Candelariella, Lepraria, Pleurococcus.*

Caloplacids ("crustaceous form" of Omura): non-leprose crustaceous lichens; examples: *Caloplaca, Lecanora, Lecidea.*

C. **Foliaceous types**

Parmeliids ("*Parmelia*-type" of Barkman): thallus incised almost as far as the centre and with more or less discrete lobes, rather loosely adhering to the substratum; examples: species of *Physcia* and *Xanthoria.*

Lobariids ("type de *Lobaria*" of Hilitzer): thallus more superficially incised (incisions not reaching as far as the centre), its lobes broad and ascending; example: *Peltigera.*

Anaptychiids ("type d'*Anaptychia*" of Hilitzer): thallus incised almost to the centre, with long and narrow, discrete, ascending or patent lobes (a transition between the groups C and D); example: *Anaptychia.*

D. **Fruticose types**

Cetrariids ("*Cetraria*-form" of Frey): thallus with rigid, erect or horizontal to pendent, flattened branches; examples: *Ramalina, Cladonia* sect. *Cladonia.*

Usneids ("*Usnea*-form" of Frey): pendulous forms with slender and repeatedly ramified branches; example: *Usnea.*

Cladoniids ("*Cladonia*-form" of Frey): erect, "suffruticose" forms with repeatedly ramified stems; examples: *Cladonia* sect. *Cladina, Cornicularia.*

N.B. The term "Caloplacids" is deliberately chosen to preclude possible confusion because, in a more restricted sense, a *Verrucaria* type, *Lecanora* type, *Lecidea* type, etc. have been described.

By surveying the growth forms system and by relating it to wall-dwelling vegetation, we find that on walls the following growth habit forms are rare or even entirely lacking: corynephorids, androsaceids, neckerids (slightly more frequent in Southern Europe), isotheciids, lobariids, usneids, and cladoniids. In this respect mural vegetation does not only distinguish itself from terrestrial vegetation (which includes e.g. many corynephorids and also buxbaumiids, polytrichids, isotheciids and cladoniids) and from epiphytic vegetation, but also from vegetation on rock surfaces (with e.g. many androsaceids, leucobryids, neckerids and lobariids).

4.6.5 Flower colour

As far as can be ascertained, very little is known of the possible relation between the abiotic environment and flower colour. SCHRÖTER (1932) devoted some attention to the differences in colour and in hue between flowers in montane zones and in the lowlands. He remarked in this conection (p. 76): *"Das Höhenklima begünstigt durch starkes Licht und niedere Temperatur die Bildung roter Farbstoffe (Rotwerden weisser Doldenblüten, Rotüberlaufen der Grasspelzen)".*

The formation of anthocyanins, especially in the vegetative parts, is probably also relatively strong in other extreme habitats with strong nocturnal heat loss through radiation, particularly in open pioneer vegetation of various kinds e.g. in halophytic and also in mural vegetation. Presumably light is one of the determining factors, because the red pigments do not develop in many species if they occur in more sheltered places. It is not at all unlikely that a study of the pigmentation of the leaves and other assimilating organs in different habitats may also yield interesting results. One has the impression that glaucous colours are of more common occurrence in physiologically drier environments (halophytes).

In the spectrum a category 'no flowers' is included to account for the contribution of the Pteridophytes, which in life form and growth form type are rather strongly reminiscent of the Spermatophytes, to mural vegetation. The spectrum of petal colours does not have the pretention of given a colour characteristic of mural vegetation that is more than a contribution to the description. In certain cases there is a distinct relation between the floral colour and the pollination mechanism (manifest in flowers with a carrion odour, and in ornithophilous forms). The data are available to workers who believe that they can put the information to good use in the field of floral biology or otherwise.

For a study of the distribution of floral colours, the following categories were distinguished: scarious (e.g. *Sagina procumbens*), white, violet, blue, green, yellow, pink or lilac, red, and others (e.g. brown). The predominant colour of the corolla was always chosen when more colours occur in one flower e.g. yellow for flowers of *Ranunculus* with yellow white-based petals. In the case of variable petal colours, the most common colour was chosen e.g. red for *Centranthus ruber* (which is sometimes white-flowered). In a number of cases only one of the possible colour variants was recorded, which points to a relation between the colour type and the environment e.g. only blue flowers of *Pulmonaria* and no red ones. Two colours are only recorded in those cases in which consistently the same two colours appear in a flower or inflorescence, as in a number of *Anthemidae* e.g. *Matricaria* and *Chrysanthemum*.

4.7 Composition

The specific composition, the average number of species and the absolute number of species present is discussed in connection with vegetational characteristics (see under 6). In chapter 9 a number of species more or less characteristic of mural vegetation are discussed in some detail.

5 Principles of classification

During the last few decades a discussion has started concerning the use of objective and subjective criteria for the classification of vegetation types and it has become quite clear that there are many degrees of subjectivity. Starting-points for a classification are, for instance, faithful species or differential combinations of species (e.g. Braun-Blanquet, 1964); presence and affinity indices (e.g. Sørensen, 1948), multivariate analyses (Williams & Lambert, 1959-1961), discriminant function (Goodall, 1953), factor analysis (Dagnelie, 1960) and, more particularly for ordination, the methods of gradient analysis according to the Wisconsin school (e.g. Bray & Curtis, 1957) and Whittaker (e.g. 1957).

Some of these methods were employed and their results compared. The extent to which the results of à mathematical evaluation based purely on qualitative vegetational characteristics (presence of species, etc.) differ from those obtained in part from quantitative data (viz. percentage of coverage) was also ascertained.

5.1 Criteria for classification and ordination

For several decades, there was a controversy between the so-called "Scandinavian School" or Uppsala School in phytosociology, to which Du Rietz was a keen contributor, and the Franco-Swiss or Zürich-Montpellier School with Braun-Blanquet as the most important protagonist. Since the Amsterdam Botanical Congress of 1935 these two schools of thought have gradually become closer and various workers have made serious attempts to combine the positive elements of both.

In the meantime several British and American workers formulated a number of principles and methods that initially differed appreciably from those of the botanists in continental Europe. In Russia also a special "school" originated. In the last 10 to 15 years the study of these various hypotheses and methods of analysis has gradually gained ground about and botanists have worked towards the development of a novel and better approach to phytosociological problems. Continental botanists have tended to introduce more exact and objective methods and the same can be said of the Anglo-American School, though initially here subjective methods of inquiry (such as Tansley's scale of abundance) were in vogue or the working hypotheses were not founded on sound postulates (e.g. certain aspects of Clements' climax theory).

The drawbacks of the Franco-Swiss School have been discussed by various workers e.g. by Poore (1955-1956), whose comments concerning a number of points show, incidentally, that he was not altogether conversant with the methods of this school (see also Moore, 1962), but some of his counter-arguments carry conviction all the same.

In the Anglo-American School a controversy developed concerning whether a classification is possible or ordination methods are to be preferred, in which workers in continental Europe became involved. The conviction is rapidly gaining ground that both methods of inquiry have their merits.

Before starting a discussion on mural vegetation from this point of view, it will serve a useful purpose to expound the author's views regarding a possible grouping of vegetation units. It is necessary to make a sharp distinction between principles applicable in studies of a single type or stand of vegetation or in investigations within a limited area (such as gradient analysis, or work done in an area considerably smaller than a phytogeographical district), and fundamentals to be used in inquiries into the mutual relationships of vegetation types occurring in a larger area. In continental Europe, studies of the latter type prevail much more than they do in Anglo-Saxon countries.

Of course not all the differences are so essential and it is necessary to recognise various gradients. However, it will be clear that, for instance, the summation of a number of small experimental areas used in frequential analyses is not comparable to the more or less complete lists of species recorded from experimental surfaces when the presence or the constancy is taken as one of the starting-points for the comparison.

The difference in the organisational level of the subject under study is one of the principal causes of mutual misunderstanding, but to a certain extent "provincialism" must have contributed to the disagreements in opinion: workers are often not very familiar with the literature from other language groups. Thus even the adepts of the Franco-Swiss School sometimes completely disregard the literature published outside their own country (compare, in this connection e.g. the survey of vegetation units published by Oberdorfer, 1967).

5.2 The ecosystem as a starting-point

One of the controversies among ecologists concerns the moot point at what level of organisation vegetation units should be recognised.

Realising that interpretations can be given at more than one level, Gams (1918) distinguished various types of *synusia*. Unfortunately his approach fell into disuse and several authors produced definitions of the concept of a synusium which at best only partially covered Gams' starting-point (compare Tuomikoski, 1942). Du Rietz (1932), for instance, related the concept with stratification and Braun-Blanquet (e.g. 1964) identified it with ecologically uniform assemblies of taxa belonging to the same category of life forms. The latter interpretation almost coincides with Gams' *"Synusien 2. Grades"*.

Gams also employed the terms *phytocoenoses* and *biocoenoses* for abstract vegetation units, but these terms were later usually applied in a much wider sense, viz. with the inclusion of Gams' concept of an isocoenosis in a broader geographical relation. It is to be regretted that plant ecology is burdened with many terms that have different meanings according to the author. This makes it necessary to define the interpretation and the meaning of terms and concepts first before discussing any ecological problem to avoid ambiguities. I do not intend to give a digest of ecological definitions which could in the least serve as a suggestion, although such a synthesis is ultimately unavoidable, because this is beyond the competence of a single worker. All definitions given in the present paper are, therefore, no more than working hypotheses. However, the use of a term which might cause semantic confusion has been avoided wherever possible.

Before starting ecological research, one must realise that the object of inquiry is a "system" or forms an integral part of a "system"; a "system" being an assembly of interrelated elements which thus forms a structurally organised whole. The structural order is, upon the whole, purposeful. Its constituting elements can in turn consist of subordinate elements, and thus the system itself of sub-systems.

Both a synusium and a biocoenosis (biotic community) can, generally speaking, be seen as a system, and ecological systems can be studied and discussed at various levels of organisation, e.g. at the level of the autecology of a single individual, of a 'biotic community' of some sort or of the whole biosphere. Even the system at the level of the habitat of a single individual may, if a higher plant is involved, comprise several subsystems including a rhizosphere and a phyllosphere. Within an individual specimen of a higher organism, systems can again be distinguished at a number of levels of integration: organ systems, organs, tissues, cells, organelles, etc. down to at least the macromolecular level, and the study of the dynamic aspects of such subordinate systems belongs to the physiologists.

Every species reacts in its own specific way on environmental factors and behaves as an autonomous entity (compare the concept *niche*). A biotic community can contain a number of subsystems such as "Schichtsynusien" (i.e. superimposed layers, see e.g. Du Rietz, 1932), epiphytic vegetation, or epilithic vegetation on rocks in a forest.

Ramensky (1925) calls such "internally" homogeneous portions of what · he considers to be a biotic community or *coenosis,* properly speaking a landscape element, *stenocoenoses.* Biotic communities and landscape elements form a part of the biosphere, the system comprising all life on earth and in its turn forming a subordinate part of systems of ultimately cosmic level.

The various levels of integration are not always sharply distinguishable and that is why it is necessary to establish beforehand at which level of organisation the object of inquiry is located, premising that biological research at any of these levels is equally useful.

Several authors recognise ecosystems at the level of the population e.g. Evans (1956). If a population is defined as an assembly of individuals of the same taxonomic species i.e. as a collective group of organisms within a vegetation stand or a landscape unit, such a "system" does not fit in an ecological series, because it is of a different nature and belongs rather to the discipline of population dynamics and genetics, even if there is a certain relation between populations and ecosystems and the regulation of the numbers of individuals or shoots. Only in monospecific stands of vegetation do these concepts coincide. One should also bear in mind that the term population is sometimes applied in the sense of vegetation stands or landscape units (compare Westhoff, 1965) or other "statistical assemblies" of elements. The necessity of recognition of the vagueness of the limits, and of a discussion concerning vegetation units and ecosystems at various 'levels' is demonstrated by the following consideration.

If a forest is regarded as a 'biotic community' (or biocoenosis), it is undoubtedly seen as an ecosystem which can be subdivided into various stenocoenoses. The question arises as to how far these stenocoenoses are interdependent and form a single system, because only if this condition is fulfilled are the stenocoenoses equivalent to subsystems. If a ravine forest (or 'Schluchtwald') with ashes and maples is compared with an oak-hornbeam wood on calcareous soil in W. or central Europe, the occurrence in the first type of forest of a number of stenocoenoses recognisable as separate structural entities (such as epilithic vegetation of mosses and lichens and chasmophytic vegetation) is rather striking, and in this type of wood the stratification is usually manifest.

The stratification is also quite obvious in several types of Boreal coniferous forests, the superimposed layers of vegetation, moreover, being more or less clearly, and sometimes to a large extent, independent of the composition of other layers. This is the most cogent argument for Lippmaa's (1934) contention that the different vegetation layers represent basic units. The more advanced integration of the elements in an oak-hornbeam wood is apparent from the vague lines of demarcation between these elements, and, moreover, the soil has a much more homogeneous structure (not only a more constant texture, but also fewer boulders present). One thus gains the impression that the stability of a landscape element (a forest) is related to soil structure, and one can visualise series or gradients along which the vegetational structure alters clinally. Generally speaking, one may expect a more close-knit integration of elements as the environment becomes less extreme. Such a clinal increase in stability can be observed more or less clearly when one goes geographically from N.- to central Europe, and orographically from the montane regions down towards the lowlands.

Factors other than macroclimatological can likewise create extreme environmental conditions e.g. strong local wind action (in coastal areas) and an excessive inclination (ravine forests). It thus becomes clear that in not so stable environments the stenocoenoses exhibit a higher degree of autonomy than in the more stable ones, and are sometimes even more or less independent of the landscape element, or only indirectly e.g. by being dependent on shading. In such cases the units are also found in other elements of the landscape e.g. *Vaccinium*-layers in various kinds of Boreal woodland and vegetation of *Phyllitis scolopendrium* and *Gymnocarpium robertianum* and other ferns outside the ravine forests on unforrested gravelly slopes at higher altitudes. Contrariwise, an oak-hornbeam wood contains a number of species that are to a very high degree 'faithful' to this type of deciduous forest. Epiphytic vegetation can likewise be regarded both as an autonomous ecosystem and as a subsystem of a tree or a forest. For simular reasons, pools and small brooks in a forest on wet soil can be regarded as stenocoenoses. All stenocoenoses of this kind, moreover, exhibit clearly independent and constant patterns of succession. They usually show an individual structure, seasonal periodicity, dissemination spectrum and metabolism.

It is practically sure that the stenocoenoses by themselves represent systems of a certain level of organisation, but it is not quite so certain if a forest in its totality may invariably be considered to constitute a system. In certain instances this may indeed be accepted, but in other cases this is not permissible or only with certain restrictions. Generally speaking, subsystems are the more autonomous, the more they are spatially isolated from one another.

Let us now consider pioneer vegetation in this light. As a rule it consists of open stands poor in species (if one considers only the macroscopic plant world). In the most extreme situation each individual plant can be more or less completely independent of all other specimens in the stand, namely, if the density is low and the plants are of a relatively small size. During a progressive development of a vegetation cover the landscape element gradually assumes the character of an initially very simply constructed system, which progressively becomes more intricate, at least if the stand is capable of changing the environment in such a way that the latter becomes less extreme (see chapter 7). Pioneer vegetation can, in the extreme case, be nothing but a sparse stand of autonomous elements and is actually not a system. What is in this case regarded as a 'vegetation unit' is functionally hardly comparable to an intricate, stable landscape element.

The question whether habitat-induced differentiations within a landscape element must be interpreted as different vegetation units (see e.g. SUKATSCHEW, 1929) or a landscape element must be seen as an entity (Braun-Blanquet and his followers), can not be answered unambiguously if one starts from the concept of an *ecosystem*. This can be defined *as any system in which organisms and the non-living environment interact to produce exchanges of materials* (compare e.g. ODUM, 1959). In this connection one must bear in mind that only in exceptional cases do we have more than a very incomplete knowledge of the functional relations between vegetational elements, so that, for the time being, it is not very realistic to make the ecosystem our starting-point, however logical this may seem to be.

As long as so little is known of the functional interrelations, we can start from the hypothesis that it is useful to refer to repetitively occurring vegetation elements characterised by a more or less constant composition and structure, and somehow hanging together in an ecological sense, as a 'vegetation type'. At the level of this unit another prerequisite must be met, viz. a minimum degree of homogeneity i.e. a minimum of overdispersion in a representative surface area (which, if need be, can be found by adding up the surface areas of vegetation fragments if the representative area is not attained within one element of the vegetation unit).

As early as 1924 RAMENSKY pointed out that the continuity of vegetation (which can of course also be considered at various topographical or geographical levels) does not permit the distinction of rigid units. Not the assemblies of elements are constant, but the fundamental causes and 'laws' to which the assemblies and structures owe their existence.

Sharp boundaries do exist, but these are almost always the result of sudden changes in the ecological conditions, or else they are artificial. Only in unstable habitats are sharp boundaries sometimes present as a result of primarity.

Applying these considerations to mural vegetation, we can first of all establish that it belongs to pioneer vegetation with a usually simple structure. The composition may include a number of 'incidental' species which probably have hardly any functional relation with the rest of the vegetation cover or none at all (this concerns the great number of species with a low presence). The recognition of horizontal patterns and the optical assessment of the degree of homogeneity or inhomogeneity usually causes few difficulties. However, the question of the possible relations of the vertical elements is not so easy. There can hardly be any doubt that the connection is negligible in a large number of cases e.g. between crustaceous lichens, bryophytes and vascular plants.

As in epiphytic vegetation (compare BARKMAN, 1958), bryophytes and lichens usually constitute different kinds of vegetation forming a mosaic pattern. Not infrequently the bryophytes are restricted in their occurrence to the joints of the wall, whereas the crustaceous lichens may occur both on the stones (or bricks) and on the mortar joints. If a well-developed stand of vascular plants is present, the bryophyte layer is mostly suppressed, and its composition at any rate but rarely differs from bryophyte vegetation encountered elsewhere on the same wall or in comparable situations. However, some vascular plants are presumably associated to some extent with a moss layer, such as *Sagina procumbens,* which shows a great ecological affinity with such species as *Ceratodon purpureus* and *Bryum argenteum* (see 6.13). The modest stature of *Sagina procumbens* presumably also plays a role. (However, a great affinity is not a sufficient indication of any functional relationship.)

In all instances both the vascular plants and the bryophytes were recorded. The degrees of correlative affinity were not only calculated between vascular plants alone and mosses alone, but also between vascular plants and mosses. In order to ascertain if this is justified, one could perhaps compare information analysis of all species inclusive, and of the mosses separately, or of stenocoenoses one believes to have recognised by their physiognomy, as the case may be.

It is likely that the composition of the moss layer under a stand of vascular plants frequently corresponds to that of moss vegetation without vascular plants and this would imply that the amount of information from the mosses is the same after data evaluation according to either method.

Conceivably such species as *Sagina procumbens* will prove to be closely associated with the moss cover; the seeds of this species, at any rate, can germinate in a moss layer.

5.3 Classification and ordination

Vegetation units can be arranged according to different principles or criteria. In recent years discussions have started, specially in the pertaining Anglo-Saxon literature, concerning the question whether such a systematic arrangement must be based on classification or on ordination. Classification presupposes the presence of discontinuities and the totality of the vegetation patterns on earth is supposed to be a varied mosaic of discontinuous landscape elements. Ordination is the arrangement of the elements in their relation to one or to more gradients and postulates a complex continuum without sharp boundaries. The existence of such a continuum was already recognised by RAMENSKY (1924), and by GLEASON (1926). The continuity is perceptible, both in a quantitive and in a qualitative sense, in space as well as in time, and we thus distinguish *ecoclines* and *chronoclines* (succession seres) which are both equally important in various categories of landscape elements such as several forms of hydro- and hygroseres. This is one of the reasons why the distinction of abstract units is much more difficult in peat bogs and marshes than in pioneer vegetation or in 'climax vegetation'. Such continuous vegetation types can not easily be moulded into a rigid system of arrangement. The continuity in space can be considered at various levels of organisation e.g. in a zonation, but also according to a topocline (WESTHOFF, 1947) e.g. when vegetation types are compared from eu-atlantic areas via subatlantic and subcontinental to continental regions, or along an orocline in the mountains. The introduction of such concepts as *topoclin and oecocline* (MEIJER DREES, 1951) and their derivates (such as *chlorocline, conductocline, acidocline, basicline, hygrocline, humidocline,* etc., compare SEGAL, 1965) indicates that followers of the Franco-Swiss School are also well aware of the occurrence of continua. In other words, workers who aim at a classification, nevertheless, give the continuity principle a central place in plant ecology. The introduction of the 'cline' concept is of course based on the analogy with Huxley's cline concept in idiosystematics which explicitly aims at a classification. In phytosociology, followers of the Zürich-Montpellier School have even applied ordination up to a point when they arranged their tables, although the procedure was admittedly heuristic as a rule.

There can be no doubt that continuity exists at various levels of organization, but the occurrence of discontinuities can not be denied either, even if we admit that the latter are at least partly of anthropogenic origin. Discontinuities usually coincide with sudden changes in environmental conditions and relatively homogeneous vegetation covers large surface areas only in extreme environments. The discontinuities are also graded and can be recognised at the level of the stenocoenoses, or between e.g. ant or termite heaps, paths and streams in a landscape element (such as a forest) and the remainder of that element, but also at the level of the boundaries between landscape elements (forests-meadows) and between formation classes on a map of the world. In the latter case the geographical isolation plays a role, in the other instances a topological or ecological isolation.

Vegetation can, therefore, be both continuous and discontinuous and the disagreement between the different phytosociological methods has been much overrated (compare MOORE, 1962; IVIMEY-COOK & PROCTOR, 1966). It is true that in the Anglo-Saxon countries the investigations were mostly aimed at very local conditions and situations, and that they were of a more analytical nature than in continental Europe, but there are gradations. For an analytical study in a small area ordination methods seem to be better suited and the results may even provide a basis for a system of classification, if this would serve a useful purpose. However, ordination methods have to contend with the complex interrelation of ecological factors, and consequently the methodological difficulties are not inconsiderable, especially if there is no obvious masterfactor or principal limiting factor and certain important environmental factors are insufficiently known. One can of course apply ordination analysis to the composition of the stand itself in such cases and this may even prove to be more effective than ordination by any available measurement of environment (WHITTAKER, 1967). According to Whittaker, a *coenocline,* despite its continuity, can be divided into community types.

If one starts from abstract units with a wide range of variability, so that the boundaries between these units are more or less vague, the difference between classification and ordination is not a very distinct one. In such a case, the use of a classification method may have some practical advantages, not only for the description of vegetation units, but also as a basis for vegetation mapping.

The classification methods current in continental Europe are mostly based on a heuristic approach. The obvious intuitive element in these methods need not be disadvantageous. Intuition is an important source of inspiration in scientific inquiry, which can not be summarily dismissed as a 'subjective' line of approach.

(In my opinion, 'intuition' is, at least partly, 'subconscious knowledge' digested by the brain so fast that the conscious mind can not keep pace with it – a process somewhat similar to data evaluation by means of a computing device with built-in "memory".)

In point of fact a classification, although based on abstract vegetation units, can indeed be an arrangement according to certain objective criteria. A table of relevés of stands of vegetation is nothing but a grouping of data which can be purposefully interpretated in a certain way as is often done in statistics: the arrangement of recorded data can be made to agree as closely as possible with some presupposed hypotheses, but the data themselves are not altered at all. That the compilation of tables can be achieved in an objective way is demonstrated by the method described by ELLENBERG (1956). The selection of the basic data is, however, crucial and may thus become the stumbling block. The question arises if the samples can be chosen in an objective way, but the answer is dependent on the problem posed and, in practice, the level of organisation that is involved, be it in a less significant way.

An objective sampling within vegetation containing homogeneous stands of considerable size does not offer great difficulties. However, in the direction of a gradient, this is already much more problematical, because the determining ecological factor (or a complex of factors) does not necessarilly show a simple (e.g. linear) relation in respect of the distance and because it remains to be seen if the effect of such factors can be measured in a linear scale. The influence of a factor varies with each species and changes, moreover, along the whole reach of the gradient. There are indications that the effect of at least certain environmental factors on certain species e.g. in connection with production, changes logarithmically along gradient stretches (where the production does not reach its optimum or suboptimum) rather than linearly. Only if analytical recordings of sufficiently restricted area have been made and in an adequate number of replicates, can a maximum of information be expected. If the surface area of the samples does not correspond with the structural complexity of the object, a spurious continuity (or discontinuity, as the case may be) may even be suggested in cases where it does not exist (Poore, 1962). Likewise, the comparison of rather similar stands of vegetation within larger regions by means of random sampling becomes more and more difficult. The various landscape elements appear to be very irregularly distributed and the essence of the problem, the degree of correspondence (and possible identity), causes a difficulty. Only if the recorded data are so numerous and from such a large area that a selection *a posteriori* can be made, and the relevés obtained at random, is there any certitude of objectivity. However, in the field one can not always meet these conditions. Some vegetation units are scarce. but possess a clearly definable composition, structure and ecology. During my studies concerning vegetation of cracks and joints in paved roads it appeared as if in sea ports a specific vegetation unit containing some halophytes is frequently met with (see 8.2.1). This could only be substantiated by recording this kind of vegetation in a number of ports and sea-side resorts. This also applies to special types of vegetation in a number of singular habitats (dunes, high mountains). In other words, it is only after the investigation has begun, that the recordings in such widely divergent habitats appear to be of particular interest. The selection of the experimental areas can not be achieved by a very simple *a priori* indication of suitable sites based on some random distribution on a topographic map. Furthermore, the potential diagnostic value of rare taxa for the recognition of vegetation units or as indicators of singular environmental factors must not be precluded.

Both classification and ordination can be meaningful and may supplement one another, and ordination methods need not necessarily be applied only to samples within a restricted area, but may also be used for vegetation studies in a large geographical region.

In a number of cases it is conceivable that data obtained by means of ordination methods can be the basis of a system of classification. If such a system is meaningful, this must show up in the results of ordination analysis. On the other hand, it may be better to apply ordination as a supplement to classification so as to place the units or the assemblies of species in a spatial model in which the mutual relationships are demonstrated (compare GREIG-SMITH, 1961; CRAWFORD & WISHART, 1966, 1967).

A classification only gives an abstracted approach to the actual condition. Summarising, we can say that the answer to the question which method or methods ought to be given preference depends on the specific inquiry and on the scale of the problem (POORE, 1962). Ordination (and also factor analysis) involves a good deal of calculation and the application of ordination methods is only recommended if the primary observational data render the mathematical evaluation efficient viz. if numerous and adequate numerical data are available of every category from which valuable results can be expected (e.g. a sufficient number of records of every ecological factor that appears to be essential for the interpretation).

ANDERSON'S (1965) contention that ordination is more appropriate to the study of structurally not very intricate vegetation fragments, and classification to other cases, is clearly too simplistic, and so is the almost opposite conclusion of VAN LEEUWEN & VAN DER MAAREL (1966) that the ordination is more suitable for the analysis of much varied vegetation rich in species and classification for more homogeneous stands of a poorer specific composition. Furthermore, the unwarranted extension of conclusions based on studies of a very local character, goes too far to say the least: the whole argument concerning the presence or the absence of continua in vegetation stands in a publication by GOODALL (1953) was based on recordings of a very limited number of species (14) in 256 experimental surfaces within an area not exceeding 640 m^2.

I wish to state most emphatically that I consider the relatively simple method of conveying scientific information to be the principal goal of a descriptive system of vegetation units. A wholly objective classification is an impossibility. A classification always contains useful information even though other arrangements can be made. A word of warning must be sounded against unduly detailed classifications which are too suggestive of the presence of communities that have more than a strictly local importance (and distribution). That is why in many cases an approach to the problem at the level of alliances and higher units is upon the whole internationally better understood and accepted as valid than an endless description of 'associations'.

Walls are structural elements which usually discontinously bound upon their surroundings, a feature they have in common with many other elements of landscapes strongly influenced by man. However, the wall surface is not always uniform and may show gradients, not only in a vertical direction (see under 4.3.2) but also in a horizontal direction. An example of the last case is provided by the round outer walls of wind-mills in the Netherlands, which form excellent objects for the study of the effect of exposure. Such gradients are not always gradual and especially in the vertical direction there are distinct discontinuities, the latter being caused, among other things, by the ecological limit to which the water is sucked up by capillary action and by the extent to which the protective action of the projecting stone slabs favours germination.

Geographical variation, on the other hand, is more or less clinal and it would certainly be worth while examining this variation by means of an ordination method.

5.4 Criteria for a classification

Of the many criteria that have been tried at one time or another, to find a systematic classification of vegetation units, the floristic criterion, based on 'fidelity', has found rather general acceptance in central, S.- and continental W. Europe, and later became better appreciated also in Scandinavia. The vegetation units are primarily recognised on the basis of their floristic composition i.e. of one composite character. Such a system may be artificial (but Westhoff, 1967, defends quite the contrary), and the opinion is gradually gaining ground that for a properly founded classification other criteria also ought to be used as much as possible e.g. structural and ecological ones.

The floristic criterion has especially been elaborated as a starting-point for a phytosociological classification by Braun-Blanquet. There was no lack of criticism of the principles and methods that came in vogue with the Franco-Swiss School. Sharp criticism was given, among others, by Gams (e.g. 1939):

"....*Spekulative Methoden, die bei Verwendung willkürlicher, autoritär festgelegter Abgrenzungen zu einer künstlichen Systematik führen...*"

and by Du Rietz (1932) and Katz (1933); a well-documented survey of all the disadvantages of the method was given by Tuomikoski (1942). Undeniable advantages are the relative simplicity of the method and the surveyability of the results. The procedure has proved to be very suitable to procure much approximate information in a short time, and it is, among other things, useful for a study on a relatively large geographical scale, especially in regions where no previous analytical recordings of vegetation have been made. The fundamental unit, the association, is not exclusively based on its floristic composition, but also on its physiognomic (perhaps rather: structural) and ecological aspects (compare Braun-Blanquet, 1928; Du Rietz, 1936).

As a rule this means that the additional aspects figure largely in the description of the units (or at least should do so); a vegetation unit is supposed to be more or less clearly distinguishable within a certain range of tolerable variation of these aspects. Still, the structural, ecological and other (e.g. geographical) aspects need not have an important diagnostic value. In phytosociological practice it is not only the purely qualitative floristic characters that are used to distinguish the various units, but also quantitative criteria and in addition such characteristics as vitality and fertility e.g. in the definition of the concept of a faithful species by WESTHOFF (1951): "A species occurring more in a certain plant community than in any other community of the area under investigation. 'More' refers in the first place to presence, but also to abundance, dominance, sociability and vitality."

For a further elaboration, the procedure of SZAFER & PAWLOWSKI (1927) is sometimes applied. The concept of the faithful species has been the pivot of the Franco-Swiss School for several decades. Gradually the restricted significance of this notion became realised. GAMS (1941) (compare TUOMIKOSKI, 1942) pointed out, for instance, that faithful species represent a special case of differentiating species, and MEIJER DREES (1952) and BECKING (1957) discussed the theoretical possibilities when the areas of a faithful species and a vegetation unit coincide only partially. Their conclusions can directly be applied to the concept of a differentiating species. Regrettably, different workers have altogether diverse ideas of what is a locally, a territorially and a regionally faithful species.

The most simple point of view is held by those authors who regard absolutely and regionally faithful species as synonymous concepts and distinguish locally faithful species only if the area of distribution of the vegetation unit does not completely coincide with the distributional area of the species, all this irrespective of the absolute size of these areas.

Meijer Drees defines a regionally faithful species as a species that is faithful in the whole of the region in which its distribution coincides with that of the vegetation unit, whereas a locally faithful species is faithful only in a part of the common area of distribution of species and vegetation unit. BECKING (1957) considers these notions to be dependent on the percentage relationship of the surface areas of vegetation unit and area of overlap. The overlap is supposed to amount to 10-50% in the case of locally faithful species, 50-90% in the case of regional ones, and 90-100% for absolutely faithful species. A considerable difficulty is that the size and the limits of such areas can usually not so easily be determined, the more so because the areas are often very incompletely and inaccurately known. Territorial faithfuls are mostly considered synonymous with regional ones.

ELLENBERG (1956) has given different definitions for the three concepts and he considers each concept to be only dependent on the size of the area in which the species is faithful. The size of that area is not sharply defined. Meijer Drees' system is the most promising in my opinion. Barkman (1958) adopted this system, but altered the definitions of a locally and of a regionally faithful species by using the meaning of workers other than Meijer Drees. The practical objection one can bring forward against his proposal to restrict the term 'locally faithful species' only to those cases in which the distribution of the species falls completely within that of the association, is that such cases are relatively rare. Barkman also introduced a new terminology for all possible cases.

The various concepts have been sharply defined by Meijer Drees and provide, to my mind, a sound basis for the distinction of the various possibilities concerning the relations between faithful species (and differentiating species) and the relevant association. A minor addition could be put up for discussion (Segal, 1965). One must raise the question of what the minimum area is within which the concepts of a locally, territorially and regionally faithful species are relevant. To this end I propose that the area of an association must not be smaller than a phytogeographical district, unless the distributional area of the faithful taxon is smaller. A locally or regionally faithful species in the sense of Meijer Drees must occur in at least a whole phytographical district such as the Dutch Wadden District [1]). Absolute faithfuls are referable to several categories. One group has a very restricted local distribution and includes, for instance, endemic forms. The term 'territorially faithful species' could be reserved for this special case. One could stipulate that the distributional area of these territorial forms must not exceed that of a phytogeographical province. A second group comprises species found in extreme habitats, and a third category includes all faithful taxa of larger phytosociological units (which usually have a fairly wide distribution). It follows that faithful species of associations only have a local significance. This does not apply so strictly to the faithful species of vegetation units of higher order, however, but such entities of a higher order can mostly be adequately described by means of physiognomic and ecological terms and they are in fact often ecological units (compare Segal & Westhoff, 1959; Ivimey-Cook & Proctor, 1966).

There is far less agreement concerning the starting-point of vegetation analysis among protagonists of a floristic system than is sometimes claimed. The basic units are no longer recognised by a great many authors

---

1) There is no agreement either concerning the interpretation of the concept of a phytogeopraphical 'district'. The 'Wadden District' sensu Van Soest (e.g., in Heukels & van Ooststroom, 1962) is only a part of the 'Atlantic district' sensu Braun-Blanquet, and others. The term 'district' is used here in the sense of Van Soest.

on the absolute degree of fidelity, but more and more on relative fidelity, the concept of a faithful species increasingly being replaced by the differentiating combination of species (BEEFTINK, 1965; SEGAL, 1965). Also the *opulent species* concept tends to replace the fidelity concept (Maas & Westhoff in DOING, 1956). An opulent species is defined as a species with a locally optimal development in a certain vegetation type, but with insufficient fidelity. Quite apart from this change in the fundamental concept, more subtle differences in terminology originated ("Kennart" besides "Charakterart", "Trennart" besides "Differentialart", compare MOORE, 1962). Moreover, the starting-point has been shifted by various workers to the ecological groups (DUVIGNEAUD, 1946; ELLENBERG, 1956) i.e. groups of species, which, in a particular set of ecological and geographical conditions, reach maximum development and vitality, form the characteristic core of the association, and usually fix its physiognomy also (POORE, 1956); SCAMONI, PASSARGE & HOFMANN (1965) have put in a plea for *sociological groups (Soziologische Artengruppen).*

One of the most important differences in starting-point concerns the meaning of the terms 'dominance' and 'presence'. It is quite clear that these are not always applied in combination with the concept of the degree of fidelity (defined e.g. according to Szafer & Pawlowski), as is evident from the use of such names as the *Poëtum annuae* or the *Polygonetum avicularis* (KNAPP, 1961; PASSARGE, 1963). The constantly present species that do not belong to the differentiating species combination have also gained more importance. According to BRAUN-BLANQUET (1959), stands with a constant composition but without faithful species are either ecologically (or choro-ecologically) determined pauperisms, or initial stages (or, alternatively, ephemerous, transitional stages of anthropogenic origin, if they are not mixtures of two, or rarely of more, associations). This interpretation only holds true if the meaning of the terms 'faithful species' and 'associations' has exactly the sense ascribed to them by this author. It also remains to be seen if it is permissible to disregard certain relevés and to 'purify' tables by omitting recordings of allegedly 'impure' (inhomogeneous) stands altogether, a procedure advocated by some students. Such a 'cleaning' process readily leads to the suggestion of the presence of discontinuities which in reality do not exist at all, does not do any justice to the possible variability, and may thus result in the loss of useful information concerning the sociological and ecological amplitudes of certain species.

Among the objections raised against the floristic approach to phytosociology the criticism regarding the concept of fidelity is the most important. Even the criteria indicated by Szafer & Pawlowski are inadequate to assess the degree of fidelity and their objectivity leaves much to be desired. Even in cases of local fidelity one must presuppose that the degree of fidelity must be related to *all* types of vegetation in the area under investigation, but it is dependent on the selection of the samples just the same. The sampling changes appreciably if, instead of recording parts of "typical" or "pure" stands, one also takes relevés of intermediate samples. If in a certain area where the vegetation cover is going to be completely described there is no question of a selection of samples reasonably approaching a method of random sampling, this brings the method into disrepute. There is a considerable likelihood of a kind of self-deceit of the workers in that they, when describing associations, seek for something rather than merely record something.

Plant ecology has moved into a transitional phase. On the one hand workers tend to look for more objective methods and a sharper analysis, but on the other hand they are trying to find an approach which leaves some room for the gradations and relativities so essential in the domain of the life sciences; sharp demarcations and discontinuities being of rare occurrence in these disciplines. In the next few decades phytosociologists must attempt to find a synthesis between these two strivings. A more objective form of inquiry will most probably automatically lead to a sufficient degree of relativity. The solution suggested here is only an opinion which, I hope, may lead to a fruitful discussion.

The starting-point must be that a system is useful as a method of conveying scientific information if it can provide a more or less accurate approximation of the situation by means of a brief circumscription (which could be a convenient name) or a combination of (standardised) appropriate symbols. Such a system can only be an artificial one, but there is no cogent reason why it could not prove to be a very close approximation of a more 'natural' classification based on the greatest possible number of floristic, physiognomic, ecological, geographical, and some dynamic, criteria.

The system based on floristic composition has proved to be suitable for the characterisation of certain landscape elements up to a point, even though the system is as yet anything but stabilised. The biological criterion is possibly the most natural one in a number of cases because it usually involves the other criteria. Among the other criteria the structure is usually of considerable significance and enables an abstraction on a world-wide scale. However, floristic criteria alone fall short of the ultimate purpose in that the possibility of drawing alternative demarcation lines remains. The various alternatives can be limited by assigning a certain sequence of importance to the various criteria (SEGAL, 1965) e.g. as follows:

1. Floristic criteria: Differential combination(s) of species and constant species or species aggregates.
2. Physiognomic criteria: Structure.
3. Ecological criteria.
4. Geographical criteria.

There is at present a tendency towards the assignment of the first place to physiognomic criteria (e.g. DOING, 1962; WHITTAKER, 1967). This has the undeniable advantage that the formations can be placed hierarchically above the other units (compare RÜBEL, 1933). The question arises, furthermore, if — at least for certain types of vegetation — the physiognomic criteria should not be given first place, even if the resulting classification is not quite compatible with an arrangement on the basis of floristic composition, as is the case when, for instance, certain types of forests and scrubs are distinguished. In this case the sequence of the categories 1 and 2 of the diverse criteria should be reversed. It is even conceivable that for certain types of vegetation the floristic criteria must prevail and in other cases the physiognomic ones (compare CAIN, 1947) e.g. for relatively stable and unstable types, respectively. This could be of great practical importance for the placing of vegetation stands poor in species. The consequent application of the floristic system would require the recognition of a large number of associations with a single faithful species and referable to higher hierarchic units, even up to the level of classes, with only a few faithful species or genera (or only a single one) – compare several proposed classes such as *Zosteretea, Ruppietea* and *Spartinetea.* A physiognomic system would set bounds to such practice. A physiognomic basis of classification could even be useful to treat the palearctic coniferous forests, considering that the composition of the layers of dwarf shrubs and of herbs seem to be determined by the structure rather than the composition of the tree layer. Of such cases GAUSSEN (1933) and LIPPMAA (1934), among others, have given some striking examples. The introduction of quantitative criteria by the Scandinavian School gave rise to the concept of the *sociations* (in the sense of the definition of the Amsterdam Botanical Congress of 1935, see DU RIETZ, 1936 i.e. types of vegetation with a dominant species in every layer). Sociation and association coincide in certain instances, as is illustrated in Fig. 4, in which the density of the lines represents the diagnostic value of different species. The greater the density, the better the diagnostic significance. For associations the differential species, but especially the faithful species have diagnostic value; for a sociation in the first place the dominant species and to a lesser extent the constant species. In the left upper corner of the figure, the concept of association and sociation coincide.

(It is unfortunate that POORE, 1962, contributed an additional definition of the association concept to the phytosociological literature which covers only partially the currently accepted definition. Before 1935 the term had been used to denote diverse concepts.)

All this is not a plea to employ such terms as sociation and association. I very much prefer to start from the general concept of a plant commu-

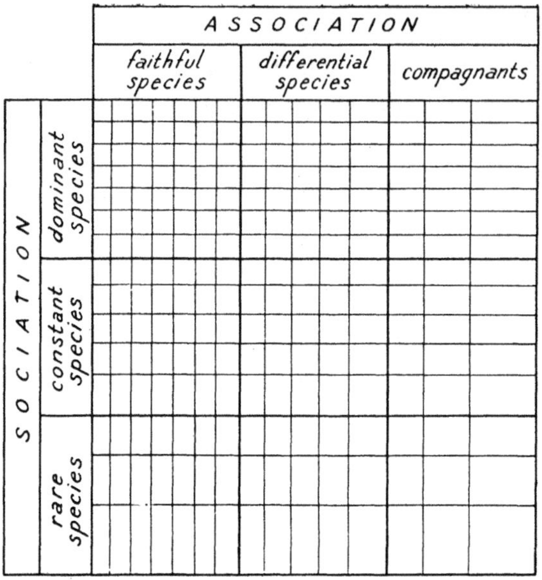

Fig. 4 Relations between associations and sociations (explanation in the text).

nity which can be used to denote a stenocoenosis of any kind as well as ecosystem at various levels of organisation up to the category we are wont to regard as a landscape element. This also is an attempt towards a synthesis of the different points of view. By doing so, one can ascertain if a given vegetation unit fits into a system of higher units in very much the same way as NORDHAGEN (1936), who regarded the sociations as subordinate elements of higher units and not necessarily of an association. If a vegetation unit is not only floristically but also physiognomically and ecologically sharply identifiable, a distinction of such a unit can be very useful. There is also some leeway to distinguish units with a constant composition but without differentiating species if this is considered to be necessary. Still, one must always bear in mind that between the units of all levels of organization sharp lines of demarcation do not exist and that the gaps in the system between certain recognised units are "filled in" by transitions and partly also by ecological series. It is also noteworthy to remember that vegetation units sometimes show about the same degree of affinity to different higher units (for an illustrative example, see BARKMAN, 1958). The fundamental concepts of the idea of an association assembly (SEGAL, 1963a, 1964) and the definition of the subassociation as the basic unit in which the differential group of species are ecological assemblies of taxa also characteristic of higher units (compare FABER, 1933; DUVIGNEAUD, 1946; SEGAL, 1963a), are all aimed at the filling of the "gaps". Examples of ecological series have been given by WHITTAKER (1967) and SEGAL (1967).

When describing units of a lower order one must always be critical when lists or tables contain a large number of 'companions'. In such relevés the tail end must be as small as possible to inspire us with confidence concerning the foundation of such a unit. And the species must, as completely as possible, be referable to ecological groups which should, generally speaking, coincide with the differential combination of species (compare DUVIGNEAUD, 1946; SEGAL & WESTHOFF, 1959), at least if one wishes to maintain a hierarchic arrangement. One can thus enlarge our insight into the variability of vegetation stands.

For the typification of vegetation units, apart from positive characteristics (their presence), negative ones (their absence) can strengthen the diagnostic value. As a basis for the evaluation of relationships of stands or of vegetation units, the further analysis can be carried out by using objective criteria such as affinity indices, either based on coefficients of affinity such as those proposed by JACCARD (1902) and SØRENSEN (1948), or on a comparative evaluation by means of χ^2 tests or of a mathematical relation in which the relative quantitative share of the various species is accounted for, such as the relations suggested by BARKMAN (1958), VASILEVICH (1962) and ORLOCI (1966). Such methods of approach will probably inspire the greatest confidence in the long run. Also on the basis of such coefficients and indexes one can attempt to establish the incidence of focal points or nodes in the sense of DUVIGNEAUD (1946: 'noyaux') or POORE (1956). An ultimate hierarchic system must not have a linear construction. However, the systems of the Franco-Swiss School likewise offer ample opportunities to arrive at a more-dimensional or ramified classification and it is even possible to arrange the units in systems of co-ordinates (compare GAMS, 1942; DUVIGNEAUD, 1946). One should, in any event, amply consider the necessity of recognising associations within higher units (such as alliances) and refrain from doing so if the continuity, be it topologically, geographically, ecologically or chronologically, is manifest, as is the case in many types of aquatic and riparian (bog and fen) vegetation. In such cases an inquiry into the clinal phenomena and their possible causes is to be preferred (compare SEGAL, 1965).

Summarising, the conclusion can be drawn that there is as yet no consensus of opinion concerning the various methods of classification and the evaluation of the recorded data (relevés, etc.) and their interpretation. More research is required before it can be decided which methods are preferable in certain instances.

If the association is accepted as a basic unit, subordinate elements can be distinguished within an association according to diverse criteria. Here, too, a certain hierarchic sequence can be attempted, in the same way as is done in the case of higher units, as follows: *subassociations,* floristically or ecologically typified, to be subdivided, on the basis of the same criteria, into *declinants* (see BARKMAN, 1958), and *variants,* geographically typified, to be subdivided by means of the same criteria into *subvariants.*

6 A survey of the various types of mural vegetation

In the following survey of vegetation units I have attempted to demonstrate the variability of mural vegetation as much as possible; the latter being considered a continuum and the units as abstractions or as clusters of certain combinations of species. It is not possible to show all the relevant documentary material in the form of tables. In the 'condensed' tables of each type usually three relevés are shown as examples. These relevés were *not* chosen for their uniformity of composition or for their approximation of a certain 'type', but rather as illustrations of the variability of mural vegetation. In addition, the following data were usually incorporated in the tables:

1. Certain data relating to the relevés viz. date (among other things, in relation to periodicity); year (the first two numbers being omitted: - (19)62, (19)64, etc.); serial number of the relevé (for documentary purposes); exposure; geographical area; percentage of coverage of the herb layer (CH); percentage of coverage of the bryophyte layer (CM).

The *exposure* is recorded as belonging to one of four classes, overshadowed walls and "exposure unknown" being referred to separate categories. The code symbols for the exposure and the geographical area have been given in 2.3 and in Appendix I.

2. Data relating to the vegetation stands.
   a. For each individual species:
      Presence in per cent. (P).
      Ultimate weights (UW) according to the estimation methods discussed in 2.2.3.
      Mean group abundance of all stands combined (MGT) according to the values mentioned in 2.2.3.
      Mean group abundance of stands where a given species is present (MGP), on the basis of the same values.
   b. For the stands as a whole:
      A combined coding for the mean number of species (MNS), the mean coverage of the herb layer (MCH) in per cent., the mean coverage of the moss layer in all stands combined (MCM) in per

cent., and, in brackets, the mean coverage of the moss layer in those stands where bryophyte vegetation is present. This is written as MNS-MCH-MCM, which can, for instance, be coded as: 13-30-4, or 16-28-31(35). In the first instance the stand is upon the average poorer in species but has a somewhat higher mean coverage of the herb layer and a markedly lower coverage of the moss layer, mosses being present in all stands in contradistinction to the second example.

In a number of cases the relationship of the various units has been estimated by means of the formula proposed by Poore (see under 2.2.5.2).

With the available material a number of other forms of data processing could be executed, but for the time being only a restricted number of calculations were chosen.

It goes without saying that the figures do not permit an accuracy expressable in per cents. From the number of stands (abbreviated to: "st.") taken into account for each vegetation unit, the accuracy can be deduced.

6.1 Vegetation with Parietaria judaica

Parietaria judaica (synonyms i.a. *P. ramiflora, P. diffusa*) has been considered as a subspecies or a variety of *P. officinalis* by several authors, or even as an ecological modification of the latter (Paclt, 1959). It has been shown, on the ground of morphological, cytogenetic, experimental taxonomic, phytogeographic and ecological differences, that *P. judaica* and *P. officinalis* are two clearly distinct and genetically separated taxa, which can be considered to be 'good' species (Mennema & Segal, 1967). The confusion originated, among other things, because forms of *P. officinalis* (which is of rare occurrence on walls) may exhibit morphological characteristics of *P. judaica,* and forms of the latter species, when occasionally growing in soil and not on a wall, may be taken for specimens of the former. *P. judaica,* if found growing in soil, almost invariably occurs along the foot of a wall on which it also grows. In such situations it is associated with species which mostly belong to the species group of the ruderal *P. officinalis* vegetation stands in W. and central Europe, thus "ecologically" resembling the latter species. *P. judaica* is originally a chasmophyte which most probably became much more widely distributed since Man started building stone (and brick) walls and thus created an environment very suitable to the requirements of this somewhat nitrophilous species. Conceivably *P. judaica* initially occurred in sites locally enriched with organic nitrogen by the excrements of birds or other animals. In the few localities where I found *P. judaica* growing in rock crevices the plant cover usually appeared to me to be quite natural: in the Trou de la Miège near Montpellier (Hérault, France), in the Gorges du Tarn (Tarn, France), near Poggio (Elba, Italy), near Ascona (Ticino, Switzerland), and near Brixham and Berry Head (Devon, Engeland). ROUX & LAHONDERE (1961) recorded it from rock crevices in Brittany and FRIEDRICH (1954) reports its occurrence on rocky and gravelly mountain slopes in the area of Lake Garda (together with *Centranthus ruber*). MOLINIER (1959) reported the occurrence of *Parietaria judaica* in natural habitats in Corsica and HORVATIC (1963) in the island of Pag (Yugoslavia), in both cases in cohabitation with *Adiantum capillus-veneris*. KNOERR (1960) found *P. judaica* growing in different vegetation types of rock crevices in some of the islands S. of Marseilles. At the present time, it is not at all unlikely that the occurrence of *P. judaica* in rock crevices is sometimes of a 'secondary' nature, the proximity of occupied sites on walls having been primary. Near Ascona, at least, this seemed the more likely considering that in this locality. *P. judaica* was associated wiht species not, or hardly ever, observed in such habitats, such as *Syringa vulgaris, Chamaerops humilis* and *Galeobdolon luteum*. Stands of mural vegetation with *P. judaica* are of common occurrence in the Mediterranean, the submediterranean, and the eu-atlantic regions, while in the subatlantic parts of Europe they are almost exclusively restricted to sites in the immediate vicinity of the streambeds of the great rivers such as the Loire, Seine, Scheldt, Meuse and Rhine, the localities petering out fast towards the north and the east. Along the geographical gradients the composition of the stands gradually alters. *P. judaica* is fairly resistent to high temperatures, but is not adapted to withstand prolonged cold spells and is, therefore, not found in more continental areas. Although long periods of drought are fatal, it can survive fairly long dry spells. Its periodicity and other biological aspects have been mentioned in a previous publication. The diaspores of *P. judaica* are, presumably, mainly dispersed by ants, and, likewise, those of a number of species usually associated with it such as *Lamium amplexicaule* and (in the Mediterranean area) species of *Veronica* (*V. cymbalaria*, etc.).

P. judaica is a calciphile and nitrophile, its nitrophilous tendencies being more pronounced in the Mediterranean area than elsewhere and in that region emphasised by the association with other more or less nitrophilous species, principally agricultural weeds and species of ruderal habitats. The variation in the composition of vegetation stands with *P. judaica* is so excessive that it is necessary to distinguish more vegetation units.

The changes in specific composition are noticeable along various geographical clines e.g.

1. in the Mediterranean area *Capparis spinosa* becomes more common as the climate becomes drier;

2. in the Mediterranean area from the lowlands to the montane zone *Asplenium ruta-muraria* and *Linaria organifolia* increase in numbers and the first species also becomes more common towards the more Atlantic parts of Europe;
3. from the Atlantic regions to the subatlantic ones *Phyllitis scolopendrium, Ceterach officinarum* and *Centranthus ruber* progressively decrease in number, whilst *Poa compressa* and *Chelidonium majus* appear more frequently;
4. the same tendency being apparent in the transitional zone from the Mediterranean region to the subatlantic zones (but *Phyllitis* not playing a role).

Other Mediterranean, Atlantic or central European species also appear or disappear along the geographical gradients and this causes the presence of a complex of communities which are more or less continuously connected by intermediate forms and which can be subdivided by means of various criteria. One could, for instance, attempt to locate an 'average' type in the transitional area between Mediterranean, Atlantic and central European vegetation units belonging to this complex and use this as the basis to describe the various geopraphically distinct units as subunits. However, stands with a sufficient representation of elements of all three geographical variants are extremely rare and a common basis can as best be found in the species combination *Parietaria judaica-Linaria cymbalaria-Tortula muralis.* Moreover, the transition from the Mediterranean region through the submediterranean to the subatlantic zone is hardly anywhere a gradual one, the mountianous regions forming barriers in many places and transitions being restricted to river valleys, e.g. the Rhone valley. That is why I have divided the communities with *P. judaica* into a number of groups according to floristic criteria which more or less clearly coincide with geographical criteria and as much as possible with structural and (naturally) ecological citeria. One group includes the Mediterranean communities and the Atlantic ones, and *Asplenium ruta-muraria* between atlantic communities. Each group embraces several units, and at least one of these is always a vegetation type in which *P. judaica* is the dominant species. Such species as *Ceterach officinarum, Umbilicus rupestris* and *Centranthus ruber* form links between the Mediterranean communities and the Atlantic ones, and *Asplenium ruta-muraria* between the Atlantic and the subatlantic variants. Stands with *Cheiranthus cheiri* occur in all regions, but in the Mediterranean area the association of this species with *P. judaica* is the rule, whereas in the Atlantic region their co-occurrence is less common to become much rarer in continental western Europe. The different variants and some other mural communities in which *P. judaica* may occur or is lacking (such as stands in which *Hedera helix* is dominant) will all be dealt with separately.

Table 48 Matrix of wall-dwelling species
See text under 6.13.

1	Parietaria judaica	[illegible]
2	Centranthus ruber	[illegible]
3	Galactites tomentosa	[illegible]
4	Bromus sterilis	[illegible]
5	Catapodium rigidum	[illegible]
6	Campanula erinus	[illegible]
7	Phagnalon sordidum	[illegible]
8	Mentha rotundifolia	[illegible]
9	Lamium amplexicaule	[illegible]
10	Mercurialis annua	[illegible]
11	Oxalis corniculata	[illegible]
12	Veronica polita	[illegible]
13	Rubus (div.) spec.	[illegible]
14	Stellaria media	[illegible]
15	Antirrhinum majus	[illegible]
16	Umbilicus rupestris	[illegible]
17	Bromus madritensis	[illegible]
18	Barbula acuta	[illegible]
19	Cheiranthus cheiri	[illegible]
20	Dactylis glomerata	[illegible]
21	Sedum album	[illegible]
22	Sedum dasyphyllum	[illegible]
23	Erigeron mucronatus	[illegible]
24	Arenaria serpyllifolia	[illegible]
25	Scorpiurium circinatum	[illegible]
26	Hedera helix	[illegible]
27	Sedum acre	[illegible]
28	Ceterach officinarum	[illegible]
29	Sonchus oleraceus	[illegible]
30	Barbula vinealis	[illegible]
31	Grimmia pulvinata	[illegible]
32	Homalothecium sericeum	[illegible]
33	Linaria cymbalaria	[illegible]
34	Poa compressa	[illegible]
35	Geranium robertianum	[illegible]
36	Hieracium lachenalii	[illegible]
37	Chelidonium majus	[illegible]
38	Schistidium apocarpum	[illegible]
39	Hypnum cupressiforme	[illegible]
40	Polypodium vulgare	[illegible]
41	Corydalis lutea	[illegible]
42	Asplenium trichomanes	[illegible]
43	Streblotrichum convolutum	[illegible]
44	Poa angustifolia	[illegible]
45	Encalypta streptocarpa	[illegible]
46	Orthotrichum anomalum	[illegible]
47	Asplenium adiantum-nigrum	[illegible]
48	Lycopus europaeus	[illegible]
49	Bryoerythrophyllum recurvirostre	[illegible]
50	Bryum argenteum	[illegible]
51	Holcus lanatus	[illegible]
52	Erigeron canadensis	[illegible]
53	Rorippa sylvestris	[illegible]
54	Epilobium roseum	[illegible]
55	Barbula unguiculata	[illegible]
56	Asplenium ruta-muraria	[illegible]
57	Mycelis muralis	[illegible]
58	Ceratodon purpureus	[illegible]
59	Poa annua	[illegible]
60	Tortula muralis	[illegible]
61	Brachythecium rutabulum	[illegible]
62	Agrostis stolonifera	[illegible]
63	Taraxacum (div.) spec.	[illegible]
64	Sagina procumbens	[illegible]
65	Urtica dioica	[illegible]
66	Amblystegium serpens	[illegible]
67	Rhynchostegium murale	[illegible]
68	Epilobium parviflorum	[illegible]
69	Cystopteris fragilis	[illegible]
70	Funaria hygrometrica	[illegible]
71	Poa pratensis	[illegible]
72	Plantago major	[illegible]
73	Athyrium filix-femina	[illegible]
74	Phyllitis scolopendrium	[illegible]
75	Bryum caespiticium	[illegible]
76	Bryum capillare	[illegible]
77	Epilobium montanum	[illegible]
78	Dryopteris filix-mas	[illegible]

Table 39 Filici-Saginetum

a. with Aspidiaceae and inops form
b. with Asplenium trichomanes
c. with Gymnocarpium robertianum
d. with Lycopus europaeus

	a.							b.							c.				
Nr C	1	2	3	P	U W	MGT	MGP	4	5	6	P	U W	MGT	MGP	7	8	9	P	U W
Date	19-11	22-8	5-8	95st				30-10	3-8	1-8	31st				19-9	27-7	29-7	39st	
Year	62	62	64					67	62	64					65	66	66		
Nr R	786	612	457					686	490	369					1227	582	594		
Ex	2	3	2					2	1	1					1	4	1		
Geo R	120	120	111					120	112	111					200	400	400		
CH	15	30	10					10	12	10					30	10	20		
CM	4	25	-					2	15	25					50	5	35		
Sagina procumbens	+p	2b	2m	91	+r-2b	3	3	2m	2m	2a	97	+r-2b	4	4	2m	+p	2m	62	+r-2b
Poa annua	+p	+r	+p	65	+r-3a	1	1	2m	2m	2m	94	+r-2a	0.8	0.9	1p	1p	1p	69	+r-1p
Tortula muralis	2m	2a	-	62	+r-4a	5	6	2a	2a	1p	84	+r-2b	4	5	-	1p	1p	76	+r-2b
Ceratodon purpureus	2m	2a	-	69	+r-4a	4	6	1p	-	2b	42	+p-2b	2	4	2b	1p	-	55	+p-2b
Dryopteris filix-mas	1p	1a	+p	79	+r-3a	2	2	+p	1a	+p	68	+r-2a	1	2	-	1p	+r	62	+r-1b
Asplenium ruta-muraria	+p	-	-	62	+r-2b	3	4	2a	1p	1p	77	+r-3b	3	3	1b	+p	-	62	+p-2b
Bryum argenteum	2m	2a	-	53	+r-2b	1	2	-	2m	-	39	+r-3b	1	3	1p	1p	+p	41	+r-2m
Linaria cymbalaria	2a	-	-	41	+p-3b	4	10	-	1b	1p	52	+p-3a	3	5	-	-	-	18	+r-2m
Bryum caespiticium	1p	1a	-	51	+p-2a	0.6	1	-	1a	+p	35	+p-2a	0.9	2	2a	2m	-	62	+p-2a
Bryum capillare	+p	-	-	29	+p-2a	1	3	2m	+r	-	52	+r-3a	2	3	1p	2m	-	28	+p-2m
Taraxacum (div.) spec.	-	-	-	19	+r-1a	<0.1	0.1	+p	+p	+r	52	+r-+a	<0.1	0.1	+p	+p	+p	31	+r-1b
Sonchus oleraceus	-	-	+p	26	+r-+p	<0.1	<0.1	+r	+r	-	29	+r-+a	<0.1	0.2	+p	+p	-	21	+r-+p
Athyrium filix-femina	+p	-	-	34	+r-3b	1	4	-	-	-	23	+r-1b	0.1	0.5	-	+p	+p	34	+r-2a
Phyllitis scolopendrium	-	+r	1a	35	+r-4a	2	7	+r	1a	+p	52	+r-2a	1	2	-	-	-	10	+p-4a
Plantago major	-	-	-	17	+r-1a	<0.1	0.1	+r	+r	+p	19	+r-+p	<0.1	<0.1	-	-	-	24	+r-+p
Epilobium roseum	-	+r	-	16	+r-1p	<0.1	0.2	-	-	+p	16	+r-2a	0.3	2	-	-	-	7	+r-2m
Epilobium montanum	-	-	-	24	+r-2m	<0.1	0.2	+r	-	+r	29	+r-1a	0.1	0.3	2m	+p	-	28	+r-2m
Poa pratensis	-	-	-	12	+r-2a	0.2	1	-	-	+r	13	+r-+p	<0.1	<0.1	-	-	-	21	+r-+p
Agrostis stolonifera	-	-	-	12	+r-2m	<0.1	0.3	-	-	-	10	+r-+p	<0.1	<0.1	-	-	-	3	+r
Dryopteris carthusiana	+p	-	-	13	+r-2a	0.2	1	-	-	-	-	-	-	-	-	-	-	-	-
Polypodium vulgare	+p	-	+p	22	+r-1b	0.2	0.7	-	1p	-	39	+r-2a	0.7	2	-	-	-	-	-
Pteridium aquilinum	-	2a	+p	15	+r-2a	0.3	2	-	-	-	16	+r-1b	0.1	0.8	-	-	-	-	-
Dryopteris dilatata	-	-	-	10	+r-3a	0.4	4	-	-	-	3	+p	<0.1	0.1	-	-	-	10	+p
Hypnum cupressiforme	-	-	-	7	+p-3a	0.6	7	-	-	-	10	+p-2b	1	12	+p	-	-	14	+r-2m
Asplenium trichomanes	-	-	-	3	+p-2m	<0.1	0.1	+r	+p	1p	87	+r-2b	2	2	+p	-	-	24	+p-2m
Streblotrichum convolutum	-	-	-	7	2m-2a	0.2	3	+p	-	-	39	+p-2a	0.9	2	-	1p	-	24	+p-2m
Asplenium adiantum-nigrum	-	-	2m	4	1p-2m	0.1	2	+p	-	-	32	+r-2m	0.3	0.9	-	-	-	-	-
Homalothecium sericeum	-	-	-	6	+p-2b	0.4	6	-	-	-	26	+p-4a	3	11	-	-	-	7	+p-1b
Epilobium parviflorum	-	-	-	8	+r-1p	<0.1	0.1	-	-	+p	16	+r-1b	0.1	0.8	-	-	-	14	+r-+p
Senecio vulgaris	-	-	-	3	+r	0	0	-	-	-	16	+r-1a	<0.1	0.3	-	-	-	3	+r
Epilobium adnatum	-	-	-	3	+r-1a	<0.1	1	-	+p	+p	16	+p	<0.1	0.1	-	-	-	-	-
Brachythecium rutabulum	-	-	-	5	+p-2a	0.2	4	-	+p	2m	13	+p-2m	0.2	1	+p	-	-	10	+p-1b
Bryum murorum	-	-	-	1	+p	<0.1	0.1	-	-	-	13	+p-1p	<0.1	0.3	-	-	-	7	1p
Epilobium lanceolatum	-	-	-	1	+p	<0.1	0.1	-	-	-	13	+r-1a	<0.1	0.4	-	-	-	-	-
Buddleja davidii	-	-	+p	7	+r-2b	0.3	4	-	-	-	10	+r-+b	0.1	1	-	-	-	-	-
Rhynchostegium murale	-	-	-	2	2m-2a	0.1	5	-	1a	+p	10	+p-1a	<0.1	0.6	2m	-	-	18	1p-2a
Chelidonium majus	-	-	-	-	-	-	-	-	-	-	10	+r-+p	<0.1	<0.1	-	+p	+a	10	+p-1a
Barbula vinealis	-	-	-	-	-	-	-	-	-	-	10	+p-2a	0.5	5	-	-	-	7	+p-2m
Cystopteris fragilis	-	-	-	5	+r-2a	0.1	3	-	-	-	-	-	-	-	2a	2m	2m	62	+r-3a
Gymnocarpium robertianum	-	-	-	-	-	-	-	-	-	-	-	-	-	-	2a	2a	-	69	+r-5
Amblystegium serpens	-	-	-	8	+p-2b	0.4	5	-	-	-	6	+p-1p	<0.1	0.3	-	2m	-	31	+p-2m
Funaria hygrometrica	-	-	-	6	+p-2m	<0.1	0.7	-	+p	-	6	+r-+p	<0.1	0.1	2a	2m	-	24	+p-2a
Bryoerytrophyllum recurvirostre	-	-	-	2	+p-2m	<0.1	1	-	-	-	6	2m	0.2	0.3	-	2m	+r	24	+r-2m
Leptobryum pyriforme	-	-	-	4	+p-2m	0.1	2	-	-	-	3	2m	0.1	3	-	2m	1p	24	+p-2m
Erigeron canadensis	-	+p	-	7	+r-1p	<0.1	0.3	-	-	-	6	+r	0	0	-	+r	-	14	+r-+p
Encalypta streptocarpa	-	-	-	-	-	-	-	-	-	-	-	-	-	-	2a	-	-	14	+p-2a
Glechoma hederacea	-	-	-	-	-	-	-	-	-	-	-	-	-	-	-	-	-	14	+r-1a
Distichium capillaceum	-	-	-	-	-	-	-	-	-	-	-	-	-	-	-	-	1p	14	1p-2m
Poa compressa	-	-	-	3	+r-+p	<0.1	<0.1	-	-	-	3	1a	<0.1	2	-	2m	2m	10	+p-2m
Didymodon rigidulus	-	-	-	-	-	-	-	-	-	-	-	-	-	-	+p	-	-	10	+p-2m
Urtica dioica	-	-	-	1	+r	0	0	-	-	+r	3	+r	0	0	-	+r	-	3	10
Marchantia polymorpha	-	+p	-	4	+p-2m	<0.1	1	-	-	2m	6	+p-2m	0.1	1	-	+r	-	3	+r
Holcus lanatus	-	-	-	4	+p-2a	<0.1	0.1	+r	-	-	3	+r	0	0	-	-	-	-	-
Lycopus europaeus	-	-	-	-	-	-	-	-	-	-	-	-	-	-	-	-	-	-	-
Rorippa sylvestris	-	-	-	1	1a	<0.1	2	-	-	-	-	-	-	-	-	-	-	-	-
Rumex crispus	-	-	-	-	-	-	-	-	-	-	-	-	-	-	-	-	-	-	-
Ranunculus sceleratus	-	-	-	-	-	-	-	-	-	-	-	-	-	-	-	-	-	-	-
Alnus glutinosa	-	-	-	-	-	-	-	-	-	-	-	-	-	-	-	-	-	-	-
Scutellaria galericulata	-	-	-	-	-	-	-	-	-	-	-	-	-	-	-	-	-	-	-
Angelica sylvestris	-	-	-	-	-	-	-	-	-	-	-	-	-	-	-	-	-	-	-
Valeriana officinalis	-	-	-	-	-	-	-	-	-	-	-	-	-	-	-	-	-	-	-
Rumex obtusifolius	-	-	-	-	-	-	-	-	-	-	-	-	-	-	-	-	-	-	-
Bidens tripartitus	-	-	-	-	-	-	-	-	-	-	-	-	-	-	-	-	-	-	-
Tussilago farfara	-	-	-	-	-	-	-	-	-	-	-	-	-	-	-	-	-	-	-
MNS - MCH - MCM		11	-	18	-	14 (18)			15	-	18	-	16			16	-	22	-

d.							total			
10	11	12	P	U W	MGT	MGP	P	U W	MGT	MGP
21-6	15-9	28-10	64st				219st			
61	60	61								
217	574	608								
3	1	1								
120	120	120								
15	35	35								
3	15	60								
2m	2m	1b	85	+r-2b	3	4	87	+r-2b	3	4
+p	1p	1a	79	+r-2a	1	2	83	+r-3a	0.6	0.7
2m	1p	1a	66	+r-2a	2	2	77	+r-4a	4	5
-	2a	3b	66	+r-3b	6	9	64	+r-4a	4	6
-	+p	-	31	+r-2m	0.3	0.9	62	+r-3a	1	2
-	-	-	28	+r-2b	0.9	3	55	+r-3b	2	4
-	2m	2m	64	+r-2b	2	3	53	+r-3b	2	2
-	1b	2b	73	+r-3a	4	6	50	+r-3b	4	7
-	1p	1p	46	+r-3a	2	4	49	+r-3a	1	2
-	-	-	39	+r-2a	1	2	36	+r-3a	1	3
+r	-	1p	52	+r-1b	0.2	0.3	36	+r-1b	<0.1	0.3
-	-	-	39	+r-2a	0.2	0.6	30	+r-2a	0.1	0.3
-	-	-	21	+r-2a	0.2	0.9	29	+r-3b	0.8	3
-	-	-	12	+r-2b	0.5	4	26	+r-4a	2	6
+r	-	+p	22	+r-1a	<0.1	0.1	25	+r-1a	<0.1	0.1
+p	+p	-	43	+r-2m	0.3	0.5	23	+r-2a	0.1	0.6
-	-	-	13	+p-2m	0.1	0.5	22	+r-2m	0.1	0.4
+p	-	-	40	+r-2m	0.3	0.6	22	+r-2a	0.2	0.7
-	+p	-	27	+r-1b	0.1	0.3	15	+r-2m	<0.1	0.3
-	-	-	-	-	-	-	5	+r-2a	0.1	1
-	-	-	1	+p	<0.1	0.1	16	+r-2a	0.2	1
-	-	-	4	+r-1a	<0.1	0.5	10	+r-2a	0.2	2
-	-	-	-	-	-	-	7	+r-3a	0.2	2
+p	-	-	3	+p-2a	0.1	4	7	+r-3a	0.5	7
-	-	-	1	2m	<0.1	3	17	+r-2b	0.2	2
-	-	2m	3	2m	0.1	3	13	+p-2a	0.3	2
-	-	-	-	-	-	-	7	+r-2m	0.1	1
-	-	-	1	2a	0.1	8	8	+p-4a	0.6	8
-	-	-	7	+r-1p	<0.1	0.1	10	+r-1b	<0.1	0.3
-	-	-	6	+r-+p	<0.1	<0.1	5	+r-1a	<0.1	0.4
-	-	-	4	+p	<0.1	0.1	5	+r-1a	<0.1	0.3
-	-	-	13	+p-3a	0.6	5	10	+p-3a	0.3	3
-	-	-	-	-	-	-	3	+p-1p	<0.1	0.3
-	-	-	-	-	-	-	2	+r-1a	<0.1	0.4
-	-	-	3	+r-+p	<0.1	0.1	5	+r-2b	0.1	3
-	-	-	-	-	-	-	5	+p-2a	0.1	3
-	-	-	1	+r	0	0	3	+r-1a	<0.1	0.3
-	-	-	-	-	-	-	2	+p-2a	0.1	4
-	-	-	4	+r-1a	<0.1	0.5	12	+r-3a	0.5	4
-	-	-	1	+p	<0.1	0.1	10	+r-5	0.9	9
-	-	-	9	+p-1p	<0.1	0.3	11	+p-2b	0.2	2
-	-	-	10	+p-2a	0.2	2	10	+r-2a	0.2	2
-	-	-	-	-	-	-	5	+r-2m	0.1	1
-	-	-	-	-	-	-	5	+p-2m	0.1	1
-	-	-	16	+r-+p	<0.1	0.1	11	+r-1p	<0.1	0.1
-	-	-	-	-	-	-	2	+p-2a	0.1	4
-	-	-	-	-	-	-	2	+r-1a	<0.1	0.8
-	-	-	-	-	-	-	2	1p-2m	<0.1	1
-	-	-	6	+r-2m	<0.1	0.7	5	+r-2m	<0.1	0.9
-	-	-	3	+p-2m	<0.1	1	2	+p-2m	<0.1	2
-	+p	+p	19	+r-1a	<0.1	0.2	8	+r-1a	<0.1	0.1
-	-	+p	15	+r-3a	0.8	5	7	+r-3a	0.3	4
+r	+p	+p	24	+r-+p	<0.1	0.1	10	+r-2a	<0.1	0.4
+p	1b	2a	85	+r-3a	4	5	26	+r-3a	1	5
+p	2a	+p	48	+r-2b	0.9	2	15	+r-2b	0.3	2
+p	-	-	38	+r-2a	0.4	1	11	+r-2a	0.1	1
-	2a	+p	37	+r-2a	0.4	1	11	+r-2a	0.1	1
-	+r	-	25	+r-2a	0.2	0.7	8	+r-2a	0.1	0.7
-	+p	-	22	+r-3a	1	5	7	+r-3a	0.3	5
+r	+p	-	22	+r-+a	0.1	0.3	7	+r-+a	<0.1	0.3
-	+r	+p	16	+r-2b	0.3	2	6	+r-2b	0.1	2
+p	+p	-	15	+r-1a	0.1	0.5	6	+r-1a	<0.1	0.5
-	-	-	13	+r-2b	0.4	3	4	+r-2b	0.1	3
-	+r	+p	10	+r-3a	0.5	5	3	+r-3a	0.2	5
	18	-	83	-	13 (18)		16	- 81	-	14 (18)

In addition in 1: Polypodium interjectum 1p; Thelypteris palustris +p; Polystichum falcatum +r; in 2: Polystichum falcatum 1a; Chamaenerion angustifolium +p; in 3; Spergularia rupicola 2a; Sagina maritima +p; Chamaenerion angustifolium, Mycelis muralis +r; Bryum caespiticium is ssp. kunzei; in 6: Polypodium interjectum, Stellaria alsine +p; Adiantum capillus-veneris, Arabidopsis thaliana, Corydalis lutea +r; Chrysanthemum parthenium, Matricaria chamomilla r; Hygroamblystegium tenax 1p; in 7: Erigeron acer 2m; Galinsoga ciliata 1a; Arenaria serpyllifolia +p; Schistidium apocarpum 2m; Orthotrichum anomalum, Camptothecium lutescens 1p; Tortella tortuosa, Fissidens taxifolius +p; in 8: Campanula rapunculoides 1a; Poa angustifolia, Artemisia vulgaris, Leontodon hispidus, Stellaria media +p; Matricaria inodora, Agrostis tenuis +r, Tortula marginata 1p; Barbula unguiculata +p; in 9: Campanula rotundifolia 1a; Poa pseudocompressa 1p; Mycelis muralis, Chamaenerion angustifolium, Cerastium fontanum +p; Hieracium lachenalii, H. cf. sabaudum +r; Trichostomum crispulum var. brevifolium 2a; Oreoweisia bruntonii +p; in 10: Rumex palustris 2a; Mentha aquatica 1p; Sonchus palustris, Ranunculus acris, Myosotis scorpioides, Sium erectum, Poa palustris +p; Cardamine pratensis, Peucedanum palustre, Hordeum murinum, Galinsoga parviflora, Galium mollugo, Festuca ovina, Carex acutiformis +r; Eucladium verticillatum 1p; in 11: Peucedanum palustre 2a; Cardamine pratensis +p; Rumex hydrolapathum, Senecio congestus, Ranúnculus repens +r; in 12: Matricaria maritima ssp. inodora 1b; Parietaria judaica 1a; Chamaenerion angustifolium, Festuca rubra, Brachythecium mildeanum +p.

LEGEND

1. Utrecht (Netherlands), wall along side Oude Gracht, facing nos 70-90.
2. Amsterdam (Netherlands), canal wall of Raamgracht, facing nos 67-69.
3. Penzance (Cornwall, England), wall above railway station in Lannoweth Terrace.
4. Muiden (Noord-Holland, Netherlands), wall in Heerengracht opposite nos 26-27.
5. Honfleur (Calvados, France), quay side wall of Bassin de l'Ouest.
6. Ottery St. Mary (Devon, England), wall along side Ridgeway opposite no. 19, facing a small stream.
7. Lofer (Salzburg, Austria), wall along road to Unken.
8. Praha-Dejvice (Czechoslovakia), wall along side of railwayline near the Starodejvicka and the Nad Bořislavkou.
9. Chřibská (Czechoslovakia), wall facing a stream in the centre of the town.
10. Blokzijl (Overijssel, Netherlands), quay side wall of Bierkade near the Noorderkade.
11. Amersfoort (Utrecht, Netherlands), canal wall of Kleine Spui in front of no. 20.
12. Dordrecht (Zuid-Holland, Netherlands), canal wall of Korte Geldersekade in front of no. 5.

6.1.1 Vegetation with Parietaria judaica and Bromus madritensis

The most common variant in the parts of the Mediterranean area that have been studied is the community with *Parietaria judaica* and *Bromus madritensis*. Roughly, two kinds of stands can in turn be distinguished within this community viz. stands with *Mercuralis annua* and stands with *Umbilicus rupestris*.

Table 4 Geographical distribution and exposure of stands of the communities of Parietaria judaica and Bromus madritensis

a. inops
b. with Mercurialis annua
c1. with Centranthus ruber and Umbilicus
c2. with Umbilicus rupestris
t. total

	exposure:	1	2	3	4	5	6	t	t%
geographical region:									
	320	1	2	3	1	-	-	7	70
a	331 - 332	1	-	-	2	1	-	3(+1)	30
	t	2	2	3	3	1	-	10(+1)	-
	t%	20	20	30	30	-	-	-	100
	310	-	1	1	-	-	-	2	4
	320	-	5	7	4	-	-	16	30
b	331 - 332	4	11	12	9	2	-	36(+2)	67
	t	4	17	20	13	2	-	54(+2)	-
	t%	7	31	37	24	-	-	-	100
	510	-	1	3	-	-	-	4	25
	320	-	-	-	-	-	-	-	-
c1	331 - 332	4	4	1	3	-	-	12	75
	t	4	5	4	3	-	-	16	-
	t%	25	31	25	19	-	-	-	100
	132	-	1	-	-	-	-	1	4
	310	-	1	2	-	-	-	3	11
	320	8	1	1	2	2	-	12(+2)	43
c2	331 - 332	2	5	3	2	1	2	12(+3)	43
	t	10	8	6	4	3	2	28(+5)	-
	t%	36	28	21	14	-	-	-	100
	132	-	1	-	-	-	-	1	1
	310	-	3	6	-	-	-	9	8
	320	9	8	11	7	2	-	34(+2)	32
t	331 - 332	11	20	16	16	4	2	63(+6)	58
	t	20	32	33	23	6	2	108(+8)	-
	t%	19	30	31	21	-	-	-	100

The community with *Mercurialis annua* (b) is more common on north-facing walls, that with *Umbilicus rupestris* (particularly c 2) on south-facing ones. In the inops form a stand can not always be identified as a depauperated or 'degenerated' stand of a particular community. In the most extreme environments, vegetation tends to be poor in species, often showing dominance of only a single species. Such depauperated stands have been refered to as **inops** (SEGAL, 1963a; this term has been suggested by Dr. V. Westhoff). In the case under discussion the exposure of the sites in which the inops form is encountered suggests that it is usually an impoverished example of the community with *Mercurialis annua,* which is not unexpected considering the more frequent occurrence of this community. The community with *Umbilicus* is connected with the community of *Parietaria judaica* and *Asplenium trichomanes* by a number of intermediate examples (see under 6.1.2), which latter community exhibits an even stronger preference for a north-facing exposure.

These communities constitute a link in the series: a1-a2-b-c (Table 3) towards the community of *Asplenium trichomanes* and *Sedum dasyphyllum* which is, generally speaking, accompanied by a shift of the south-facing exposure to a north-facing one. The community of *Parietaria judaica* and *Bromus madritensis* has a large number of differentiating species in respect of other types of mural vegetation. A high degree of presence is shown by e.g. *Bromus madritensis, Lamium amplexicaule, Campanula erinus, Mercurialis annua, Veronica cymbalaria* and *Barbula acuta. Veronica cymbalaria* possibly attains optimum development in such situations, but the ecology of this species is insufficiently known. Walls supporting stands of this community often teem with ants and with wall lizards. The inops stands average three species, the maximum being five. *Parietaria judaica* was present in all relevés, the MGP amounting to 40%. *Linaria cymbalaria* was recorded in six cases (P = 55%) with an MGP of 10%. Three species were each recorded twice viz. *Antirrhinum majus* (MGP = 2%), *Umbilicus rupestris* (MGP = 0.3%) and *Sonchus oleraceus* (MGP < 0.1%). In addition, 9 species were recorded only once with +p or +r representation, and one *(Antirrhinum latifolium)* with 2a representation.

The remainder of the vegetation types with *Bromus madritensis* is shown in Table 3, in which also two columns are included relating to the community of *Asplenium trichomanes* and *Sedum dasyphyllum,* and corresponding with those of Table 31. In the table the optimum development of a number of species among this group of communities is manifest. Apart from *Parietaria judaica,* especially *Linaria cymbalaria, Centranthus ruber, Sonchus oleraceus, Ficus carica, Bromus madritensis, Lamium amplexicaule, Veronica cymbalaria, Tortula muralis* and *Barbula acuta* are frequently represented. The relatively larger number of species of somewhat ruderal and open sites is striking. *Centranthus ruber* prefers less steep walls and particularly ledges. The species is too common in different communities with *Parietaria judaica* to justify the recognition of a separate vegetation type, although the two species have somewhat different requirements, the former usually appearing in a later successional phase and its settlement apparently being favoured by the accumulation of humus and a not very high pH. *Centranthus ruber* (and *Parietaria judaica* likewise) is sometimes dominant in ruderal vegetation (e.g. on rocky slopes near Brixham in Devonshire, England, with *Urtica dioica*). *Centranthus ruber* was also found growing in rock crevices, namely near Masera in northern Italy (albeit exclusively near the village, which suggests a secondary settlement). The original natural habitat of this species is presumably rocky slopes. On such slopes, for instance near the coast in Dorset and Devon (near Torquay, southern England), it is found in association with, among other species, *Crithmum maritimum. Sonchus oleraceus* occurs in numerous kinds of mural vegetation and may presumably be taken to be a differentiating species against the communities of true chasmogams in rock crevices. However, the individuals found growing on walls usually remain small and sterile, only rosettes being formed as a rule. *Ficus carica,* on the other hand, links mural vegetation with chasmogamic vegetation. This species, which is also encountered in other Mediterranean wall-dwelling communities with *Parietaria judaica,* usually only appears late in the succession, but on decomposed walls it develops into trees several metres high which bear fruit freely.

MEIER & BRAUN-BLANQUET (1934) considered the occurrence of *Phagnalon sordidum* an indication of a relationship with chasmophytic communities with *Asplenium glandulosum* (order *Asplenion glandulosi*), but this relationship appears to be very remote. *Phagnalon* also appears in a later successional stage and is especially encountered in stands already rich in species. *Barbula acuta* is of general occurrence in numerous communities of more or less open habitats in the Mediterranean area. This species is by no means restricted to the so-called *Onobrychidi-Barbuletum* of open sites, and not to the alliance of the *Thero-Brachypodion* either (compare BRAUN-BLANQUET et al., 1952). It is also quite common in vegetation with *Tillaea muscosa. Barbula acuta* can withstand appreciable fluctuations in humidity and also prolonged periods of drought. In the subunit with *Mercurialis annua, Bromus sterilis* and *Fumaria capreolata* are rather conspicuous.

These species, like many other species of the community (particularly the therophytes) develop principally in the spring and die off before the summer.

Parietaria communities of the Mediterranean area show a certain periodicity. *Mercurialis annua* is usually represented by its ssp. *huetii* which is more or less characteristic for rock crevices and walls (in contradistinction to the 'typical' ssp. *annua* which is normally only found on arable land and in ruderal habitats). Relevé no. 3 of Table 3 shows a transition to the community with *Parietaria lusitanica* (compare 6.1.4). The subunit with *Umbilicus rupestris* has a relatively large number of differentiating species in respect of the subunit with *Mercurialis annua,* although only a few of them are completely lacking in the latter subunit. The distinction between the two is, accordingly, partly of a quantitative nature, but it is also determined by the frequent co-occurrence of the differentiating species, the most important being *Campanula erinus, Geranium rotundifolium, Veronica polita, Senecio vulgaris, Euphorbia peplis, Lepidium graminifolium, Arenaria serpyllifolia, Catapodium rigidum, Cerastium glomeratum, Saxifraga tridactylites, Ceterach officinarum, Umbilicus rupestris, Sedum dasyphyllum, Asplenium trichomanes, Scorpiurium circinatum, Homalothecium sericeum* and *Streblotrichum convolutum.* A considerable proportion of these differentiating species are therophytes, chasmophytes or mosses. It is interesting to observe how many of these therophytes (which include a number of winter annuals) when they occur in higher latitudes or at a greater height above sea-level, tend to prefer more southern exposures or more horizontal substrata, such as tops of walls, and dunes. When in the hotter Mediterranean region, however, they seek some shelter. Stands with *Centranthus ruber* usually abound in species. A number of the above-mentioned differentiating species clearly attain their optimum development in these stands e.g. *Campanula erinus* and *Geranium rotundifolium*. However, also the (annual) *Bromus madritensis* and *Lamium amplexicaule* show the highest degree of presence. We may not assume that stands of the type a2-2 are intermediate between a2-1 and b, even though a number of species show a neat series of increasing degrees of presence in the sequence a1–a2-1–a2-2–b–c of Table 3, in which direction also a mounting preference for north-facing walls can be noticed. In this sequence, for instance, *Ceterach officinarum* and *Sedum dasyphyllum* progressively increase in number, whereas *Tortula muralis*, among other ones, exhibits a maximum of presence in type b. In the stands of type a2-2 the therophytes are conspicuous, and in a2-1 stands hemicryptophytes and chamaephytes are relatively important elements e.g. apart from *Centranthus ruber, Phagnalon sordidum, Sedum sediforme, Reichardia picroides, Antirrhinum majus* and *A. latifolium*, but also some summer annuals viz. *Geranium molle* and *Catapodium rigidum*. This environment is presumably not subjected to such extreme fluctuations in temperature or is more stable in other respects, possibly on account of a more advanced successional stage. *Antirrhinum latifolium* was recorded from a number of localities in southern France E. of Toulon. According to ROTHMALER (1956), who considers it to be a subspecies of *A. majus*, the distributional area of *A. latifolium* is restricted to S.E. France and N.W. Italy, but the latter was also found at Mourèze (Hérault, France) and is regularly encountered in Tuscany (Italy) e.g. in the island of Elba.

In stands with *Bromus madritensis* the qualitative contribution of therophytic forms is always large, but in the inops type of stands it is considerably less important (15% as against 39-45% in the richer stands, unweighted values, 0.1% and 5-9%, respectively, weighted values). Dominance of *Parietaria judaica* is, generally speaking, concomitant with a disappearance of the therophytes. Seasonal therophytes are of frequent occurrence in communities with *Umbilicus rupestris* (the score being U 10% and W 12% as against U 2% and W 1% in the stands with *Mercurialis annua*). Myrmecochory is always very important (the scores being 20-44% and 51-88%, respectively). In stands with *Mercurialis annua* and with *Umbilicus rupestris* many agricultural weeds occur, but their quantitative contribution remains small (their score: 11-12% U and 0.4-1% W, respectively). In a qualitative respect the number of species with Mediterranean distribution are more important than those with a Mediterranean-Atlantic distribution, but in a quantitative respect the reverse is the case (the scores being 20-26% and 8-9%, respectively, for Mediterranean species, and 16-22% and 63-78%, respectively for Mediterranean-Atlantic taxa; for inops stands these values are 9% and 2%, and 49% and 86%, respectively).

6.1.2 Mediterranean vegetation with Parietaria judaica and Asplenium trichomanes

Stands with *Parietaria judaica* and *Asplenium trichomanes* principally differ from the communities with *P. judaica* and *Bromus madritensis* in the absence of a number of taxa characteristic of the latter communities, such as *Phagnalon sordidum, Bromus madritensis, Lamium amplexicaule, Mercurialis annua, Veronica cymbalaria* and *Catapodium rigidum*. On the other hand, *Poa bulbosa, Asplenium trichomanes, A. ruta-muraria, Polypodium vulgare, P. australe, Sedum album, Scorpiurium circinatum, Homalothecium sericeum* and *Barbula revoluta* have a relatively high degree of presence. A number of these taxa also occur in the community of *Asplenium trichomanes* and *Sedum dasyphyllum* (see under 6.1.1).

Table 5 Geographical distribution and exposure of stands of the community of Parietaria judaica and Asplenium trichomanes

exposure:	1	2	3	4	5	t	t%
geographical region:							
132	-	-	-	1	-	1	3
310	3	3	1	1	2	8(+2)	27
320	2	-	1	-	-	3	10
331 - 332	8	5	-	5	1	18(+1)	60
t	13	8	2	7	3	30(+3)	-
t%	43	27	7	23	-	-	100

If the degree of presence of the diverse communities on north-facing walls is compared, a clearly progressive sequence can be observed:

	Type of community	Percentage on northern exposure	Percentage on southern exposure
a1	Community of Parietaria judaica and Bromus madritensis with Mercurialis annua	7	37
a2-1	Community of Parietaria judaica and Bromus madritensis with Centranthus ruber and Umbilicus rupestris	25	25
a2-2	Community of Parietaria judaica and Bromus madritensis with Umbilicus rupestris	36	21
b	Community of Parietaria judaica and Asplenium trichomanes	43	7
c	Community of Asplenium trichomanes and Sedum dasyphyllum	54	0

The occurrence of community type b on S.-facing walls is, moreover, restricted to higher altitudes. The only record in area 132 is from

Carcasonne (Aude, France) in the transitional area of the submediterranean and the Atlantic flora. It should also be borne in mind that the sites placed in the category with exposure 5 (overshadowed walls), even if they are S.-facing, almost always resemble N.-facing walls phytosociologically speaking. The community of *Parietaria judaica* and *Asplenium trichomanes* is not sharply distinguishable from the community with *Asplenium trichomanes* and *Sedum dasyphyllum* either. The latter is still poorer in species and, more particularly, lacks a number of the more typical taxa of the *Parietaria* communities, such as *Centranthus ruber, Sonchus oleraceus* and *Ficus carica,* whereas e.g. *Ceterach officinarum, Asplenium trichomanes, A. ruta-muraria, Sedum dasyphyllum,* and the hypnid mosses *Homalothecium sericeum, Hypnum cupressiforme, Rhynchostegiella tenella* and *Plasteurhynchium meridionale* (in 3 stands = 9%) attain the highest degree of presence. Furthermore, a number of nitrophilous species (such as *Senecio vulgaris* and *Bryum caespiticium*) are especially lacking in this community. The relative presence of *Parietaria judaica* is still considerable, but the abundance and the coverage are low. In the vicinity of Hyères (Var, France) *Cheilanthes fragrans* was recorded several times. At Trans-en-Provence (Var, France), in the transitional area from a submediterranean to a central European flora, stands were met within which also *Cystopteris fragilis, Bryum argenteum* and *Tortella tortuosa* occur and which apparently form a transition towards the central European *Asplenium trichomanes* communities (see under 6.6.4).

The stands were distinguished from those with *Bromus madritensis,* especially in a quantitative way, by the relatively low number of therophytes (score: 16% U, 0.4% W).

6.1.3 Vegetation with Adiantum capillus-veneris

Relevés of mural vegetation with *Adiantum capillus-veneris* were made in the areas 132, 310, 320, 331 and 332; all of damp or sheltered walls e.g. in a well at Carcasonne (Aude, France), on walls near rivulets or on walls along which water was trickling down. The composition of the stands is shown in Table 6.

By the presence in these stands of *Eucladium verticillatum, Samolus valerandi* and *Pellia endiviaefolia* they resemble stands with *Adiantum* communities of natural habitats (compare Braun-Blanquet et al., 1952: class *Adiantetea).* Occasionally a transition towards the community with *Parietaria judaica* and *Bromus madritensis* (Table 6, no. 1) was encountered, and once a transition towards the Atlantic *Asplenium trichomanes* communities (no. 3, compare 6.6.2). In one of the stands *Carex flacca* and *Schoenus nigricans* were found growing on a wall. At Valmary (no. 1) a zonation was found on a wall alongside a stream, with *Adiantum* vegetation at the water-level to about 0.3 m above, and above this vegetation of *Parietaria judaica* and *Bromus madritensis*. The combination of *Adiantum capillus-veneris* and *Parietaria judaica* (with, in addition, *Centranthus ruber* and *Ficus carica*) was recorded by FRIEDRICH (1954) from the area of Lake Garda in Italy, and from natural habitats by MOLINIER (1959) and HORVATIC (1963).

Table 6 Community of Adiantum capillus-veneris

Nr C	1	2	3	P	U W	MGT	MGP
Date	7-4	12-4	27-4	*10st*			
Year	62	62	62				
Nr R	053	081	228				
Ex	1w	3w	5w				
Geo R	310	332	132				
CH	30	30	45				
CM	30	2	10				
Adiantum capillus-veneris	2a	2a	2b	*100*	1b-4b	16	16
Parietaria judaica	2a	2b	1a	*60*	1a-2b	5	8
Asplenium trichomanes	1a	-	2b	*40*	+r-2b	2	5
Samolus valerandi	+p	1p	-	*40*	+r-1p	<0.1	0.1
Eucladium verticillatum	2a	2m	-	*30*	2m-3b	5	18
Schistidium apocarpum	1p	+p	-	*30*	+r-1p	<0.1	0.2
Ficus carica	+r	-	-	*30*	+r-+p	<0.1	<0.1
Sedum dasyphyllum	+r	-	-	*30*	+r-+p	<0.1	<0.1
Scorpiurium circinatum	2b	-	-	*20*	+p-2b	2	9
Amblystegium serpens	-	-	2a	*20*	+p-2a	0.8	4
Linaria cymbalaria	2a	+r	-	*20*	+r-2a	0.8	4
Centranthus ruber	-	+r	-	*20*	+r-2a	0.8	4
Ceterach officinarum	1a	-	1a	*20*	1a	0.3	2
Bryum caespiticium	1p	-	1p	*20*	1p	<0.1	0.4
Oxyrrhynchium swartzii	-	-	-	*20*	1p	<0.1	0.4
Bryum capillare	+p	-	-	*20*	+p	<0.1	0.1
Pellia endiviaefolia	+p	-	-	*20*	+p	<0.1	0.1
Tortula muralis	+p	-	-	*20*	+r-+p	<0.1	<0.1
Cystopteris fragilis	-	-	+r	*20*	+r	<0.1	<0.1
Umbilicus rupestris	+r	-	-	*20*	+r	<0.1	<0.1
MNS - MCH - MCM		*12*	-	*28*	-	*16*	

In addition in 1: Erigeron mucronatus 1a; Bromus madritensis, Campanula erinus, Polypodium vulgare, Taraxacum spec., Dactylis glomerata, Carex mairei +p; Sonchus oleraceus, Calamintha nepeta, Cardamine hirsuta, Origanum vulgare, Lepidium graminifolium +r; Barbula vinealis 2m; Pellia endiviaefolia +p; in 2: Hedera helix +p; in 3: Asplenium adiantum-nigrum 1b; Phyllitis scolopendrium 1a; Pteridium aquilinum 1a; Agrostis stolonifera +p; Dryopteris filix-mas +r; Barbula unguiculata 1p; Brachythecium rutabulum, Ceratodon purpureus +p.

LEGEND

1. Valmarie (Hérault, France), wall alongside a rivulet between two water mills; 0-0.5 m above water surface.
2. Castries (Hérault, France), wall behind the church, along which water was trickling down.
3. Carcasonne (Aude, France), old water well of the Château Comtal.

In communities with *Adiantum capillus-veneris* the number of hygrophytes is appreciable (score 23% U, 55% W). The representation of species with a Mediterranean-Atlantic distribution is rather high (score: 17% U, 29% W).

6.1.4 Vegetation with Parietaria lusitanica and/or Anogramma leptophylla

Mural vegetation with *Parietaria lusitanica* and/or *Anogramma leptophylla* was recorded in southern France (Department Var) and in Italy (Tuscany, especially Elba), but has also been found elsewhere in the Mediterranean region e.g. stands with *Anogramma* at Domodossola, N. Italy (TERRETAZ, 1964). At such sites is also found *Selaginella denticulata,* and mural vegetation with this species has been observed by

BRAUN-BLANQUET (1966) in northern Portugal. The exposure is shown in Table 7.

Table 7 Geographical distribution and exposure of stands of the communities of Parietaria judaica and Parietaria lusitanica and/or Anogramma leptophylla

exposure:	1	2	3	4	t	t%
geographical region:						
320	3	2	1	-	6	50
331	3	3	-	-	6	50
t	6	5	1	-	12	-
t%	50	42	8	-	-	100

A marked preference to N.- and E.-facing sites is quite apparent. The stands were always found on walls that remained more or less moist, at least during the winter season and in the spring, due to water trikling down, or to moisture from the soil (such walls being often built against heaped soil) or to capillary suction (low walls). *Anogramma leptophylla* is a thermophilous fern with a chiefly tropical and subtropical distribution that can not withstand low winter temperatures. According to TERRETAZ (1964) the sites where it occurs in N. Italy have, accordingly, a tropical microclimate: they are constantly humid and warm, the winter temperatures only occasionally dropping below 8° C. In the sites studied by him, warm spring water is responsible for the maintenance of the relatively high winter temperatures. These findings agree with those of MORTON & GAMS (1925). In the island of Elba I encountered the species repeatedly on step inclines and in rock crevices near streams, usually in association with *Selaginella denticulata, Targionia hypophylla, Rhynchostegiella tenella* and *Timmiella anomala.*

This community shows some resemblance to the *'association à Selaginella denticulata et Grammitis leptophylla'* (*Grammitis = Anogramma*), described by MOLINIER (1937) on the basis of a single relevé made on an island near Hyères (S. France), and to the relevés made by BRAUN-BLANQUET (1966) on Cambrian schists near Pessegeiro in the Basque country. The most important point of resemblance with those relevés is the co-occurrence of *Anogramma* and *Selaginella denticulata.* As early as 1915 BRAUN-BLANQUET recorded *Anogramma leptophylla* on the basal parts of walls in the Cevennes (S. France).

I encountered the combination *Anogramma-Selaginella* on walls on three occasions (i.e. in 25% of the relevés) and always in association with *Parietaria lusitanica* (which was presented in combination with *Anogramma* in 42% of the relevés). *Parietaria lusitanica* is undoubtedly more thermophilous than *P. judaica,* but is more 'poikilohygric' than *Anogramma leptophylla.* RIOUX & QUEZEL (1951) consider this species to be a faithful species of the association of *Phagnalon sordidum* and *Asplenium glandulosum* in the Montpellier area in S. France. In my opinion, *Asplenium glandulosum* is likewise a poikilohygric species. The composition of the community is shown in Table 8.

Table 8 Communities of Parietaria judaica and P. lusitanica and/or Anogramma leptophylla

Nr C	1	2	3		U W	MGT	MGP
Date	24-4	21-4	18-4				
Year	65	62	65				
Nr R	059	162	034				
Ex	1	2	3				
Geo R	320	331	320				
CH	30	50	30				
CM	40	80	s				
Parietaria judaica	2b	1a	1b	92	+p-3b	13	14
Umbilicus rupestris	1a	1b	-	83	+p-2a	2	3
Parietaria lusitanica	+p	1a	2b	75	+p-2b	3	4
Anogramma leptophylla	2m	3a	-	67	+r-3a	3	5
Fumaria capreolata	+p	-	-	58	+r-+p	<0.1	0.1
Tortula muralis	2m	3b	-	58	1a-3b	6	10
Polypodium australe	2a	+r	-	58	+r-2b	5	8
Campanula erinus	+p	-	2a	50	+p-2a	0.7	2
Geranium rotundifolium	-	+p	-	50	+r-+p	<0.1	0.1
Scorpiurium circinatum	-	2b	-	42	1b-3b	6	14
Catapodium rigidum	+p	-	-	42	+p-1p	0.1	0.4
Homalothecium sericeum	3a	-	-	33	+r-3a	3	10
Barbula acuta	2a	-	-	33	2m-2a	1	4
Sedum dasyphyllum	2m	-	+r	33	+r-2m	0.5	1
Orthotrichum diaphanum	+p	-	-	33	+p-2m	0.3	0.8
Asplenium trichomanes	2m	-	-	33	+r-2m	0.3	0.8
Stellaria media	+p	-	-	33	+r-1p	0.1	0.2
Polypodium vulgare	-	-	-	33	+r-+p	<0.1	0.1
Hyoseris radicata	+r	-	+p	33	+r-+p	<0.1	<0.1
Cerastium glomeratum	+p	-	-	33	+p	<0.1	0.1
Linaria cymbalaria	-	-	-	25	1a-2b	2	7
Bryum capillare	-	-	-	25	+p-2b	2	6
Selaginella denticulata	2a	-	-	25	2m-2a	0.7	6
Ceterach officinarum	-	-	-	25	1p-1a	0.3	1
Bryum caespiticium	2m	-	-	25	+p-2m	0.3	1
Hypnum cupressiforme	2m	-	-	25	+p-2m	0.3	1
Mercurialis annua	-	-	-	25	+r-+b	0.1	1
Hedera helix	-	-	-	25	+r-+a	0.1	0.5
Veronica hederifolia	+p	-	-	25	+p-1a	0.1	0.7
Geranium sanguineum	+p	-	-	25	+p	<0.1	0.1
Lamium amplexicaule	+p	-	-	25	+r-+p	<0.1	0.1
Rubus (div.) spec.	+p	-	-	25	+r-+p	<0.1	0.1
Sonchus oleraceus	-	-	+p	25	+r-+p	<0.1	<0.1
Urtica atrovirens	+r	-	-	25	+r	0	0
Barbula vinealis	-	3a	-	17	2m-3a	3	17
Polypodium interjectum	-	-	-	17	+p-2a	0.7	4
Streblotrichum convolutum	-	-	-	17	1p-2m	0.2	2
Bryum pallens	-	+p	-	17	+p-2m	0.2	1
Geranium pyrenaicum	-	+a	-	17	+p-1a	0.1	0.8
Dactylis glomerata	-	+a	-	17	+r-1a	0.1	0.8
Erophila verna	-	-	-	17	+p-1p	<0.1	0.3
Targionia hypophylla	-	1p	-	17	+p-1p	<0.1	0.3
Bromus sterilis	+p	-	-	17	+p	<0.1	0.1
Cardamine flexuosa	+p	-	-	17	+p	<0.1	0.1
Geranium lucidum	-	+p	-	17	+p	<0.1	0.1
Phagnalon sordidum	-	-	-	17	+r-+p	<0.1	0.1
Avena barbata	-	-	-	17	+r-+p	<0.1	0.1
Bromus madritensis	-	-	+p	17	+r-+p	<0.1	0.1
Mentha rotundifolia	+p	-	-	17	+r-+p	<0.1	0.1
Veronica polita	+r	-	-	17	+r	0	0
MNS - MCH - MCM		19	-	32	-	20	

In addition in 1: Asplenium adiantum-nigrum 1p; Lamium maculatum, Erophila verna, Sedum reflexum, Ranunculus spec. +p; Pteridium aquilinum, Vaillantia muralis, Senecio lividus +r; in 2: Senecio vulgaris, Pellia endiviaefolia +r; in 3: Veronica cymbalaria 1a; Geranium molle +p.

LEGEND

1. Ramatuelle (Var, France), foot of wall near Rue des Moulins Roux.
2. Poggio (Elba, Italy), wall alongside the road to Marciana Marina.
3. Volterraio (Elba, Italy), wall of a haybarn alongside the road from Portoferraio to Rio Marina.

The community differs from other kinds of mural vegetation in the occurrence, or the relatively high degree of presence, of a fairly large number of species. Apart from *Parietaria lusitanica, Anogramma leptophylla* and *Selaginella denticulata,* the following taxa are important as differentiating species: *Polypodium australe* and *Geranium sanguineum.* The community shows, to some extent, a resemblance with that of *Parietaria judaica* and *Bromus madritensis,* especially with its variant with *Umbilicus rupestris,* and also, on account of the high degree of presence of e.g. *Polypodium australe, P. vulgare, Scorpiurium circinatum* and *Homalothecium sericeum,* with the community of *Parietaria judaica* and *Asplenium trichomanes.* Stands with *Anogramma leptophylla* contain a high percentage of winter annuals. *Anogramma leptophylla* itself, and also *Selaginella denticulata,* can be regarded as winter annuals. The latter species tends to be perennial in colder regions, however (NEGRE, 1966). Whether the stands with *Anogramma* and those with *Parietaria lusitanica* can be considered to belong to the same community or not remains to be seen and this would, at any rate, require a more thorough study substantiated by more material. That they represent at least one distinct vegetation units is, to my mind, certain.

The question of how to classify communities with *Anogramma leptophylla* in the system of the Franco-Swiss School is not so easily answered, however. This fern does not occur only in chasmophytic vegetation and the assignment of the communities in question to the *Asplenio-Sedion* by BRAUN-BLANQUET (1966) seems to be decidedly premature, quite apart from the consideration that this alliance is not based on very solid evidence (even its name is obscure!). The assignment of the *Selaginello-Grammitetum leptophylli* Molinier 1937 to the *Isoetion* by BRAUN-BLANQUET et. al. (1952) does not seem to be a very fortunate suggestion either.

The scores of the therophytes are high (43% of the unweighted values and 16% of the weighted ones) and so are those of the hygrophytes (U 21% and W 15%). In these communities, too, myrmecochory is important (the score being U 27%, W 51%). Many of the species have a distributional area including the Mediterranean region (their score being U 26% and W 29%) or a Mediterranean-Atlantic one (score: 18% U, 46% W).

6.1.5 Vegetation with Parietaria judaica and Capparis spinosa

Mural vegetation with *Parietaria judaica* and *Capparis spinosa* has only been studied by means of relevés in Tuscany (Italy), but was also recorded in S.E. France and Portugal. ARENES (1929) recorded *Capparis spinosa* from a wall in the Provence and BRAUN-BLANQUET (1966) mentions an association of the two species with *Antirrhinum barrelieri* on the city walls of Tarragona in Spain. In Italy two groups of relevés can be distinguished, one with a more frequent occurrence of *Centranthus ruber* and the other with a high presence of *Ficus carica.*

Table 9 Exposure of stands of communities of Parietaria judaica and Capparis spinosa

a. with Ficus carica
b. with Centranthus ruber

	exposure:	1	2	3	4	t	t%
a		-	5	9	-	14	52
b		3	6	3	1	13	48
t		3	11	12	1	27	-
t%		11	41	45	4	-	100

In Italy the community shows a preference for S.- and E.-facing walls over N.- and W.-facing ones. In central Europe, in contrast, vegetation of E.-facing walls corresponds more often with that of N.-facing walls than with that of S.-facing ones (see under 6.6.4). Possibly this difference is to be ascribed to the drier climate in Tuscany, the prevailing west and north winds increasing the ecological resemblance between S.- and E.-facing walls. *Capparis spinosa* is well-adapted to periods of prolonged drought because of its coriaceous leaves. The composition of the stands is indicated in Table 10.

Table 10 Communities of Parietaria judaica and Capparis spinosa
a. with Ficus carica
b. with Centranthus ruber

	a							b							total			
Nr C	1	2	3	P	U W	MGT	MGP	4	5	6	P	U W	MGT	MGP	P	U W	MGT	MGP
Date	3-5	28-4	4-5	14st				5-5	28-4	27-4	13st				27st			
Year	65	65	65					65	65	65								
Nr R	120	083	131					139	087	070								
Ex	3	3	3					1	3	1								
CH	20	65	20					20	45	10								
CM	s	-	-					s	-	-								
Capparis spinosa	2a	2a	2b	100	2a-2b	12	12	2b	3a	2a	100	+r-3a	9	9	100	+r-3a	10	10
Parietaria judaica	2b	4a	+r	93	+r-4a	17	19	2a	2b	+p	100	+p-3a	9	9	96	+r-4a	14	14
Sonchus oleraceus	+p	-	+p	86	+r-1b	0.1	0.1	+p	+p	-	46	+p-1a	0.2	0.3	67	+r-1a	0.1	0.2
Mentha rotundifolia	1b	+p	+p	64	+r-1p	0.1	0.2	1p	-	+p	46	+r-1p	<0.1	0.1	56	+r-1p	<0.1	0.1
Bromus madritensis	1p	-	1p	43	+p-1p	0.1	0.3	+p	-	-	31	+r-+p	<0.1	0.1	37	+r-1p	<0.1	0.1
Linaria cymbalaria	-	-	+p	50	+p-2a	2	3	1p	+p	-	93	+p-2a	2	3	70	+p-2a	2	3
Helichrysum italicum	1a	-	-	29	+p-1a	0.1	0.5	-	-	+p	38	+r-+p	<0.1	0.1	33	+r-1a	<0.1	2
Oxalis corniculata	-	+p	+p	36	+p-1a	0.2	0.4	-	-	-	15	+r	0	0	26	+r-1a	0.1	0.3
Tortula muralis	-	-	+p	21	+p	<0.1	0.1	1p	-	-	15	1p	0.1	0.4	18	+p-1p	<0.1	0.2
Ficus carica	+p	+p	+a	50	+r-+b	0.6	1	-	-	-	-	-	-	-	26	+r-+b	0.3	1
Bromus sterilis	-	2m	-	21	1p-2m	0.2	1	-	-	-	7	+p	<0.1	0.1	15	+p-2m	0.1	0.9
Veronica cymbalaria	-	-	-	21	+p	<0.1	0.1	-	-	-	-	-	-	-	11	+p	<0.1	0.1
Hordeum murinum	-	-	1p	14	1p	0.1	0.4	-	-	-	-	-	-	-	7	1p	<0.1	0.4
Veronica arvensis	-	1p	-	14	+p-1p	<0.1	0.3	-	-	-	-	-	-	-	7	+p-1p	<0.1	0.3
Hypericum perforatum	+p	-	-	14	+p	<0.1	0.1	-	-	-	-	-	-	-	7	+p	<0.1	0.1
Mercurialis annua	-	-	-	14	+p	<0.1	0.1	-	-	-	-	-	-	-	7	+p	<0.1	0.1
Inula viscosa	-	-	+p	14	+r-+p	<0.1	0.1	-	-	-	-	-	-	-	7	+r-+p	<0.1	0.1
Veroñica hederifolia	-	-	-	7	+r	0	0	+r	-	+p	77	+r-1p	0.1	0.2	41	+r-1p	0.1	0.1
Avena barbata	-	-	-	7	+r	0	0	-	-	+p	23	+r-+p	<0.1	<0.1	15	+r-+p	<0.1	<0.1
Reichardia picroides	-	-	-	7	+p	<0.1	0.1	-	-	-	15	+r-+p	<0.1	0.1	11	+r-+p	<0.1	0.1
Sedum dasyphyllum	-	-	-	-	-	-	-	1p	-	1p	62	1p	0.2	0.4	30	1p	0.1	0.4
Centranthus ruber	-	-	-	-	-	-	-	+r	-	-	54	+r-2a	0.8	1	26	+r-2a	0.4	1
Catapodium rigidum	-	-	-	-	-	-	-	+p	-	+p	54	+r-+p	<0.1	0.1	26	+r-+p	<0.1	0.1
Antirrhinum latifolium	-	-	-	-	-	-	-	+r	1a	-	31	+r-1a	0.1	0.4	15	+r-1a	<0.1	0.4
Umbilicus rupestris	-	-	-	-	-	-	-	+r	-	+p	31	+r-1p	<0.1	0.2	15	+r-1p	<0.1	0.2
Geranium molle	-	-	-	-	-	-	-	-	-	+r	23	+r-+p	<0.1	<0.1	11	+r-+p	<0.1	<0.1
Lobularia maritima	-	-	-	-	-	-	-	-	-	+p	23	+r-+p	<0.1	0.1	11	+r-+p	<0.1	0.1
Antirrhinum majus	-	-	-	-	-	-	-	-	1b	-	15	+p-1b	0.3	2	7	+p-1b	0.1	2
Dactylis glomerata	-	-	-	-	-	-	-	-	-	1p	15	+p-1p	<0.1	0.3	7	+r-1p	<0.1	0.3
Sedum reflexum	-	-	-	-	-	-	-	-	-	(+p)	15	+p	<0.1	0.1	7	+p	<0.1	0.1
Papaver rhoeas	+p	-	-	14	+p	<0.1	0.1	-	-	-		-	-	-	7	+p	<0.1	0.1
MNS - MCH - MCM		8	-	28	-	0 (s)			10	-	19	-	0 (s)		8	- 24	-	0

In addition in 1: Campanula erinus, Melica ciliata +p; in 2: Urtica atrovirens, Stellaria media +p; in 6: Arum italicum +r.

LEGEND

1. Siena, wall facing Strada Esterna di Fontebranda, outside of the Porta Fontebranda.
2. Fucecchio, city wall.
3. Poggibonsi, wall along high road nr 2 opposite km-sign V 252.
4. Florence, wall near Forte di Belvedere at the beginning of the Via di S. Leonardo.
5. Certaldo, city wall along Vicolo dell'Osteria near nr 32.
6. Pisa, wall facing the Dome square.

The community is most closely related to the community of *Parietaria judaica* and *Bromus madritensis*, but is differentiated from the latter by the presence of *Capparis spinosa* and by high degrees of presence of *Helichrysum italicum, Mentha rotundifolia* and *Veronica hederifolia* as against the absence or low degrees of presence of e.g. *Phagnalon sordidum, Rubus* spec., *Ceterach officinarum* and numerous therophytes and mosses (such as *Mercurialis annua, Lamium amplexi-*

caule, Geranium rotundifolium, species of *Barbula* and *Bryum,* and *Hypnaceae).* The average number of species per relevé is 9 and this is much lower than that of stands with *Bromus madritensis.* The structure also differs considerably.

Horvatic (1963) described an association found on old walls and rubble heaps in the island of Pag (Yugoslavia). This community he called *Asplenio-Cotyledonetum horizontalis,* and he presumed *Capparis spinosa, Theligonum cynocrambe* and *Cheilanthes fragrans* to be faithful species, with *Parietaria judaica* (referred to as '*P. vulgaris* Hill.') as an important companion. This vegetation type is probably closely allied to Tuscan *Capparis spinosa* stands, and the latter may even be an impoverished representation of the *Asplenio-Cotyledonetum horizontalis.*

In stands with *Capparis spinosa* many therophytes occur (their score: 26-29% U, 1-2% W) and microphanerophytes are also important (their score: 11-19% U and 37-41% W). Seasonal xerophytes are qualitatively important in stands with *Centranthus ruber* (their score: 11% U and 1% W), but they are altogether absent in the stands with *Ficus carica.* Myrmecochory is important in both types of stands (the score of the myrmecochores being U 29% and W 45%, and U 35% and W 56%, respectively). The constituting taxa mostly have either a Mediterranean distribution (score: U 28-30% and W 37-41%) or a Mediterranean-Atlantic one (score: U 21-26% and W 45-54%).

6.1.6 Vegetation with Parietaria judaica and Hyoscyamus albus

Stands of mural vegetation with *Parietaria judaica* and *Hyoscyamus albus* agree with stands having *P. judaica* and *Capparis spinosa* in so far that they are poor in species and that the bryophyte layer is usually poorly developed (although upon the whole somewhat better than in the latter community). The resemblance with the community of *P. judaica* and *Bromus madritensis* is considerably greater, but the differences are nevertheless appreciable. The community seems to be less common in areas where *Capparis spinosa* occurs.

Table 11 Geographical distribution and exposure of stands of the community of Parietaria judaica and Hyoscyamus albus

exposure:	1	2	3	4	t	t%
geographical region:						
310	-	-	1	-	1	9
320	-	-	1	-	1	9
331 - 332	-	6	-	3	9	82
t	-	6	2	3	11	-
t%	-	55	18	27	-	100

The most frequently occurring exposure is E. whilst exposures to the N. are entirely lacking. *Hyoscyamus albus* withstands long dry spells reasonably well. The floral composition of the community can be found in Table 12.

Table 12 Community of Parietaria judaica and Hyoscyamus albus

Mr C	1	2	3	P	U W	MGT	MGP
Date	12-4	20-4	26-4	11et			
Year	62	62	62				
Nr R	085	148	223				
Ex	2	2	4				
Geo R	332	331	332				
CH	30	50	12				
CM	-	-	7				
Parietaria judaica	2b	3b	2a	100	1b-5a	27	27
Hyoscyamus albus	2b	1b	1a	100	+a-2b	5	5
Sonchus oleraceus	-	+r	-	45	+r-1p	0.1	0.2
Tortula muralis	-	-	2m	36	1p-2m	0.6	2
Veronica polita	+r	-	1p	36	+r-1p	0.1	0.2
Euphorbia helioscopia	+r	-	+r	36	+r	0	0
Barbula acuta	-	-	2a	27	1p-2a	1	4
Bromus sterilis	-	-	+r	27	+p-2m	0.2	0.8
Mercurialis annua	-	-	-	27	+r-1p	0.1	0.2
Dactylis glomerata	+r	-	-	27	+r	0	0
Fumaria capreolata	-	-	-	18	1a-2a	0.9	5
Oxalis corniculata	-	-	-	18	1p-2m	0.1	2
Senecio vulgaris	-	-	1p	18	+r-1p	0.1	0.2
Lepidium graminifolium	-	+r	-	18	+r	0	0
Phagnalon sordidum	-	+r	-	18	+r	0	0
MNS - MCH - MCM		9	-	34	-	1 (3)	

In addition in 2: Linaria cymbalaria 2a; Stellaria media 1p; Sonchus asper, Bromus spec. +p; Inula viscosa +r; in 3: Capsella bursa-pastoris, Sisymbrium irio 1p; Lamium amplexicaule +p; Erophila verna +r.

LEGEND
1. Castries (Hérault, France), wall near the church.
2. Hyères (Var, France), wall near Rue Lamalgue.
3. Montbrun (Aude, France), garden wall.

Floristically the community is rather clearly differentiated compared to the community with *Bromus madritensis* by the occurrence of *Hyoscyamus albus,* and less so by the presence of *Euphorbia helioscopia* and *Veronica polita,* but it is mainly negatively characterised by the absence of the same group that are lacking in the stands with *Capparis spinosa* (see under 6.1.5). Presumably the latter community is more thermophilous and possibly also less nitrophilous; the community with *Hyoscyamus albus* is floristically differentiated against it by the presence of *Phagnalon sordidum* and *Poa bulbosa,* in addition to the above-mentioned species, and it is, furthermore, characterised by the absence (or low presence) of e.g. *Capparis spinosa, Helichrysum italicum, Ficus carica, Mentha rotundifolia* and *Veronica hederifolia.*

Oberdorfer (1954) described three relevés from Greece of *Parietaria judaica* vegetation with *Hyoscyamus albus* in which *Sisymbrium irio* appeared twice and which he classified among ruderal vegetation of the alliance *Chenopodion muralis.* The community he describes, in my opinion, links up with the community described in this sub-chapter.

Arenes (1929) recorded *Hyoscyamus albus* from a wall in the Provence (S. France). In one relevé *Cheiranthus cheiri* was encountered (with the representation +a).

Therophytes are quantitatively important in the community with *Hyoscyamus albus* (their score being U 42%, W 5%). Myrmecochory is a very important form of dispersal (the score being U 44%, W 94%). Many species have a Mediterranean distribution (their score: 32% U, 14% W), or a Mediterranean-Atlantic one (their score: 18% U, 81% W).

As early as 1915 BRAUN-BLANQUET described mural vegetation from the Cevennes, but a clear insight was not obtained until later. The description of a *Filicetum murale* and a *Parietarietum murale* from the Provence (S. France) by ARENES (1929), who contributed only a few lists of species that give a not very homotonous [1]) picture of such vegetation, did not increase the insight appreciably. His *Filicetum murale* shows some resemblance to the community of *Asplenium trichomanes* and *Sedum dasyphyllum* (to be discussed under 6.6.1) and a few of his relevés on which the *Parietarietum murale* was based are related to communities of *Parietaria judaica* with *Bromus madritensis* and of *P. judaica* with *Asplenium trichomanes*. Later, BRAUN-BLANQUET (1931) emended the *Parietarietum murale* of Arènes as "Association à *Parietaria ramiflora* et *Oxalis corniculata*" without giving a clear description with pertaining relevés. However, a survey was given by BRAUN-BLANQUET et al. (1952). The available data were compiled by NICKLFELD & MEIER (1962) who distinguished:

1. a *Parietaria ramiflora* facies (corresponding with the initial stages of the succession series);
2. a *Phagnalon sordidum* facies, not found on north-facing walls (BRAUN-BLANQUET et al., 1952: 'variant with *Phagnalon sordidum*'), and
3. a *Ceterach officinarum* facies without *Parietaria* but with *Asplenium trichomanes* and species of *Sedum*.

The material is not very homotonous either. Their facies (2) agrees roughly with communities of *P. judaica* and *Bromus madritensis*, and of *P. judaica* and *Asplenium trichomanes*, and (3) with the community of *Asplenium trichomanes* and *Sedum dasyphyllum*. Up to that time .these communities were included in the *Asplenion glandulosi*, *Phagnalon sordidum* being supposed to be indicative of their relationships, but in 1966 BRAUN-BLANQUET referred the *Oxalido-Parietarietum ramiflorae* to the alliance of the *Parietario-Galion murale* of Martínez 1955 and to the order *Parietarietalia murale* Martínez 1955. A description of these units can be found in MARTINEZ (1960), who mentions as faithful species of the order: *Parietaria judaica, Linaria cymbalaria, Ficus carica, Mercurialis annua* and *Vaillantia muralis*. However, *Ficus carica* frequently occurs in chasmophytic vegetation, *Mercurialis annua* (at least ssp. *annua*) exhibits optimum development in ruderal sites, and *Vaillantia muralis* is only rarely found growing as a chasmophyte but more commonly in vegetation on tops of walls and on open rocky slopes. Martínez distinguished two alliances viz. the *Parietario-Galion murale* and the *Parietario-Centranthion rubri*, of which he supposed the former to be present in southern Spain in areas where *Quercus ilex* vegetation (*Quercetalia ilicis*) predominates and to be characterised by *Antirrhinum hispanicum* var. *hispanicum, Oryzopsis miliacea, Parietaria mauretanica, P. lusitanica* and *Galium murale*, and the latter in oceanic regions where non-deciduous hardwood forests (*Querco-Fagetea*) prevail with the faithful species *Hypericum hircinum, Centranthus ruber* and *Daucus gummifer*. However, *Galium murale* and *Daucus gummifer* attain - presumably also in Spain - optimum development in other habitats:
G. murale prefers the same sites as *Vaillantia muralis*, and *Daucus gummifer* is frequently found growing on coastal rocks (often in association with *Crithmum maritimum*). The wide-spread occurrence of *Centranthus ruber* in Mediterranean

[1]) Compare NORDHAGEN (1954): homogeneity of the relevés in a table.

areas where *Quercus ilex* is endemic and where non-deciduous forests are wanting, at least in France and in Italy, does not support the suggestions made by Martínez. The almost complete, or total absence, of the characteristic species of the alliance in the *Oxalido-Parietarietum ramiflorae* in Braun-Blanquet (1966) and in the relevés of central European(!) mural vegetation in Oberdorfer (1967) does not make it very plausible that the vegetation types under discussion belong to an alliance "*Parietario-Galion murale*". For the time being it is not clear what the meaning is of the two alliances described by Martínez. Oberdorfer proposed a class for mural vegetation: '*Cymbalario-Parietarietea diffusae*', mentioning as characteristic wall-dwelling species all those enumerated in Segal (1962b): *Linaria cymbalaria, Parietaria judaica, Corydalis lutea, Antirrhinum majus, Centranthus ruber* and *Erigeron mucronatus.* Oberdorfer's proposal is not accepted by the present author, not only because many stands of mural vegetation with *Linaria cymbalaria*, especially in W. and central Europe, do not contain *Parietaria judaica*, but also because it is thought that the classification of mural vegetation with other types of chasmophytic vegetation in the class *Asplenietea rupestris* is more appropriate, both on floristic and on structural-ecological grounds.

Pignatti (1953) distinguished an association of *Linaria cymbalaria* and *Parietaria ramiflora* in N. Italy, in which *Campanula pyramidalis* (an endemic species of the eastern Alps) and *Sedum album* appear in a high degree of presence (III = 40-60%). The combination of taxa present indicates a community which is closely related to that of *Parietaria judaica* and *Asplenium trichomanes*, and can presumably best be considered to represent a sub-unity of the latter. Lorenzoni (1961) described a subassociation with *Ceterach officinarum* which does not contain *Parietaria judaica*, however, and rather belongs to the community of *Asplenium trichomanes* and *Sedum dasyphyllum.* Both Italian workers referred these communities to the alliance of the *Potentillion caulescentis.*

Braun-Blanquet (1966) distinguished a Spanish subassociation of the *Parietaria judaica* community rich in mosses, which he named *Oxalido-Parietaretum ramiflorae homalothecietosum.* Its species combination shows quite clearly that this supposed subassociation in fact represents the community of *Parietaria judaica* and *Asplenium trichomanes.* The *Parietaria judaica* communities studied are nearly all connected by transitional cases and an attempt has been made here to construct a picture of the situation without emphasising any discontinuities which are non-existent anyway. The classification proposed here is based on abstractions, but the delimitations are not altogether picked at random but are based on careful considerations. It is necessary to propose a system for pragmatic reasons, in the first place to enable a simple and all-embracing exchange of views in an efficient way. Two alternative possibilities out of a much greater number have been selected for proposal, but the present author decidedly prefers the first solution.

Class: *Asplenietea rupestris* Meier & Braun-Blanquet 1934

Order: *Tortulo-Cymbalarietalia*

Alliance: *Parietarion judaicae*

A s s o c i a t i o n s:

1.	Bromo-Parietarietum judaicae	Oxalido-Parietarietum judaicae Br.-Bl. 1931
	a. mercurialetosum annuae	brometosum madritense
	b. umbilicetosum	—
2.	Sedo-Parietarietum judaicae	homalothecietosum
	a. campanuletosum pyramidalis	—
3.	Adianto-Parietarietum judaicae	(adiantetosum)
4.	Anogrammo-Parietarietum	(parietarietosum lusitanicae)
5.	Capparidi-Parietarietum judaicae	(capparidetosum spinosae)
	a. ficetosum caricae	—
	b. centranthetosum rubrae	—
6.	Hyoscyamo-Parietarietum judaicae	(hyoscyametosum albi)

Connected by transitional stands are especially the pairs 1-2, 1-5, 1-6, 2-3 and 2-4. In addition, there are transitions towards Atlantic and central European communities with *Parietaria judaica* and *Cheiranthus cheiri* to be discussed presently, but also towards vegetation of trampled sites with *Poa annua* (more particularly with types occurring near wall bases) in which in the Mediterranean region also *Cynodon dactylon* is usually represented. An example of the *Bromo-Parietarietum poetosum annuae* follows:

No. 65121, 1-5-1965: Siena (Italy), near base of wall, 0.1 x 10 m²; exposure: S; herb layer: 40% coverage.

Parietaria judaica	2a	Cynodon dactylon	2a
Oxalis corniculata	2a	Mentha rotundifolia	1a
Poa annua	2a	Sonchus oleraceus	+p

Vegetation type no. 1a can be regarded as a *Parietaria* community with a relatively high percentage of ruderal taxa, and no. 1b as a transition towards the alliance of the *Cymbalario-Asplenion* to be discussed in the following pages, or to chasmophytic communities of that same alliance. The vegetation no. 2a presumably represents a transition towards *Sedo-Scleranthetea* communities; the nos. 5a and 5b are roughly comparable to the nos. 1a and 1b, respectively. The initial stages of the various communities and the impoverished inops variants have been disregarded in this short survey.

6.1.8 Vegetation with Cheiranthus cheiri

The question whether the wallflower is truly indigenous in S.W.-, W.- and central Europe is still undecided. Some workers are of the opinion that it was originally endemic in southern Greece and the Greek Archipelago (Rouy & Foucauld, 1893; *Flora Europaea*); many others report an initially "Mediterranean" origin. Hegi, for instance, is of the opinion that this species occurs in central Europe rather frequently as a garden escape and is now firmly established in lime-rich sites. However, the distributional data clearly point to a different interpretation: *Cheiranthus cheiri* is chiefly encountered in the valleys of the great rivers and in the coastal regions of W.- and central Europe and it has, most probably, followed the same migration route as comparable species, such as *Parietaria judaica, Corydalis lutea* and possibly also *Linaria cymbalaria,* in addition to a number of species of ferns (compare 3.3.1). In central Europe the wallflower is regularly encountered on rocky slopes, but upon the whole only in the immediate vicinity of human settlements where enrichment of the substratum with nitrogen compounds seems likely. *Cheiranthus cheiri* prefers a not excessive inclination of its substratum and mostly occurs with *Sedum album* and *Poa compressa* (e.g. in the area of Dinant, Belgium), in a community which has so far not been described and presumably is the subsequent phase of succession after communities of exposed rocky slopes *(Sedo-Scleranthetea).* In S. England (Dorset) the species under discussion was found growing on rocky coastal cliffs, associated with *Lobularia maritima* and *Matthiola incana.* In northern Brittany (Plévenon) *Cheiranthus* occurred with *Parietaria* in rock crevices where *Asplenium marinum* was dominant and also *Armeria maritima* occurred, but also in this locality the sites are in the proximity of human settlements where the wall flower was growing on walls, so that in this area its occurrence on rocks need not be a primary one. Communities with *Cheiranthus cheiri* in S.-, W.- and central Europe differ appreciably in their specific composition, but also in this case there are transitions which can most clearly be observed in the areas linking the Mediterranean and the Atlantic regions, the Loire basin apparently being the most important geographical link and the walls of the numerous old castles along that river having provided a secondary environment for the plant during its migration. *Cheiranthus cheiri* is a nitrophilous species which shows a dislike for very steep substrata and, accordingly, hardly ever acts as a pioneer of mural vegetation. It usually appears only after the wall is strongly decomposed and much fine-grained sediment and humus have accumulated. In the various geographical regions wall vegetation with *Cheiranthus* is preceded by a different type of pioneer community. In southern Europe the pioneer stands are usually poorly developed stands of *Parietaria judaica* communities, whilst in W.- and central Europe they are, as a rule, stands of the *Filici-Saginetum* (see 6.9.1) or a pioneer community of *Linaria cymbalaria* and *Asplenium ruta-muraria.*

In central Europe sometimes an association with *Hieracium amplexicaule* (see under 6.4) represents this stage. Now and then the *Cheiranthus* stands develop better on the upper portions of the wall than the pioneer community and thus form a zonation which is also the result of the stronger decomposition of the upper parts of the wall. The author saw a very striking example near Volterra (Tuscany, Italy) with *Parietaria judaica* vegetation below the zone formed by *Cheiranthus* vegetation.) If wallflower vegetation is preceded by *Parietaria* vegetation, it is very probable that as long as the plant cover is rather sparse, *Parietaria judaica* and *Linaria cymbalaria* can still increase after *Cheiranthus* has become established but are ultimately ousted out by *Cheiranthus* and by *Centranthus ruber*. This could be deduced by a comparative study, in a number of places, of the stands of vegetation on walls of corresponding structure, composition, exposure etc., but of different age. In Cahors (Lot, France), for instance, two parts of a wall alongside the river Lot were compared (on 28-4-1962) and the differences between the stands they supported presumably agree with successional stages. Relevé b is the one made of an older wall:

	a	b
Number relevé	62233	62344
Inclination	90°	88°-90°
CH	35(50)	50(70)
CM	5	15
Number of species	12	16
Parietaria judaica	2b	2a
Linaria cymbalaria	2b	1b
Centranthus ruber	1a	2b
Cheiranthus cheiri	+r	2b
Asplenium trichomanes	+r	—
Barbula vinealis	+p	2a
(and other ones)		

The subsequent development after a precursory phase consisting of *Linaria cymbalaria - Asplenium ruta-muraria* vegetation is almost invariably associated with a decrease in numbers of these two species as the wallflower becomes more numerous (see under 6.1.11).

Stands of *Cheiranthus cheiri* most probably represent local climax vegetation in many instances. In Caen (W. France) transitions towards mural scrub vegetation were noticed, with representation of such species as *Acer pseudoplatanus, Fraxinus excelsior, Clematis vitalba,* and *Hedera helix*. The trees attain a height of 4.5 m. This community was also rich in animal life and contained e.g. various snails, ants *(Lasius niger* and *L. flavus),* wall lizards, and wasps nesting in the wall.

6.1.8.1 Vegetation with Cheiranthus cheiri and Sedum dasyphyllum

In the parts of the Mediterranean region visited by the author the wallflower is not very common and the impression was gained that the species becomes more numerous towards higher altitudes and towards the Atlantic coast.

Table 13 Geographical distribution and exposure of stands of the community of Cheiranthus cheiri and Sedum dasyphyllum

exposure:	1	2	3	4	t	t%
geographical region:						
132	1	-	-	-	1	7
310	-	1	4	-	5	33
320	1	1	1	-	3	20
331-332	4	-	1	1	6	4
t	6	2	6	1	15	-
t%	40	13	40	7	-	100

In the same direction a shift of the exposition of the stands can be observed. In the eu-mediterranean zone the species prefers N.-facing sites, but in submediterranean regions it is mostly seen on S.-facing walls. *Cheiranthus cheiri* most probably can not withstand prolonged periods of droughts and high temperatures or strong fluctuations in the temperature as well as *Parietaria judaica* does. The one record of region 132 in the table (near Carcasonne, S. France) is situated in the border area between that region and area 310. The composition of the stands is indicated in Table 14, and it shows that this community is somewhat allied to the *Sedo-Parietarietum,* from which it mainly differs in the relatively high degree of presence of *Cheiranthus cheiri, Centranthus ruber* and *Bromus sterilis,* and in the absence or relatively low degree of presence of *Asplenium trichomanes, A. ruta-muraria, Polypodium australe, Homalothecium sericeum, Barbula revoluta,* and other species. In respect of the W.- and central European wallflower communities, the Mediterranean stands are differentiated, among other things, by the relatively high degree of presence of *Sedum dasyphyllum, Ceterach officinarum, Umbilicus rupestris, Barbula acuta* and several other species with a dominantly Mediterranean distribution.

6.1.8.2 Vegetation with Cheiranthus cheiri and Asplenium ruta-muraria

The predominantly extra-mediterranean communities of wall-flower and wall-rue are well differentiated by the species enumerated in the aforegoing paragraph. The correspondence between stands of Atlantic and of subatlantic areas is fairly strong and the many transitional cases do not render the classification of these stands an easy matter. Three groups of communities are more or less evident viz.

1. community with *Centranthus ruber* and *Sedum album,*
2. community with *Poa compressa* and *Parietaria judaica,* and
3. community with *Poa compressa* and *Ceratodon purpureus.*

Apart from these, inops variants occur which are not easily referable to any of the three types or seem to be facies expressions of initial stages of the communities. A small group of relevés could not be classified in one of these four categories. These groups of stands differ to a degree

Table 14 Communities of Cheiranthus cheiri

a. with Sedum dasyphyllum
b. with Asplenium ruta-muraria
b1. and Centranthus ruber and Sedum acre
b2 and Poa compressa and Parietaria judaica
b3 and Poa compressa and Ceratodon purpureus

	a.							b1.					
Nr C	1	2	3	P	U W	MGT	MGP	4	5	6	P	U W	MGT
Date	20-4	25-4	25-4	15st				6-8	3-8	27-9	22st		
Year	62	62	62					62	64	61			
Nr R	144	218	213					526	403	538			
Ex	1	3	2					4	1	4			
Geo R	331	310	310					112	111	111			
CH	40	30	30					17	20	15			
CM	s	2	-					7	40	2			
Cheiranthus cheiri	2a	2a	2b	100	1a-2b	10	10	2a	2a	2a	100	1p-2b	15
Linaria cymbalaria	+r	-	1a	40	+r-2a	2	4	1b	1b	+r	82	+r-2a	3
Centranthus ruber	2b	-	-	40	+a-2b	3	8	+p	+a	1a	95	+r-4b	7
Antirrhinum majus	-	-	1a	20	+p-1a	0.2	1	-	+p	1p	27	+r-1p	0.1
Sonchus oleraceus	+p	+p	-	47	+r-+p	<0.1	0.1	-	-	+r	41	+r-1a	0.1
Asplenium trichomanes	-	-	1b	27	+r-2a	1	5	-	+p	+r	23	+r-2m	0.2
Bromus sterilis	+r	+r	-	27	+r-+p	<0.1	0.1	-	-	+r	18	+r	0
Hedera helix	-		-	13	+p-+a	0.1	0.8	-	+p	-	32	+r-+b	0.3
Parietaria judaica	1a	2b	-	100	+p-2b	8	8	2a	-	1b	73	+r-3b	9
Polypodium vulgare	-	-	-	7	+p	<0.1	0.1	+p	1b	+p	36	+p-2a	0.6
Catapodium rigidum	-	-	-	13	+r-+p	<0.1	0.1	-	-	-	27	+p-1p	0.1
Taraxacum (div.) spec.	1b	-	-	27	+r-1b	0.3	1	-	-	+r	14	+r-+p	<0.1
Tortula muralis	+r	2m	-	53	+r-2m	0.4	0.7	1a	2m	1p	95	+p-2b	3
Ficus carica	-	-	-	13	+r-+a	0.1	0.8	-	-	-	-	-	-
Saxifraga tridactylites	-	-	+r	20	+r-+p	<0.1	0.1	-	-	-	5	2m	0.1
Ceterach officinarum	+r	+p	1b	60	+r-1b	<0.1	2	-	-	-	-	-	-
Mercurialis annua	+r	+r	-	27	+r-+p	<0.1	0.1	-	-	-	-	-	-
Sedum dasyphyllum	+r	+p	+p	80	+r-1b	0.4	0.5	-	-	-	-	-	-
Umbilicus rupestris	-	+p	+p	40	+r-+p	<0.1	0.1	-	-	-	9	+r-1p	<0.1
Barbula acuta	+r	-	-	33	+r-2m	0.2	0.6	-	-	-	5	+p	<0.1
Scorpiurium circinatum	-	-	-	20	1a-2b	3	13	-	-	-	9	1b-2a	0.5
Chaenorrhinum origanifolium	-	-	+p	13	+p-1p	<0.1	0.3	-	-	-	-	-	-
Poa bulbosa	-	+r	-	27	+r-+p	<0.1	<0.1	-	-	-	5	+r	0
Geranium rotundifolium	+r	-	-	27	+r	0	0	-	-	-	-	-	-
Oryzopsis miliacea	1p	-	-	20	1p-1a	0.2	0.8	-	-	-	-	-	-
Mentha rotundifolia	+r	-	-	13	+r	0	0	-	-	-	5	+r	0
Antirrhinum latifolium	2a	-	-	13	1b-2a	0.8	6	-	-	-	-	-	-
Geranium lucidum	-	-	-	13	+r	0	0	-	-	-	-	-	-
Orthotrichum cupulatum	-	-	-	13	+r	0	0	-	-	-	5	2m	0.1
Hornungia petraea	-	-	-	13	+p	<0.1	0.1	-	-	-	-	-	-
Brachypodium ramosum	+p	-	-	13	+r-+p	<0.1	0.1	-	-	-	-	-	-
Senecio vulgaris	+r	-	-	13	+r	0	0	-	-	-	-	-	-
Arenaria serpyllifolia	-	-	-	7	+p	<0.1	0.1	-	-	-	18	+p-1p	<0.1
Dactylis glomerata	-	-	-	7	+p	<0.1	0.1	-	-	-	41	+r-+p	<0.1
Asplenium ruta-muraria	-	-	-	-	-	-	-	-	-	1p	27	+r-1p	<0.1
Barbula vinealis	-	-	-	-	-	-	-	1p	2m	-	64	+p-3b	4
Streblotrichum convolutum	-	-	-	-	-	-	-	+r	2m	-	45	+r-2m	0.5
Oxalis corniculata	-	-	-	13	+r-2a	0.5	4	-	-	-	-	-	-
Phagnalon sordidum	+r	-	-	20	+r-1p	<0.1	<0.1	-	-	-	-	-	-
Bryum capillare	-	-	-	-	-	-	-	+p	+p	-	45	+r-2m	0.3
Bryum argenteum	-	-	-	-	-	-	-	+r	-	+p	45	+r-1p	0.1
Bryum caespiticium	-	-	-	-	-	-	-	-	-	-	27	+p-2m	0.2
Homalothecium sericeum	-	-	-	7	2b	1	18	1b	2b	-	64	+r-3b	5
Grimmia pulvinata	-	-	-	-	-	-	-	-	-	1p	18	+p-2m	0.1
Poa annua	-	-	-	-	-	-	-	-	+p	-	14	+r-+p	<0.1
Sedum acre	-	-	-	-	-	-	-	-	2m	1a	32	+r-2a	0.5
Poa pratensis	-	-	-	-	-	-	-	-	-	+p	23	+p-1a	0.1
Petrorhaga prolifera	-	-	-	-	-	-	-	-	-	-	23	+r-2b	0.8
Tortula intermedia	-	-	-	-	-	-	-	1a	-	+p	27	+p-1a	0.2
Hypnum cupressiforme	-	-	-	-	-	-	-	-	-	-	23	+p-2a	0.8
Sagina procumbens	-	-	-	-	-	-	-	-	-	-	14	+r-2m	0.1
Poa compressa	-	-	-	-	-	-	-	-	-	-	5	1p	<0.1
Chelidonium majus	-	-	-	-	-	-	-	-	-	-	-	-	-
Ceratodon purpureus	-	-	-	-	-	-	-	-	-	+p	18	+p-1p	<0.1
Poa angustifolia	-	-	-	-	-	-	-	-	-	-	5	+r	0
Campanula rotundifolia	-	-	-	-	-	-	-	-	-	-	-	-	-
Artemisia absinthium	-	-	-	-	-	-	-	-	-	-	-	-	-
Sedum album	-	+r	-	7	+r	0	0	-	-	-	5	+p	<0.1
Tortula ruralis	-	-	-	-	-	-	-	-	-	-	5	+p	<0.1
Hieracium amplexicaule	-	-	-	-	-	-	-	-	-	-	-	-	-
Festuca rubra	-	-	-	-	-	-	-	-	-	-	14	1p-1a	0.2
MNS - MCH - MCM		13	-	30	-	4			16	-	28	-	31 (35)

In addition in 1: Melica minuta +p; Verbascum sinuatum +r; in 2: Veronica polita, Cardamine hirsuta, Urtica dioica, Capsella bursa-pastoris +r; in 4: Plantago lanceolata, Spergularia rubra +r; in 5: Camptothecium lutescens, Rhynchostegium confertum 2m; in 6: Erigeron mucronatus, Sagina apetala +r; Barbula revoluta 2p; Didymodon rigidulus, D. trifarius 1p; in 9: Hypericum perforatum, Erigeron acer, Clematis vitalba, Cardaminopsis arenosa, Verbascum cf. lychnitis +r; in 10: Plantago lanceolata +r; in 11: Chrysanthemum leucanthemum +p; Cirsium arvense r; Brachythecium rutabulum 3a; in 12: Veronica chamaedrys, Poa chaixii +r; Senecio jacobea r; in 13: Saxifraga rosacea ssp. sponhemica, Cardaminopsis arenosa, Barbula unguiculata 1p; Bryoerythrophyllum recurvirostre, Brachythecium rutabulum +p.

LEGEND

1. Ollioules (Var, France), wall alongside the river Reppe.
2. Rivière-sur-Tarn (Aveyron, France), wall alongside road nr N 407.
3. Ste. Enimie (Lozère, France), wall of a house.
4. Granville (Manche, France), city wall near Panorama Tour du Roc.
5. Totnes (Devon, England), wall alongside Station Road near Fore Street.
6. Topsham (Devon, England), wall alongside mouth of the river Exe.
7. Carcasonne (Aude, France), wall at Western side of the Cité, above a nursery.
8. Candes-St.-Martin (Indre-et-Loire, France), wall behind church.
9. Luxembourg (town), wall on rock Um Bock.
10. Avallon (Yonne, France), rampart wall under Rue de la Fontaine-Neuve.
11. Culemborg (Gelderland, Netherlands), city wall opposite Rozenstraat.
12. Bad Wimpfen (Württemberg, Western Germany), castle wall near Burgstaffel.
13. Bouillon (Luxembourg, Belgium), schistaceous slopes below the castle, along the river Semois; inclination 50-95°, mean inclination 85°.

in their geographical distribution as will be clear from Table 15, in which table also a summary is given of the various exposures of the sites with stands of *Cheiranthus cheiri* and *Sedum dasyphyllum* so as to facilitate their mutual comparability.

Table 15 Geographical distribution and exposure of stands of communities with Cheiranthus cheiri

a. with Sedum dasyphyllum
b. with Asplenium ruta-muraria
b1. and Centranthus ruber and Sedum album
b2. and Poa compressa and Parietaria judaica
b3. and Poa compressa and Ceratodon purpureus
b4. inops and other vegetation types

	exposure:	1	2	3	4	6	t	t%
geographical region:								
b1	111 - 112	4	4	3	8	-	19	86
	132	-	1	-	2	-	3	14
	t	4	5	3	10	-	22	-
	t%	18	23	14	45	-	-	100
b2	131	-	1	2	1	-	4	29
	132	-	3	2	5	-	10	71
	t	-	4	4	6	-	14	-
	t%	-	29	29	43	-	-	100
b3	111	-	1	-	3	-	4	10
	120	2	-	1	-	-	3	8
	131	6	5	3	4	1	18 (+1)	47
	132	2	1	6	4	-	13	34
	t	10	7	10	11	1	38 (+1)	-
	t%	26	18	26	29	-	-	99
b4	111 - 112	-	3	1	2	-	6	-
	120	-	-	-	1	-	1	-
	132	-	-	2	-	-	2	-
	t	-	3	3	3	-	9	-
b	111 - 112	34	8	4	13	-	29	35
	120	2	-	1	1	-	4	5
	131	6	6	5	5	1	22	27
	132	2	5	10	11	-	28	34
	t	14	19	20	30	1	83	-
	t%	18	23	24	36	-	-	100
a	132 - 332	6	2	6	1	-	15	-
	t%	40	13	40	7	-	-	100

Vegetation with *Asplenium ruta-muraria* is predominantly encountered on W.- and E.-facing walls and but rarely on N.-facing ones. Its occurrence on W.-facing walls prevails in the eu-atlantic areas and in 132, but in 132 (and likewise in the adjoing area 310) also numerous S.-facing sites are found. In area 131 (and presumably also in area 120) there seems to be no special preference to a particular exposure. Of the three principal types, (1) appears to be most closely allied to the community of *Cheiranthus* and *Sedum dasyphyllum,* and (3) shows the least degree of relationship. The affinity coefficients according to Poore are 0.40 and 0.31, respectively. The mutual affinities between the groups according to this coefficient are as follows:

Affinity between (1) and (2): 0.69
„ „ (2) „ (3): 0.68
„ „ (1) „ (3): 0.51.

For all these groups differentially linking species connecting them two by two can be indicated viz. *Dactylis glomerata* for (1) and (2), *Poa compressa* for (2) and (3), and *Polypodium vulgare* s.s. for (1) and (3). Stands of (1) are characterised by e.g. their relatively high degrees of presence of *Sedum album* and of *Tortula intermedia,* and by the absence of *Poa compressa, Chelidonium majus* and *Campanula rotundifolia*; stands of (2) by the presence of *Ficus carica* and *Artemisia absinthium,* and the low degree of representation (or total absence) of *Asplenium trichomanes* and *Tortula intermedia;* and stands of (3) by the presence of *Ceratodon purpureus* and *Sedum album,* and the absence of *Parietaria judaica.*

Stands of (1) are predominantly met with in eu-atlantic regions. *Petrorhaga prolifera* was recorded occasionally in N.W. France (Loire basin and Granville). At Carcasonne (Aude, France) stands were recorded which showed a fairly manifest transition towards the communities of *Cheiranthus cheiri* and *Sedum dasyphyllum* by the occurrence of *Umbilicus rupestris, Scorpiurium circinatum* and some other species with a mainly Mediterranean distribution. Similar transitional stands are intermediate between those of group (2) and those of the community of *Cheiranthus cheiri* and *Sedum dasyphyllum,* and it is particularly in stands of this kind that *Ficus carica* and *Ceterach officinarum* are consistently encountered (relevés from Carcasonne and from Cahors, Lot, France). An example is given in Table 14, no. 7. Group (2) is restricted in its occurrence to areas 131 and 132; group (3) likewise predominantly occurs in 131 and 132, but it extends farther, across areas 120 and 111, but in area 111 it is usually only found in regions not immediately bordering upon the English Channel (e.g. near Canterbury, S.E. of London). *Hieracium amplexicaule* was only rarely recorded in *Cheiranthus* vegetation (Netherlands: Maestricht and Valkenburg; Belgium: Tongeren; W. Germany: Bad Wimpfen). The total absence of *Parietaria judaica* is especially characteristic of the stands of this group. In area 131 the community is rather strongly restricted to the valleys of the great rivers (especially those of the rivers Rhine and Meuse and their tributaries, and this holds also for the stands of group (2). The most northernly situated locality of stands with *Centranthus ruber* is Namur in central Belgium. In the Rhine valley this species does not grow wild. Near Bouillon (S.E. Belgium) a closely allied community was found on steep rock faces alongside the river Semois, in the immediate vicinity of the town. A relevé from this area (no. 13) is included in Table 14.

In a few localities in Brittany *Cheiranthus* was found associated with *Erigeron mucronatus* (e.g. at Concarneau), in other places it occurred together with *Crithmum maritmum* and *Spergularia rupicola,* and, accordingly, transitions towards communities with these species do occur. Another species sometimes occurring with *Cheiranthus* on walls is *Elytrigia pungens.* In Brittany this community also shows affinities towards mural *Crithmum maritimum* vegetation. Some examples are shown in

Table 16 Communities of Cheiranthus cheiri in coastal regions

Nr C	1	2	3
Date	7-8	7-8	20-9
Year	62	62	61
Nr R	553	550	492
Ex	3	2	4
Geo R	112	112	120
CH	37	22	50
CM	-	-	3
Cheiranthus cheiri	3a	2b	3b
Parietaria judaica	+p	1a	-
Dactylis glomerata	+r	1a	-
Crithmum maritimum	1b	-	-
Spergularia rupicola	1a	+r	-
Elytrigia pungens	-	1p	1b
Daucus gummifer	-	+p	-
Beta maritima	-	+r	-
Armoracia rusticana	-	+r	-
Matthiola incana	-	+p	-
Sonchus oleraceus	-	+r	-
Asplenium ruta-muraria	-	-	+p
Erigeron canadensis	-	-	+p
Tortula muralis	-	-	1p
Barbula vinealis	-	-	2m
Streblotrichum convolutum	-	-	1p
Ceratodon purpureus	-	-	1p

LEGEND

1. Plévenon (Côtes-du-Nord, France), castle wall Fort de la Latte.
2. Plévenon (Côtes-du-Nord, France), wall of the bridge of Fort de la Latte.
3. Rithem (Zeeland, Netherlands), wall of Fort Rammekens.

Table 16. Most probably the establishment of *Elytrigia* on walls is favoured by aerial transport of soil particles by the action of wind. Finally, it is noteworthy that the wallflower also grows on the flat tops of walls and that in such sites it is often associated with winter annuals and with species of *Sedum* (particularly with *S. acre*). In such situations, as is the case with vegetation of exposed rock surfaces, pioneer stands of the *Sedo-Scleranthetea* precede the settlement of *Cheiranthus cheiri*. Well-developed stands of such vegetation on wall tops were seen at e.g. Exeter (Devon, England) and Eijsden (Limburg, Netherlands).

Stands of *Cheiranthus cheiri* communities show an increasing quantitative contribution of hemicryptophytes in the sequence: stands with *Poa compressa* (27%); with *P. compressa* and *Parietaria judaica* (43%); and with *Centranthus ruber* and *Parietaria judaica* (58%), but at the same time their qualitative contribution falls off a little (60%, 52% and 51%, respectively). Furthermore, a decrease of the quantitative contribution of the chamaephytic elements can be observed (from 69% to 56% and again to 37%, weighted values 26%, 23% and 23%, respectively). In the inops stands the contribution of the therophytes is relatively small (score: 8% U, 0.3 W, as against 12-21% and 1-4%, respectively, in all other communities with *Cheiranthus*). In the communities with *Sedum dasyphyllum*, seasonal xerophytes play a relatively important part as far as their qualitative contribution is concerned (the score being 11% U, 2% W, as against 2-5% and 0.3-2%, respectively, in all other communities with *Cheiranthus*). Myrmecochory is particularly important in communities with *Parietaria judaica*.

In stands with *Poa compressa* many species of dry, open habitats occur, also in combinations with *Parietaria judaica* (the score being

U 16% and 17%, respectively, and W 10% and 3%, respectively). In all communities with *Cheiranthus* the scores of species with a Mediterranean-Atlantic distribution are always high, viz. U 19-43% and W 53-82%.

Like the communities of *Cheiranthus cheiri* and *Sedum dasyphyllum*, communities of wallflower and wall-rue can develop from pioneer stands with *Parietaria judaica*. *Centranthus ruber* usually becomes established in a later successional stage. In the areas 120 and 131 *Parietaria judaica* is relatively rare, however, and in these areas wallflower communities may develop by way of pioneer stands allied to the *Filici-Saginetum* and/or the *Sedo-Scleranthetea*. The pioneer stands are initially poor in species and consist usually of a community of wall-rue, ivy-leaved toadflax and the moss *Tortula muralis*; *Asplenium ruta-muraria* usually being the first to settle. This inops kind of stand can also act as the initial prophase of other communities with e.g. *Corydalis lutea* or *Parietaria*.

It is not clear whether the *Cymbalarietum muralis* Görs mentioned by OBERDORFER (1967) agrees with these inops stands. The distinction of such inops associations within a floristic system is not to be recommended because it only leads to the raising of numerous facies types and initial (pioneer) stands of vegetation without their own faithful species to the rank of an association (compare the remarks regarding the *'Poetum annuae'* under 7.2).

6.1.8.3 A survey of Cheiranthus cheiri communities

In 1954 OBERDORFER described a *"Cheiranthus-Parietaria ramiflora-association"* which he originally included in the alliance of ruderal communities called the *Arction*. He published 5 relevés with *Parietaria judaica*, of which 2 contained *Cheiranthus cheiri* and of these 2 there was one (viz., a relevé made by Sissingh, from Nijmegen, Netherlands) borrowed from a "Rundbrief" [1]). In Sissingh's relevé *Erysimum cheiranthoides* was almost certainly mistaken for *Cheiranthus cheiri* [2]). In subsequent papers OBERDORFER (1956, 1957) maintained the *Cheirantho-Parietarietum ("ramiflorae")* as a *"Mauer*-Unkrautgesellschaft". From his descriptions (1957) it is quite obvious that his relevés refer to stands of a distinctly ruderal character such as those found at the base of walls. Apparently central European communities with *Parietaria judaica* but without wallflowers were included in his concept of the *Cheirantho-Parietarie-*

[1]) 11. Rundbrief der Zentralstelle für Vegetationskartierung (The "Rundbriefe" were stencilled issues containing information from the „Gross-Deutsche Reich" during the Second World War).

[2]) The locality (Nijmegen) was already known to Dutch botanists of the 19th century and the site is (or was) well-known to several very active naturalists of Nijmegen who never recorded *Cheiranthus*. However, *Erysimum cheiranthoides* still grows there. The improbability of the occurrence of *Cheiranthus* and the likelihood of representation of *Erysimum* follows also from Sissingh's list of species. Thus only a single relevé in Oberdorfer's table remains with "Cheiranthus +"!

tum. Later (OBERDORFER, 1967) referred the *"Cheirantho-Parietarietum diffusae"*, together with the *"Cymbalarietum muralis"*, to the alliance *Galio-Parietarion* Martínez 1960 [1]) (order *Parietarietalia muralis* Martínez 1960, class *Cymbalario-Parietarietea diffusae*).

HÜBSCHMANN (1967) gave a table based on 15 relevés with *Parietaria judaica* but without *Cheiranthus* from the Moselle valley, which he all referred to the *Cheirantho-Parietarietum*. These relevés link up with the present author's data from central Europe (see under 6.1.11).
Some possible classifications of *Cheiranthus* communities are the following:

Alliance : *Parietarion judaicae*

Associations:

1. Sedo-Cheiranthetum cheiri	1. Sedo-Cheiranthetum cheiri
a.	sedetosum dasyphylli
2. Asplenio-Cheiranthetum cheiri	asplenietosum rutae-murariae
b. (?) sedetosum acris	(sedetosum acris)
c. (?) poëtosum compressae	(poëtosum compressae)

Transitional cases towards ruderal vegetation can be united into a different subassociation, etc. The variants (b) and (c) both show a distinct affinity to the *Sedo-Scleranthetea*, but (c) somewhat more to the *Artemisietea*, although the differences are only gradual.

6.1.9 Vegetation with Crithmum maritimum

On walls in coastal areas, close to the shore, principally in Normandy, Brittany and S. England and usually on quay-sides, *Crithmum maritimum* is sometimes encountered. In the areas mentioned, this species is very common on coastal cliffs and its appearance on walls is not at all unexpected, but more striking is its association with typical wall-dwellers such as *Linaria cymbalaria* and *Parietaria judaica*. In Brixham (Devon, England) the combination of *Crithmum* and *Parietaria judaica* is also found in crevices of coastal rocks.

KNOERR (1960) recorded the combination from rock crevices near the coast in some islands S. of Marseilles, and HORVATIC (1963) from the island of Pag (Yugoslavia). *Crithmum maritimum* has also been recorded from quay-sides at Hendaye and Bidassoa (in the S.W. corner of France; ALLORGE, 1941).

Relevés of mural stands are shown in Table 17. In this table all those species which are common in, or restricted to, coastal areas and are rare or absent on walls in the interior, are indicated with the symbol d. Other coastal forms were once or twice recorded from walls e.g. *Limonium binervosum* and *L. lychnidifolium*, which on rocky cliffs are usually associated with *Crithmum*. In S. England 6 stands and in France 3

---

[1]) It appears likely, also considering the erroneous citation of the date of publication (not "1960", but 1955: fide Martínez), that Oberdorfer did not consult the paper by Martínez or at least did not study it thoroughly.

stands were recorded, of which 5 where found on E.-facing walls, 3 on W.-facing ones and one on a S.-facing wall surface.

Table 17 Community of Crithmum maritimum and Linaria cymbalaria

Nr C	1	2	3	P	U W	MGT	MGP
Date	29-9	5-8	8-8	%st			
Year	61	64	62				
Nr R	556	454	519				
Ex	4	4	3				
Geo R	111	111	112				
CH	90	40	35				
CM	15	2	3				
d Crithmum maritimum	4a	2a	2b	*100*	1a-4a	14	14
Linaria cymbalaria	3a	-	-	*78*	1a-3a	9	11
Parietaria judaica	-	1b	-	*67*	+p-1b	1	2
Centranthus ruber	1a	-	+r	*56*	+r-1a	0.1	0.9
d Tortula muralis	-	2m	-	*44*	+p-2m	0.5	1
Sonchus oleraceus	-	+p	+r	*44*	+r-+p	<0.1	<0.1
d Matricaria maritima ssp. maritima	+p	2b	-	*44*	+p-2b	2	6
d Plantago coronopus	-	2a	+p	*33*	+p-2a	0.9	3
d Sagina maritima	-	-	+p	*33*	+r-+p	<0.1	0.1
Bryum argenteum	2a	-	-	*22*	2m-2a	1	5
Bryum caespiticium	2a	-	+p	*22*	+p-2a	1	4
d Festuca rubra	-	2m	-	*22*	+p-2m	0.3	1
d Spergularia rupicola	-	+r	-	*22*	+r-1a	0.2	0.8
Poa annua	-	+p	-	*22*	+p-1p	0.1	0.3
Poa pratensis	-	-	-	*22*	+p-1p	0.1	0.3
Streblotrichum convolutum	-	-	1p	*22*	+p-1p	0.1	0.3
Catapodium rigidum	-	-	+r	*22*	+r-1p	<0.1	0.2
d Senecio jacobea	-	+p	+p	*22*	.+p	<0.1	0.1
Sedum acre	-	-	-	*22*	+p	<0.1	0.1
d Catapodium marinum	-	-	+p	*22*	+p	<0.1	0.1
Asplenium ruta-muraria	-	-	-	*22*	+r-+p	<0.1	0.1
Taraxacum (div.) spec.	-	+p	+r	*22*	+r-+p	<0.1	0.1
Plantago lanceolata	-	+p	+r	*22*	+r-+p	<0.1	0.1
Rumex crispus	-	-	+p	*22*	+r-+p	<0.1	0.1
Erysimum cheiranthoides	-	-	+p	*22*	+r-+p	<0.1	0.1
MNS - MCH - MCM		*11*	-	*33*	-	*3 (5)*	

In addition in 2: Phyllitis scolopendrium, Polypodium vulgare 1a; Dactylis glomerata +p; in 3: Medicago lupulina 1b; Rubus spec. 1a; Salvia pratensis 1p; Sedum acre, d Limonium lychnidifolium +p; d Daucus gummifer, Crepis vesicaria ssp. taraxacifolia, Calendula arvensis, Elytrigia repens, Senecio vulgaris, Erigeron mucronatus +r, Bryoerytrophyllum recurvirostre 2m; Didymodon trifarius +p.

LEGEND

1. Brixham (Devon, England), quay side wall.
2. Porthleven (Cornwall, England), quay side wall.
3. Carteret (Manche, France), rampart wall above harbour.

The community can be included in the *Parietarion judaicae* as a special coastal association which, if this is deemed recommendable, can be referred to by the name of *Linario-Crithmetum*.

In the community with *Crithmum maritimum* the contribution of species with a Mediterranean-Atlantic distribution is considerable, viz. 30% (U) and 53% (W).

6.1.10 Vegetation with Erigeron mucronatus

Communities with *Erigeron mucronatus* have been encountered in both the Mediterranean and the eu-atlantic parts of Europe. This species, native in N. America, and cultivated as an ornamental in Europe, has run wild and now behaves like a well-established neophyte with a marked preference for walls but occasionally (e.g. in the area of Lake Como) also encountered in rocky sites. The stands in southern Europe and in the Atlantic belt have a number of species in common, but they can clearly be distinguished in:

a. communities with *Ficus carica* (S. Europe), and
b. communities with *Asplenium adiantum-nigrum* (W. Europe).

Table 18 Geographical distribution and exposure of stands of communities with Erigeron mucronatus
a. with Ficus carica
b. with Asplenium adiantum-nigrum

	exposure:	1	2	3	4	6	t	t%
geographical region:								
a	111	-	1	4	1	-	6	40
a	112	1	4	2	2	-	9	60
a	t	1	5	6	3	-	15	-
a	t%	7	33	44	20	-	-	100
b	310	1	1	1	1	-	4	40
b	320	1	1	1	1	-	4	40
b	331	-	1	-	1	1	2(+1)	20
b	t	2	3	2	3	(1)	10(+1)	-
b	t%	20	30	20	30	-	-	100

Apparently the communities found in S. Europe show no pronounced preference to a certain type of exposure, but those occurring in W. Europe are more frequently encountered on S.-facing walls and but rarely on N.-facing ones. *Erigeron mucronatus* seems to avoid very dry habitats and is, therefore, often associated with species that do not survive (or become dormant during) periods of prolonged drought, such as *Asplenium trichomanes* and *Ceterach officinarum.* The impression was gained that this alien species is frequently met with in areas where granites or schists occur and are used as building stones. A coincidence of occurrence with *Asplenium adiantum-nigrum* is more commonly encountered than cohabitation with *A. ruta-muraria* and this might point to a certain dislike of *Erigeron mucronatus* of walls built of lime-stone or brick, but both kinds of walls are of the mortar-jointed type and the mortar is always rich in lime. If the composition of the mortar is not the same in the different areas (and this needs looking into), certain processes may be involved which lower the pH of the surface of the joints more rapidly in walls built of acid rocks than in lime-stone or brick walls (compare 6.6.2).

The stands of type (a) are floristically differentiated by a number of Mediterranean forms such as *Ficus carica, Phagnalon sordidum* and *Mentha rotundifolia*, and those of type (b) by e.g. the ferns *Asplenium adiantum-nigrum, A. ruta-muraria* and *Phyllitis scolopendrium.*

Both types have a certain affinity to other communities with *Asplenium trichomanes.* The affinity index (Poore's index) of (a) and (b) is 0.53. Stands of (a) are upon the whole richer in species than those of (b), but the percentage of coverage of their herb layer is lower as a rule. Community (a) shows some affinity to both the *Bromo-* and the *Sedo-Parietarietum.* In a single instance *Parietaria lusitanica* was recorded (Table 19, no. 3). Community (b) is allied to a number of vegetation types, more particularly with several Atlantic communities with *Asplenium trichomanes* and with *Parietaria judaica.* Sometimes transitions

Table 19 Communities of Erigeron mucronatus and Linaria cymbalaria
a. with Ficus carica,
b. with Asplenium adiantum-nigrum

	a.							b.							total			
Nr C	1	2	3	P	U W	MGT	MGP	4	5	6	P	U W	MGT	MGP	P	U W	MGT	MGP
Date	26-4	21-4	9-5	11st				4-8	9-8	8-8	15st				26st			
Year	65	62	65					64	62	62								
Mr R	068	166	156					433	575	561								
Ex	2	6	3					3	3	4								
Geo R	320	331	310					111	112	112								
CH	50	30	25					55	40	25								
CM	2	-	-					5	-	5								
Erigeron mucronatus	3a	2b	2b	100	1a-3b	16	16	4a	3b	2b	100	1b-4a	30	30	100	1a-4a	24	24
Linaria cymbalaria	1a	-	2a	64	+p-2a	2	4	2a	-	-	73	+p-2a	3	4	68	+p-2a	3	3
Sonchus oleraceus	+p	1p	-	55	+r-1p	0.1	0.1	+p	-	-	60	+r-1p	0.1	0.1	58	+r-+p	0.1	0.1
Tortula muralis	2m	-	-	45	2m-2a	2	4	2m	-	+p	60	+p-2m	0.8	1	60	+p-2a	1	2
Parietaria judaica	2b	2a	-	64	+p-3a	8	12	-	+p	-	40	+p-2b	2	5	50	+p-3a	5	10
Asplenium trichomanes	-	-	2a	45	+r-2a	1	2	2a	+p	+r	47	+r-2a	0.6	1	46	+r-2a	0.8	2
Centranthus ruber	-	-	1b	36	+r-1b	0.4	1	+p	+p	-	57	+p-2a	1	2	46	+r-2a	0.8	2
Homalothecium sericeum	-	-	-	18	1p-2m	0.3	1	1p	-	1b	33	+p-1b	0.7	2	27	+p-2m	0.5	2
Umbilicus rupestris	-	+p	-	18	+p-1a	0.1	0.8	-	-	-	20	+p-2m	0.2	1	19	+p-2m	0.2	0.9
Ceterach officinarum	-	-	1p	45	+p-2m	0.3	0.7	1p	-	-	13	+p	<0.1	0.1	27	+p-2m	0.1	0.5
Catapodium rigidum	1p	-	-	27	+r-1p	<0.1	0.2	1p	-	-	13	1p	0.1	0.4	19	+r-1p	0.1	0.3
Mercurialis annua	-	+p	-	27	+r-+p	<0.1	<0.1	-	-	-	13	+r-+p	<0.1	0.1	19	+r-+p	<0.1	<0.1
Euphorbia peplus	-	1a	-	18	+r-1a	0.1	0.8	-	-	+r	13	+p-+p	<0.1	0.1	15	+r-1a	0.1	0.4
Ficus carica	-	-	+b	55	+r-+b	0.5	0.8	-	-	-	7	+r	0	0	27	+r-+b	0.2	0.7
Mentha rotundifolia	2m	+p	-	36	+p-2m	0.3	0.8	-	-	-	-	-	-	-	15	+p-2m	0.1	0.8
Phagnalon sordidum	-	+r	-	36	+r-1a	0.1	0.5	-	-	-	-	-	-	-	12	+r-1a	<0.1	0.5
Sedum dasyphyllum	-	-	1a	27	1p-2a	0.9	3	-	-	-	-	-	-	-	12	1p-2a	0.4	3
Campanula erinus	-	1a	-	27	+p-2m	0.4	1	-	-	-	-	-	-	-	12	+r-2m	0.2	1
Bromus madritensis	2m	+p	-	27	+p-2m	0.2	0.9	-	-	-	-	-	-	-	12	+p-2m	0.1	0.9
Barbula acuta	1p	-	-	18	1p-2m	0.3	1	-	-	-	-	-	-	-	8	1p-2m	0.1	1
Fumaria capreolata	2m	+r	-	18	+r-2m	0.2	1	-	-	-	-	-	-	-	8	+r-2m	0.1	1
Crepis vesicaria ssp. taraxacifolia	-	+r	-	18	+r-1a	0.1	1	-	-	-	-	-	-	-	8	+r-1a	0.1	0.8
Avena barbata	-	1p	-	18	1p	0.1	0.4	-	-	-	-	-	-	-	8	1p	<0.1	0.4
Veronica polita	1p	-	-	18	+p-1p	<0.1	0.3	-	-	-	-	-	-	-	8	+p-1p	<0.1	0.3
Bromus sterilis	+p	-	-	18	+p	<0.1	0.1	-	-	-	-	-	-	-	8	+p	<0.1	0.1
Sedum sediforme	-	+p	-	18	+p	<0.1	0.1	-	-	-	-	-	-	-	8	+p	<0.1	0.1
Inula viscosa	-	-	-	18	+r	0	0	-	-	-	-	-	-	-	8	+r	0	0
Galactites tomentosa	-	+r	-	18	+r	0	0	-	-	-	-	-	-	-	8	+r	0	0
Euphorbia helioscopia	-	+r	-	18	+r	0	0	-	-	-	-	-	-	-	8	+r	0	0
Barbula vinealis	-	-	-	9	2m	0.2	3	2a	-	-	13	2m-2a	0.7	5	12	2m-2a	0.5	4
Oxalis corniculata	-	-	-	9	+r	0	0	-	-	-	13	+r-+p	<0.1	0.1	12	+r-+p	<0.1	<0.1
Hedera helix	-	-	+b	9	+b	0.3	4	-	+p	-	20	+r-+b	0.5	2	15	+p-1b	0.4	3
Poa annua	-	-	-	9	1p	<0.1	0.4	-	-	1a	20	+r-1a	0.1	0.5	15	+r-1a	0.1	0.5
Asplenium ruta-muraria	-	-	-	9	2m	0.2	3	-	-	-	33	+r-2m	0.2	0.5	23	+r-2m	0.2	0.9
Taraxacum (div.) spec.	-	-	-	9	+r	0	0	+p	-	+r	33	+r-+p	<0.1	0.1	23	+r-+p	<0.1	0.1
Asplenium adiantum-nigrum	-	-	-	-	-	-	-	-	-	1b	60	+r-2m	1	2	35	+r-2m	0.6	2
Polypodium vulgare	-	-	-	-	-	-	-	+r	-	-	47	+r-1p	0.1	0.1	27	+r-1p	<0.1	0.1
Dactylis glomerata	-	-	-	-	-	-	-	+p	+r	-	33	+r-1p	<0.1	0.1	19	+r-1p	<0.1	0.1
Streblotrichum convolutum	-	-	-	-	-	-	-	+p	-	-	20	+p-2m	0.2	0.9	12	+p-2m	0.1	0.9
Bryum capillare	-	-	-	-	-	-	-	-	-	-	20	+p-2m	0.2	1	12	+p-2m	0.1	1
Rubus (div.) spec.	-	-	-	-	-	-	-	-	+p	-	20	+p-+a	0.1	0.5	12	+p-+a	0.1	6
Buddleja davidii	-	-	-	-	-	-	-	+p	-	-	20	+p	<0.1	0.1	12	+p	<0.1	0.1
Phyllitis scolopendrium	-	-	-	-	-	-	-	-	-	-	20	+r-+p	<0.1	<0.1	12	+r-+p	<0.1	<0.1
Bromus erectus	-	-	-	-	-	-	-	-	-	-	13	+p-2m	0.2	1	8	+p-2m	0.1	0.9
Bryoerytrophyllum recurvirostre	-	-	-	-	-	-	-	-	-	2m	13	2p-2m	0.2	1	8	2p-2m	0.1	1
Barbula unguiculata	-	-	-	-	-	-	-	-	-	+p	13	+p-2m	0.1	0.8	8	+p-2m	0.1	0.8
Orthotrichum anomalum	-	-	-	-	-	-	-	1p	-	-	13	+p-1p	<0.1	0.3	8	+p-1p	<0.1	0.3
Epilobium parviflorum	-	-	-	-	-	-	-	-	-	+p	13	+r-+p	<0.1	0.1	8	+r-+p	<0.1	0.1
Bryum murorum	-	-	-	-	-	-	-	-	-	+p	13	+p	<0.1	0.1	8	+p	<0.1	0.1
Bryum argenteum	-	-	-	-	-	-	-	-	-	-	13	+r-+p	<0.1	0.1	8	+r-+p	<0.1	0.1
Agrostis stolonifera	-	-	-	-	-	-	-	-	-	+r	13	+r	0	0	8	+r	0	0
MNS - MCH - MCM		15	-	31	-	2 (5)			13	-	36	-	3 (4)		14	-	34	- 3 (5)

In addition in 1: Parietaria lusitanica 2m; Arabidopsis thaliana, Euphorbia exigua 1p; Veronica cymbalaria, Fumaria officinalis, Reichardia picroides, Hyoseris radicata, Sherardia arvensis, Geranium sanguineum, G. molle, Sonchus arvensis +p; in 2: Matthiola incana 1b; Scrophularia peregrina, Antirrhinum majus +p; Centaurea aspera, Apium graveolens, Muscari neglectum +r; in 3: Sempervivum tectorum +r; in 5: Mycelis muralis +r; in 6: Sagina procumbens 1a; Dryopteris filix-mas, Pteridium aquilinum +p.

LEGEND

1. Parrana San Giusto (Toscana, Italy), wall alongside road to Siena.
2. Ramatuelle (Var, France), wall alongside Rue Victor Leon.
3. Ascona (Ticino, Zwitserland), wall alongside road to Brissago.
4. Yealmpton (Cornwall, England), wall alongside high road above the church.
5. Quimperlé (Finistère, France), wall alongside road N 783 near the Mairie.
6. Ploueden (Finistère, France), wall around church.

towards the *Filici-Saginetum* were noted (Table 19, no. 6), in a single case *Soleirolia soleirolii* appeared in the relevé and twice a few specimens of *Cheiranthus cheiri.* Presumably the initial stage of communities of type (a) is mostly a community with *Parietaria judaica,* of those of type (b) rather a community with *Linaria cymbalaria* and *Asplenium ruta-muraria.* Both type (a) and type (b) can be regarded as *Erigeron mucronatus* facies of various communities, but the frequently high percentage of coverage of this species and the special ecological conditions plead in favour of a separate treatment.

The communities can be classified in the *Parietarion judaicae* as follows:

1. Association Fico-Erigeronetum mucronati	Linario-Erigeronetum mucronati ficetosum caricae
2. Association Polypodio-Erigeronetum mucronati	polypodietosum vulgaris

In stands with *Erigeron mucronatus* phanerophytes are rather conspicuous, at least qualitatively, in the stands with *Asplenium adiantum-nigrum* being represented with a score of 10% and with *Ficus carica* of 7%. In the stands with *F. carica,* however, therophytes are more numerous (score: 30% U and 5% W as against 11% and 1%, respectively). Myrmecochory is important in stands with *F. carica* and is hardly (or not at all) of importance in stands with *A. adiantum-nigrum* (the respective scores being U: 25% and 8%, W: 30% and 7%).

6.1.11 Vegetation with Parietaria judaica and Asplenium ruta-muraria

Parietaria judaica communities occurring in W.- and central Europe are easily distinguished from Mediterranean types, the resemblance with those of eu-atlantic regions (especially of area 112) being the greatest. Roughly speaking at least three categories can be distinguished:

a. communities with *Centranthus ruber* and/or *Phyllitis scolopendrium*
b. communitics with *Poa compressa,* and
c. communities with *Lycopus europaeus*

Table 20 Geographical distribution and exposure of stands of Parietaria judaica and Asplenium ruta-muraria
a. with Centranthus ruber and/or Phyllitis scolopendrium
b. with Poa compressa
c. with Lycopus europaeus

	exposure:	1	2	3	4	t	t%
	geographical region:						
a	111	2	2	8	3	15	40
a	112	5	4	1	5	15	40
a	120	1	-	-	1	2	5
a	132	3	1	2	-	6	15
a	t	11	7	11	9	38	-
a	t%	29	18	29	24	-	100
b	111	-	1	-	-	1	3
b	112	-	-	1	-	1	3
b	120	2	3	7	1	13	44
b	131	-	2	4	2	8	28
b	132	1	-	3	-	4	14
b	310	1	-	-	-	1	3
b	332	-	-	-	1	1	3
b	t	4	6	15	4	29	-
b	t%	14	21	52	14	-	100
c	111	2	-	-	-	2	29
c	120	-	1	-	3	4	57
c	132	-	1	-	-	1	14
c	t	2	2	-	3	7	-
c	t%	29	29	-	43	-	100

Type (a) was predominantly encountered in the areas 111 and 112, (b) in the areas 120 and 131, and (c) in area 120 especially.

Eight mural sites of community type (a) i.e. 20%, were facing a body of water (four in area 111, two in area 112, one in area 120 and one in area 132), six of community type (b) i.e. 18%, and all of community type (c). There is apparently no marked preference for any special exposure in types (a) and (c) but in type (b) the distribution of the exposure is very irregular, the highest number of the relevés being found on S.-facing walls. Type (a) shows a greater affinity to the *Sedo-Parietarietum* than to the *Bromo-Parietarietum* (Poore's index: 0.45 and 0.35, respectively). The resemblances are primarily determined by the species common to the higher phytosociological categories: *Parietaria, Centranthus* (alliance), *Linaria cymbalaria, Antirrhinum majus, Tortula muralis* (order), *Asplenium trichomanes, Umbilicus rupestris* (class), etc. The communities are well separated, not only floristically, but also structurally, ecologically, and geographically. In stands referable to type (a) *Parietaria judaica* is not invariably represented and sometimes only *Centranthus ruber* is found (see Table 21, no. 2). The latter is more common in W. France than in S. England. Transitional stands linking the community to the W. European *Asplenium trichomanes* communities to be discussed presently (Table 21, no. 3) and to the *Filici-Saginetum* (*do.*, no. 1) occur, but species which are frequently present in these communities are mostly also found growing together in *Parietaria judaica* communities (the greatest affinity thus being with vegetation of *Asplenium trichomanes* and *Poa annua*, see 6.6.3). Occasionally *Erigeron mucronatus, Chrysanthemum parthenium* and *Soleirolia soleirolii* were noted. As species which differentiate it from type (b), *Centranthus ruber, Phyllitis scolopendrium, Asplenium trichomanes Polypodium vulgare* and *Umbilicus rupestris* may be especially mentioned. Much rarer are e.g. *Linaria purpurea, Buddleja davidii* and *Bryum murorum* (each of them represented as (P = 8%), but they may nevertheless be considered to be important differentiating species. The most important differentiating species of community type (b) is *Poa compressa.* Type (b) is clearly poorer in species, but the percentages of coverage of the herb layer is upon the average higher than in stands of type (a). The richest in species is type (c), which is widely scattered and is, for instance, very well developed in sites in several Flemish towns such as Bruges and Lierre [1]). It forms a transition towards communities with *Sagina procumbens* and *Lycopus europaeus* (the *Filici-Saginetum,* see under 6.9.1.4,

---

[1]) In the Table, the P values of species recorded only once in relevés of stands of type (c) are put in brackets.

Table 21 Communities of Parietaria judaica and Asplenium ruta-muraria

a. with Centranthus ruber and/or Phyllitis scolopendrium
b. with Poa compressa
b1. with Poa compressa and Sedum album
c. with Lycopus europaeus
d. inops

	a.							b.						
Nr C	1	2	3	*P*	U W	MGT	MGP	4	5	6	*P*	U W	MGT	MGP
Date	30-9	5-8	9-8	*38st*				15-10	30-10	28-9	*29st*			
Year	61	62	62					60	60	65				
Nr R	577	523	577					393	450	1273				
Ex	1	3	1					3	1	3				
Geo R	111	112	112					120	131	132				
CH	25	55	65					25	70	55				
CL	35	3	4					2	2	3				
Parietaria judaica	2b	1b	2b	*81*	1a-4a	15	18	2b	4a	3b	*100*	2a-4b	29	29
Linaria cymbalaria	1b	+r	2b	*74*	+r-2b	7	10	1b	2a	2b	*80*	+p-4a	8	10
Tortula muralis	2a	1p	1a	*68*	+p-2a	2	3	2m	2m	2m	*83*	+r-3a	3	4
Asplenium ruta-muraria	+r	1p	-	*34*	+r-1a	0.2	0.5	2m	2m	2m	*62*	+r-2m	0.6	1
Ceratodon purpureus	2m	-	-	*26*	+p-2b	1	5	2m	1p	-	*48*	+r-2a	0.9	2
Bryum capillare	-	-	+p	*18*	+p-2a	0.4	2	-	-	-	*34*	+p-2m	0.2	0.5
Bryum caespiticium	+p	+p	-	*37*	+r-2m	0.3	0.7	+p	-	-	*18*	+r-+p	<0.1	0.1
Bryum argenteum	-	+p	-	*13*	+r-1p	0.1	0.2	+p	-	-	*28*	+r-2m	0.3	1
Homalothecium sericeum	+r	-	-	*34*	+r-3a	2	6	-	+p	1b	*41*	+r-2m	0.5	1
Sonchus oleraceus	+r	+p	+r	*57*	+r-1p	<0.1	0.1	-	-	-	*41*	+r-1p	<0.1	0.1
Taraxacus (div.) spec.	+p	-	-	*29*	+r-+p	<0.1	<0.1	-	-	+r	*28*	+r-+p	<0.1	<0.1
Poa annua	1p	-	-	*15*	+r-2m	0.1	0.6	+r	-	-	*21*	+r-+p	<0.1	0.1
Poa pratensis	-	-	-	*15*	+p-1p	<0.1	0.2	+r	-	-	*14*	+r-1p	<0.1	0.1
Dactylis glomerata	-	-	+r	*15*	+r-1a	<0.1	0.3	-	-	-	*3*	+p	<0.1	0.1
Sagina procumbens	1a	-	-	*21*	+r-2a	0.3	2	-	-	-	*10*	+r-+p	<0.1	<0.1
Streblotrichum convolutum	2m	+p	-	*21*	+p-2a	0.6	3	-	2m	-	*21*	+r-2m	0.3	1
Hypnum cupressiforme	1a	-	-	*18*	+p-1a	0.1	0.5	-	-	-	*10*	+r-+p	<0.1	0.1
Poa angustifolia	-	-	2m	*18*	+p-2m	0.1	0.7	-	-	-	*10*	+r-+p	<0.1	0.1
Amblystegium serpens	-	-	-	*13*	+p-2b	0.5	4	-	-	-	*10*	+r-+p	<0.1	0.1
Chelidonium majus	-	-	-	*10*	+r-1a	<0.1	0.4	-	+p	-	*21*	+r-3a	1	7
Phyllitis scolopendrium	+p	-	-	*34*	+r-2m	0.4	1	-	-	-	-	-	-	-
Dryopteris filix-mas	1p	-	-	*15*	+r-1p	<0.1	0.1	-	-	-	-	-	-	-
Centranthus ruber	-	4a	2b	*60*	+a-4a	6	10	-	-	-	-	-	-	-
Asplenium trichomanes	·	1p	+p	*50*	+r-2b	1	3	-	-	-	*7*	+p-1a	0.1	0.8
Polypodium vulgare	1p	()	+p	*26*	+r-1b	0.3	1	-	-	-	*3*	+p	<0.1	0.1
Barbula vinealis	-	+p	-	*26*	+p-2a	0.6	2	-	-	-	*7*	+p-2m	0.1	1
Hedera helix	-	-	+p	*18*	+r-+b	0.1	0.5	-	-	-	*7*	+p-+b	0.1	2
Rubus (div.) spec.	-	-	+p	*18*	+r-+a	<0.1	0.3	-	-	-	*3*	+r	0	0
Antirrhinum majus	-	+r	-	*15*	+r-2a	0.2	1	-	-	-	*3*	+r	0	0
Tortula intermedia	-	2m	-	*15*	+r-2m	0.2	1	-	-	-	*3*	2a	0.3	8
Umbilicus rupestris	-	-	+r	*13*	+r-1a	0.1	0.4	-	-	-	-	-	-	-
Asplenium adiantum-nigrum	-	-	+r	*10*	+r-1a	<0.1	0.4	-	-	-	-	-	-	-
Poa compressa	-	-	-	*3*	1p	<0.1	0.4	+p	2m	2m	*83*	+r-2m	0.8	1
Sedum album	-	-	-	*8*	+p-2b	0.7	3	-	-	+p	*10*	+p-2b	0.6	6
Reseda lutea	-	-	-	-	-	-	-	-	-	-	-	-	-	-
Artemisia vulgaris	-	-	-	-	-	-	-	-	-	-	*3*	+p	<0.1	0.1
Ballota nigra	-	-	-	-	-	-	-	-	-	-	*3*	+p	<0.1	0.1
Convolvulus arvensis	-	-	-	-	-	-	-	-	-	-	*3*	+p	<0.1	0.1
Lycopus europaeus	-	-	-	-	-	-	-	-	-	-	-	-	-	-
Epilobium roseum	+p	-	-	*3*	+p	<0.1	<0.1	-	-	-	-	-	-	-
Alnus glutinosa	-	-	-	-	-	-	-	-	-	-	-	-	-	-
Brachythecium rutabulum	+r	-	-	*5*	+r-+p	<0.1	0.1	-	-	-	-	-	-	-
Athyrium filix-femina	-	-	-	*3*	1b	0.1	4	-	-	-	-	-	-	-
Rumex obtusifolius	-	-	-	-	-	-	-	-	-	-	-	-	-	-
Angelica archangelica	-	-	-	-	-	-	-	-	-	-	-	-	-	-
Epilobium parviflorum	+r	-	-	*3*	+r	<0.1	<0.1	-	-	-	-	-	-	-
Lolium perenne	-	-	-	-	-	-	-	-	-	-	-	-	-	-
Marchantia polymorpha	-	-	-	-	-	-	-	-	-	-	-	-	-	-
Plantago major	-	-	-	-	-	-	-	-	-	-	-	-	-	-
Erysimum cheiranthoides	-	-	-	*5*	1p	<0.1	0.4	-	-	-	*7*	1a-1b	0.2	3
MNS - MCM - MCM		*13*	-	*30*	-	*9*	*(10)*		*10*	-	*42*	- *6*	*(7)*	

In addition in 1: Epilobium montanum +p; Taxus baccata, Epilobium adnatum, Senecio vulgaris +r; Barbula unguiculata 2b; Marchantia polymorpha +p; in 2: Sedum acre +r; in 3: Linaria purpurea +p, Erigeron mucronatus +r; Bryoerythrophyllum recurvirostre 2m; in 4: Solanum nigrum +r; in 7: Eupatorium cannabinum 2a; Lythrum salicaria, Arrhenatherum elatius +p, Impatiens glandulifera, Urtica dioica +r; in 8: Valeriana officinalis, Galium mollugo ssp. erectum +r; in 9: Matricaria matricarioides 1b; M. maritima ssp. inodora, Rorippa sylvestris 1p; Chamaenerion angustifolium, Capsella bursa-pastoris +p; Plantago lanceolata, Tussilago farfara +r; Oxyrrhynchium swartzii 2b; Drepanocladus aduncus +p; in 13: Acer pseudoplatanus +a.

LEGEND

1. Exeter (Devon, England), town wall near staircase along Bartholomew Street East.
2. Granville (Manche, France), rampart wall near the Haute Ville.
3. Quimperlé (Finistère, France), wall alongside road nr N 783 opposite Rue de Poradec.
4. Buren (Gelderland, Netherlands), rampart wall alongside the river Korne near Peperstraat.
5. Laon (Aisne, France), rampart wall.
6. Avallon (Yonne, France), rampart wall below Rue de la Fontaine-Neuve.
7. Kamp Bornhofen (Rheinland-Pfalz, Western Germany), wall alongside railway siding.
8. Moselkern (Rheinland-Pfalz, Western Germany), wall of a vineyard near the river Mosel.
9. Lay (Rheinland-Pfalz, Western Germany), wall of a vineyard 3 km E. of the village.
10. Exeter (Devon, England), wall of Exe river branch between Commercial Road and Edmund Street.
11. Loches (Indre-et-Loire, France), wall above river near stadium.
12. Dordrecht (Zuid-Holland, Netherlands), canal wall of Wijnhaven behind Voorstraat.
13. Exeter (Devon, England), wall behind Bartholomew Street near Mint.
14. Geertruidenberg (Noord-Brabant, Netherlands), St. Geertruidskerk (church).
15. Brughes (Belgium), canal wall Groene Rei near bridge to Paardenstraat.

where its characteristic species combination and ecology is discussed in detail). Type (c) is particularly well-developed on the lower parts of walls arising from a mass of water (canals, rivers, smaller streams, lakes), upon which by the capillary suction of water in the substratum, a number of riparian forms are enabled to grow. In a few localities in the Dutch estuary area of the great rivers (Dordrecht, Kralingse Veer) *Angelica archangelica* was found growing in this community (Table 21, no. 9) and sometimes *Oenanthe aquatica,* both of which are forms adapted to appreciable fluctuations in the water table. Community type (b) likewise shows a certain affinity towards the *Filici-Saginetum,* but transitions towards *Sedo-Scleranthetea* communities and ruderal communities *(Artemisietea* and *Chenopodietea)* are also encountered. Stands transitional towards *Sedo-Scleranthetea* often contain *Sedum album* (e.g. relevé no. 6 of Table 21) and/or *Barbula fallax,* and they are best developed on more or less slanting walls. Walls with a marked slant, and more particularly those that have been enriched with nutrients (such as the walls surrounding vineyards, which receive rain water with leached-out fertilisers or fertile soil penetrating into their cracks and crevices), carry stands with a strong admixture of species of ruderal habitats, such as *Artemisia vulgaris, Ballota nigra, Reseda lutea* and *Erysimum cheiranthoides,* which may even become dominant and aspect-forming. Hübschmann (1957) tabulated 12 relevés from the Moselle valley, which he included in the *"Cheirantho-Parietarietum"* of Oberdorfer and which most probably agree with Oberdorfer's concept of this community (see under 6.1.8.3). Three examples of relevés from the Moselle valley (the nos. 7, 8 and 9 of Table 21) are included in the present publication. In this area, *Rumex scutatus* is frequently met with in mural vegetation of walls of vineyards (see under 6.3). Mural vegetation with *Sedum album* is also found in the areas 111 and 112, but vegetation containing this species (or one or several different representative(s) of the genus, more particularly *S. forsteranum* and *S. acre*), is usually better developed on wall tops, and as a rule includes *Poa compressa* with e.g. *Arenaria serpyllifolia, Cerastium semidecandrum, Erophila verna, Saxifraga tridactylites, Barbula fallax* and *B. hornschuchiana, Tortula ruralis, Cladonia fimbriata, C. coniocraea* and *C. pyxidata* var. *chlorophaea.* Such stands have up to now not been adequately described. A species differentiating the *Sedum*-containing stands under discussion from central European *Sedum* vegetation is presumably *Catapodium rigidum.* In such stands in W. Europe *Centranthus ruber* and *Linaria cymbalaria* sometimes appear.

Between the three types (a), (b) and (c) there are some intermediate and 'transitional' stands linking them. Poore's index of affinity between (a) and (b) is 0.68 and this is fairly high. Type (c) shows a greater affinity to (b) (affinity index 0.53) than to (a)(affinity index 0.46).

Inops-types of stands can not always with any degree of certainty be referred to either (a) or (b) (e.g. the nos. 14 and 15 of the Table).

Parietaria judaica is nitrophilous to some extent and the nutrients it requires are sometimes brought in by rainwater trickling down or by the action of wind, but presumably the source is quite often the soil against which the wall is built up; at any rate *P. judaica* is of frequent occurrence on walls built against earth walls, ramparts or terraced arable land (vineyards, etc.) which include walls of old fortresses. A concomitant factor is that such walls do not become desiccated so quickly as comparable free-standing walls.

The preference shown for S.-facing sites in area 120 could be studied in some places where walls of a corresponding type of masonry are situated close together. The influence of the exposure could very clearly be observed on the round wall of a windmill at Buren (Gelderland, Netherlands). Here optimal development of the community with *Parietaria* had taken place on the part facing S. to S.-W. and the stand became gradually more poorly developed towards both sides to disappear completely on the parts facing W., N. and E. (where *Asplenium ruta-muraria* and *Polypodium vulgare* were present as well as on the S.-facing portions).

The initial phase of the *Parietaria judaica* communities consists in most cases of a community of ivy-leaved toadflax and wall-rue, the latter being usually the first to become established. When *Parietaria judaica* appears, *Linaria cymbalaria* can continue to increase in numbers and in coverage, but as a rule *Asplenium ruta-muraria* gradually decreases. As long as the rate of coverage of *Parietaria* remains at a low level, the coverage of *Linaria cymbalaria* may attain values of more than 60 per cent., but as soon as *Parietaria* is well on the increase *Linaria* starts declining. Such species as *Centranthus ruber* and *Poa compressa* usually become established later on. This sequence of developmental

Table 22 Stands of Parietaria judaica vegetation on parts of S-facing rampart walls of increasing age at Buren (Netherlands), recorded on 5-10-1960

Nr C	1	2	3	4	5	6	7
CH	3	1	8	20	30	50	80
Maximum height of herbaceous forms (in dm)	0.1	0.2	0.3	1.4	2	3	6
CM	-	•	2	3	3	3	5
Asplenium ruta-muraria	+p	1p	2b	1b	+p	-	-
Linaria cymbalaria	-	1p	1b	2a	3a	3a	2a
Tortula muralis	-	1p	1a	2m	2m	1b	1p
Streblotrichum convolutum	-	+p	1p	1p	+r	-	-
Parietaria judaica	-	-	+p	1a	1b	2b	4b
Ceratodon purpureus	-	-	1p	1a	+p	1p	+p
Bryum argenteum	-	-	1p	+p	-	+p	-
Poa annua	-	-	+r	+r	-	-	-
Sagina procumbens	-	-	+r	-	-	-	-
Solanum nigrum	-	-	+r	-	-	-	-
Grimmia pulvinata	-	-	+r	-	-	-	-
Poa compressa	-	-	-	-	+p	2a	2a
Homalothecium sericeum	-	-	-	-	+r	2b	+p
Taraxacum spec.	-	-	-	-	+p	-	-
Bryum caespiticium	-	-	-	-	+p	-	-
Rumex acetosa	-	-	-	-	-	+r	-

stages can be deduced from a study of walls of corresponding structure and exposure but of a different age. Some localities in England (such as Canterbury) and the Netherlands (especially Buren) offered suitable sites for the purpose. At Buren different parts of the old fortress wall had been restored or renovated with interval of tens or scores of years up to the recent time, so that from year to year the development could be traced. Table 22 is a survey of relevés made of a number of S.-facing walls of different ages at Buren. Although this series of relevés may strictly speaking not be taken for a successional sequence, it seems to be a very close approximation of the course of development of the community At Buren the regeneration of the stands on cleaned or not completely renovated parts of walls could also be studied. On such walls a marked initial increase of *Linaria cymbalaria* is soon followed by a rapid increase of *Parietaria* and the ousting of wall-rue in sites where viable remains of *Parietaria* had survived in parts of that wall.

A tentative classification of the communities with *Parietaria* under discussion in the *Parietarion judaicae* follows:

1. Association: *Asplenio-Parietarietum judaicae*
    - a. *asplenietosum trichomanis*
    - b. *poetosum compressae*
    - c. variant with *Sedum album*
    - d. *artemisietosum vulgaris*
    - e. *lycopetosum*

(a) shows transitions mainly towards eu-atlantic *Asplenium trichomanes* and chasmophytic communities; (e) towards riparian vegetation; (b) and particularly (c) towards *Sedo-Scleranthetea* communities, and (d) towards ruderal associations and communities of agricultural weeds.

Many species found in (d) also occur in (b). Vegetation on the surfaces of wall tops and ledges with *Centranthus ruber* and/or *Linaria cymbalaria* can best be regarded as a subassociation of one of the *Sedo-Scleranthetea* communities (presumably a community with *Poa compressa*; compare Lebrun et al., 1949: "Groupement à *Poa compressa* et *Saxifraga tridactylites*").

Stands with *Lycopus europaeus* are distinguished from all other communities with *Parietaria judaica* and *Asplenium ruta-muraria* by the qualitatively smaller contribution of chamaephytic forms (6% against 13-18%), by the scarcity of hemixerophytes (score: 3% U, 0.0% W, as against U 14-16% and W 2-5%), and, naturally, by the high scores of riparian forms and marsh plants (9% and 12%, respectively). Myrmecochory is not so important as it is in the Mediterranean *Parietaria judaica* communities if one relates it to the number of myrmecochorous taxa, but quantitatively this mode of dispersal is about equally important (13-32% and 51-82%, respectively).

6.2 Vegetation with Hedera helix

The ivy is of common occurrence on walls, even if all walls on which ivy is purposely grown are disregarded. *Hedera* is capable of overgrowing a wall from a stem rooting in the soil near the foot of that wall, but it can also grow in cracks and joints of walls and occasionally on top of a wall (in the latter case it develops laterally and often downwards). The first settlement of *Hedera* usually takes place in communities with *Linaria cymbalaria* or, as is the rule in S. and W. Europe, with *Parietaria judaica.* Especially if *Hedera* becomes dominant, such stands are mostly not clearly referable to a type of *Parietaria* vegetation, but in some cases show affinity to the communities of *P. judaica* with *Sedum dasyphyllum, Asplenium ruta-muraria* or *Anogramma leptophylla,* respectively, or to communities with *Asplenium trichomanes* without *Parietaria.* Stands with ivy could, accordingly, be regarded as facies types of diverse communities, if the vegetational structure did not differ so much from all other types of mural vegetation.

Stands of mural vegetation with *Hedera helix* have been recorded in all areas included in the present study. Towards the more continental parts of Europe they become poorer in species and the more typical wall-dwelling forms gradually disappear. The richest stands occur in the eu-atlantic regions, and particularly in S. England fine examples of mural vegetation with ivy and *Linaria cymbalaria* were noted.

Table 23 Geographical distribution and exposure of stands of Hedera helix and Linaria cymbalaria

exposure:	1	2	3	4	5	t	t%
geographical region:							
111	-	2	4	2	1	8	32
112	1	-	-	-	-	1	4
120	1	1	2	1	1	6	24
131	-	-	-	1	1	2	8
132	-	1	1	1	-	3	12
320	-	1	-	-	-	1	4
331 - 332	1	1	-	1	1	4	16
t						25	100

Table 24 Community of Hedera helix and Linaria cymbalaria

Nr C	1	2	3	P %st	U W	MGT	MGP
Date	12-4	28-4	29-9				
Year	62	62	61				
Nr R	091	239	571				
Ex	4	4	4				
Geo R	332	132	111				
CH	75	85	80				
CM	-	30	2				
Hedera helix	4b	4b	4a	100	2b-5c	67	67
Tortula muralis	-	+p	2m	56	+p-2b	1	3
Linaria cymbalaria	+r	1a	2b	52	+p-3b	3	7
Parietaria judaica	+p	-	+r	44	+r-2a	2	4
Asplenium trichomanes	-	-	+p	32	+p-2b	1	4
Asplenium ruta-muraria	-	-	2a	32	+r-2a	0.2	0.8
Centranthus ruber	1a	+p	-	28	+p-2b	1	4
Orobanche he_erae	+p	1b	-	24	+p-1b	0.3	1
Umbilicus rupestris	-	-	-	24	+r-1a	0.2	0.9
Ceterach officinarum	-	1a	-	24	+r-1a	0.1	0.6
Homalothecium sericeum	-	3a	-	20	+p-3a	2	11
Ceratodon purpureus	-	-	-	20	+r-1p	0.1	0.3
Poa compressa	-	+r	+p	16	+r-1b	0.3	2
Polypodium vulgare	-	-	-	16	+r-+p	<0.1	0.1
Hypnum cupressiforme	-	-	-	12	+r-2b	0.7	6
Brachythecium rutabulum	-	-	-	12	1a-2b	0.2	1
Streblotrichum convolutum	-	-	1p	12	+r-2m	0.1	0.7
Bryum caespiticium	-	-	-	12	+p-1p	<0.1	0.3
Dactylis glomerata	-	-	-	12	+p-1p	<0.1	0.3
Sonchus oleraceus	-	-	+p	12	+r-+p	<0.1	<0.1
Geranium robertianum	-	-	+r	12	+r-+p	<0.1	<0.1
MNS - MCH - MCM		9	-	78	-	9 (16)	

In addition in 2: Sedum album 1b; Stellaria media var. apetala, Saxifraga tridactylites +p; Poa pratensis, Reichardia picroides, Lagoseris sancta, Senecio vulgaris, Poa bulbosa, Geranium molle, Lamium purpureum +r; in 3: Holcus lanatus +r.

LEGEND

1. Sommières (Gard, France), city wall.
2. Cahors (Lot, France), garden wall near Place Lucterius.
3. Ashburton (Devon, England), wall behind church.

No definite conclusion can be drawn as regards the exposure of the stands (see Table 23). In S. Europe *Hedera* seems to avoid S.-facing walls, but in W. Europe it occurs rather often on S.-exposed substrata. A summary of the vegetation analysis is shown in Table 24.

The stands in S. Europe differ to some extent from those in W. Europe by the occurrence of e.g. *Veronica cymbalaria,* but great differences can hardly be apparent because the total number of species is not usually large enough and the majority of those that are present occurs in both areas with a high degree of presence. The only species which is restricted to this type of mural vegetation is the parasitic *Orobanche hederae,* which seems to require well-developed, dense stands of ivy. These are found not only on walls, but also on rocky slopes and in escarpment forests of *Fraxinus excelsior* and *Acer pseudoplatanus* with *Hedera* facies as noted in S. England. HEPPER (1954) recorded *Orobanche hederae* from dense *Hedera* vegetation of fixed and mobile dunes at Priory Bay on Caldy Island (Pembrokeshire, Wales). In the majority of the cases this broomrape is found in open country, but sometimes in shady places in forests and it is not clear whether it requires light for the germination of its seeds.

Hedera vegetation on rocky slopes was recorded in e.g. the area of Torquay and Berry's Head (Devon, England) and contained such species as *Centranthus ruber, Umbilicus rupestris* and *Asplenium* spec. div. Such stands closely resemble those found on walls. The "Hederaie" described from W. France by GEHU & GEHU-FRANCK (1961) also shows affinity by the appearance of *Orobanche hederae* and *Umbilicus rupestris,* but it differs in the high presence of particularly *Asplenium lanceolatum, Pteridium aquilinum, Teucrium scorodonia* and *Festuca rubra* and this suggests that it is characteristic of sites on a more acid subsoil, which agrees with the pH values and rather low calcium content found by these authors.

Hedera helix is by no means rare in diverse deciduous forest types throughout Europe and in forest edge- and scrub vegetation. In how far *Orobanche hederae* is presented in all these habitats is not known to me. In sites where mural stands of a *Hedera helix-Linaria cymbalaria* community develop from an initial stage of *Linaria cymbalaria-Asplenium ruta-muraria* and/or *A. adiantum-nigrum* vegetation, *Asplenium trichomanes* apparently finds a favourable environment in the not too dense stands of *Hedera* where the microclimate is considerably more suitable as a habitat for this fern than it is in more open vegetation on walls.

The community of *Hedera* and *Linaria cymbalaria* with *Orobanche hederae* can not clearly be regarded as a facies form of any other but a rock-dwelling or mural type of vegetation, and it is here provisionally included in the *Parietarion judaicae* as the *Asplenio-Hederetum*, but it could, alternatively, be placed in the order *Tortulo-Cymbalarietalia* without further indication. It might also be worth considering if communities with *Hedera* showing affinities towards chasmophytic vegetation types could be combined into a separate unit within the class *Asplenietea rupestris* without assigning a special place to them (unless one decides to propose a special higher category as was done in the case of communities with *Polypodium*, see under 6.8).

The *Asplenio-Hederetum* could be divided into a subassociation *cymbalarietosum*, a subassociation *pteridietosum*, etc.

6.3 Vegetation with Rumex scutatus

Rumex scutatus prefers gravelly slopes as a habitat in S. and central Europe and occurs especially in places where the gravel or rock debris is very coarse and not very stable tending to shift. Details relating to this type of environment were given by KUHN (1937), who also recorded the occurrence of the species in mural vegetation. During the present investigation, the species under discussion was most frequently encountered on mortar-jointed schist walls in the valleys of the rivers Meuse and Moselle, but also on mortar-jointed walls built of granite, sandstone and limestone.

Table 25 Geographical distribution and exposure of stands of Rumex scutatus and Linaria cymbalaria

exposure:	1	2	3	4	6	t	t%
geographical region:							
131	2	1	1	3	1	8	80
200	-	1	-	-	-	1	10
310	-	1	-	-	-	1	10
t	2	3	1	3	(1)	9(+1)	-
t%	22	33	11	33	-	-	100

Table 26 Community of Rumex scutatus and Linaria cymbalaria

Nr C	1	2	3	P 10st	U W	MGT	MGP
Date	11-9	4-9	11-7				
Year	62	62	62				
Nr R	674	637	446				
Ex	4	6	3				
Geo R	131	131	131				
CH	25	25	15				
CM	2	3	5				
Rumex scutatus	2b	1b	2a	100	1b-4a	14	14
Sonchus oleraceus	+p	+r	+p	70	+p-1a	0.4	0.5
Linaria cymbalaria	1b	2b	-	60	1b-2b	6	11
Tortula muralis	1p	1p	2m	60	+p-2m	1	2
Poa compressa	+r	+r	+p	60	+r-2m	0.5	0.3
Asplenium ruta-muraria	2m	+r	+p	50	+r-2m	0.4	0.8
Bryum capillare	+r	-	+p	50	+r-2m	0.3	0.5
Poa angustifolia	-	+p	1a	50	+p-1a	0.2	0.5
Chelidonium majus	+r	-	-	40	+r-+p	<0.1	<0.1
Streblotrichum convolutum	+r	-	-	40	+r-+p	<0.1	<0.1
Homalothecium sericeum	+p	+p	-	30	+p-2b	2	6
Asplenium trichomanes	-	1p	-	30	+p-2m	0.3	1
Bromus sterilis	-	-	+p	30	+p-1p	0.1	0.2
Dactylis glomerata	-	-	-	30	+r-1p	<0.1	0.1
Galium mollugo ssp. erectum	-	-	+p	30	+p	<0.1	0.1
Arenaria serpyllifolia	-	-	+p	30	+p	<0.1	0.1
Taraxacum (div.) spec.	+r	-	-	30	+r-+p	<0.1	0.1
Senecio jacobaea	+r	-	+p	30	+r-+p	<0.1	<0.1
Geranium robertianum	-	+r	-	20	+r-2a	0.8	4
Bryum argenteum	-	-	2m	20	2m	0.5	3
Arabidopsis thaliana	-	-	-	20	+r-1a	0.2	0.8
Urtica dioica	-	-	-	20	+p	<0.1	0.1
Tanacetum vulgare	-	-	+p	20	+r-+p	<0.1	0.1
Mercurialis annua	-	-	+r	20	+r	0	0
Erigeron canadensis	-	-	+p	20	+r-+p	<0.1	0.1
MNS - MCH - MCM		14	-	13	-	7(9)	

In addition in 1: Epilobium montanum +p; Taxus baccata +r; in 2: Poa pseudocompressa, Parietaria judaica +p; Polypodium vulgare, Mycelis muralis, Sonchus asper +r; Barbula unguiculata 2m; Leptobryum pyriforme 2p; Bryum caespiticium 1p; in 3: Diplotaxis tenuifolius 1a; Melica ciliata, Vulpia myuros, Isatis tinctoria, Artemisia vulgaris, Verbascum lychnitis, Erysimum cheiranthoides +p; Sedum acre, Dipsacus pilosus, Lactuca serriola, Achillea millefolium +r.

LEGEND

1. Yvoir (Namur, Belgium), wall along the river Meuse along road to Namur.
2. Cochem (Rheinland-Pfalz, Western Germany), wall underneath castle along road to Trier.
3. Kamp-Bornhofen (Rheinland-Pfalz, Western Germany), wall along the river Rhine, S. of the village.

Outside of the area 131, mural vegetation with *Rumex scutatus* was noted at Briançon (Hautes-Alpes, France) and Cavigliano (N. Italy).

In the latter locality, *Sedum dasyphyllum, Veronica cymbalaria, Erigeron mucronatus* and *Bromus madritensis* appear in the relevés and conceivably there is a Mediterranean variant or type which occurs in N. Italy and/or S. Switzerland. In the Moselle valley *Rumex scutatus* was encountered on a few occasions (represented with a low percentage of coverage) in stands of *Parietaria judaica* vegetation (also in a relevé made by HÜBSCHMANN, 1967). A clear-cut preference to a particular exposure was not observed, but the impression was gained that *Rumex scutatus* tends to avoid excessive insolation. It is not infrequently found growing on walls of vineyards and possibly a certain degree of instability of the environment plays a part: frequent treating and cultivation of the soil above the wall, scattering of soil particles, etc.

The contribution of typical wall-dwellers to the stands is not very large and it is a question whether the community with *Rumex scutatus* is indeed referable to the assembly of mural vegetation types (order *Tortulo-Cymbalarietalia)* or if they represent an extreme case of a special *Rumex scutatus* community (to be called *Rumicetum scutati cymbalarietosum*). The second alternative seems to be the best solution, at least for the time being.

In stands with *Rumex scutatus* species of open and dry habitats are well represented (score: U 19%, W 3%). The contribution of taxa with a Mediterranean-central European distribution is quantitatively important (score: U 9%, W 57%).

6.4 Vegetation with Hieracium amplexicaule or other species of Hieracium

Hieracium amplexicaule is a chasmophyte especially common in the Mediterranean region but extending to W.- and central Europe as far as Scotland, the Jura Mts. and the Black Forest. According to VAN SOEST (1934), its subspecies *amplexicaule* occurs in Britain, and also on walls at Tongeren (Belgium), its ssp. *speluncarum* at e.g. Maestricht (Netherlands), Wimpfen and Kassel (W. Germany), and its ssp. *pulmonarioides* at e.g. Jena (E. Germany) and Altona near Hamburg (W. Germany). According to HEGI's Flora, the species must be regarded as a garden escape in central Europe where it occurs on city walls and castles. Van Soest is of the opinion that the pappus-bearing achenes of this plant are capable of bridging great distances and that there is, accordingly, no cogent reason to accept the idea of a secondary introduction. An additional argument pleading in favour of a "natural" dispersal is the pattern of distribution along the great rivers and their tributaries (Meuse, Rhine, Elbe, Weser, and also Danube and Inn: ssp. *speluncarum* at Salzburg in Austria). Stands with *H. amplexicaule* were recorded at Maestricht, Tongeren, Bad Wimpfen and Salzburg. This species, like e.g. *Cheiranthus cheiri,* shows a marked preference for not so steeply inclined walls.

At Tongeren it was almost exclusively found on the tops of walls and on ledges. It is also found in the community with *Cheiranthus cheiri* (at Maestricht and at Wimpfen), and there is undoubtedly a certain degree of affinity between that community and stands of the *Hieracium amplexicaule* community. However, stands with *Hieracium* occurring in central Europe never contain *Parietaria judaica,* not even when the hawk-weed is associated with *Cheiranthus.*

At Wimpfen, Maestricht and Valkenburg (Limburg, Netherlands), stands on walls of a different age could be compared. It seems likely that as a rule *Hieracium amplexicaule* communities develop from an initial stand of *Linaria cymbalaria* and *Asplenium ruta-muraria,* and that ultimately a *Cheiranthus cheiri* community succeeds the *Hieracium* community. Accordingly, the community with hawk-weed can be regarded as a variant of the initial community of ivy-leaved toadflax and wall-rue, the more so because the abundance of *Hieracium* hardly ever attains a considerable value. On more slanting walls the number of species increases appreciably and the succession series also shows a progressively higher number of taxa, which increase is concomitant with a higher percentage of coverage of the layer of herbaceous forms. On three walls in Maestricht (a, b and c in order of increasing age), built of marl blocks, situated close together and all S.-facing, the following data were recorded:

	Number of species	PC of herb layer	PC of moss layer
wall a	8	15	2
wall b	9	35	1
wall c	14	50	20

(N.B.: Wall c is several centuries old).

Our knowledge of the ecology of *Hieracium amplexicaule* is too scanty to make it plausible why this species is so rare in central Europe. The scattered localities are isolated 'frontier posts' of its area of distribution, but at first sight many more sites seem to fulfil all requirements for its occurrence. The most important point of agreement between the sites supporting vegetation with *Hieracium* is that they consist of age-old walls built of soft materials such as marl.

Other hawk-weeds have also been recorded from walls e.g. *Hieracium murorum, H. maculatum, H. praecox* and *H. lachenalii.*

In contradistinction to *H. amplexicaule,* these species are not typical chasmophytes or wall-dwellers. *H. murorum* is a species of deciduous forests, but it is sometimes found in stands of mural vegetation that are surprisingly similar to those with *H. amplexicaule* (Table 27, b). Stands with *H. murorum* usually have an exposure towards the W. and the N. and apparently they avoid strong insolation or excessive heat (or great fluctuations in the temperature). Stands with other species of *Hieracium* are also rather similar (e.g., Table 27, c), but nearly always the contribution of the hawk-weeds is not substantial. They are especially encountered in initial phases of vegetation development in communities of *Linaria cymbalaria* and *Asplenium ruta-muraria,* often associated with *Chelidonium majus* and *Poa compressa,* or in communities with *Cheiranthus*

Table 27 Community of Hieracium amplexicaule and Asplenium ruta-muraria (a), community of Hieracium murorum and Asplenium ruta-muraria (b) and community of Hieracium maculatum and Asplenium ruta-muraria (c)

	a							b		c
Nr C	1	2	3	P abs dst	U W	MGT	MGP	4	P abs dst	5
Date	16-11	10-7	16-7					14-9		5-9
Year	60	62	66					64		62
Nr R	506	434	461					672		638
Ex	1	1	4					1		4
Geo R	131	131	200					131		131
CH	50	50	20					20		10
CM	15	20	3					6		5
Hieracium amplexicaule	2b	3b	2a	6	2a-3b	21	21	-	-	-
Linaria cymbalaria	-	1a	1b	5	1a-2b	5	6	-	2	-
Tortula muralis	2a	-	1p	5	1p-2a	2	3	2a	4	+p
Asplenium ruta-muraria	2a	+r	+p	5	+r-2a	1	2	2m	4	2m
Poa compressa	2b	2m	-	4	1b-2b	4	7	-	2	2m
Asplenium trichomanes	+r	1a	2m	3	+r-2m	0.7	1	+p	3	-
Dryopteris filix-mas	2b	-	-	2	+a-2b	3	10	+p	1	-
Mercurialis annua	-	-	-	2	+a	0.5	2	+p	1	-
Hedera helix	-	+r	-	2	+r-+b	0.6	2	-	-	-
Bryum capillare	2m	-	-	2	1p-2m	0.5	1	-	1	-
Ceratodon purpureus	2m	-	-	2	+p-2m	0.4	1	2m	1	-
Streblotrichum convolutum	1p	-	-	2	1p	0.1	0.4	-	1	-
Chelidonium majus	-	+p	+r	2	+r-+p	<0.1	0.1	1a	3	-
Epilobium montanum	-	-	-	2	+r-+p	<0.1	0.1	+p	1	-
Taraxacum (div.) spec.	+r	+r	-	2	+r	0	0	1a	2	-
Hieracium murorum	-	-	-	-	-	-	-	2a	4	-
Homalothecium sericeum	+a	-	-	1	+a	0.4	2	-	2	-
Mycelis muralis	-	-	-	-	-	-	-	+p	2	-
Hieracium maculatum	-	-	-	-	-	-	-	-	1	1b
Hieracium umbellatum	-	-	-	-	-	-	-	-		+p
Campanula rotundifolia	-	-	2a	1	2a	1	8	1b	1	-
Poa angustifolia	-	+p	-	1	+p	<0.1	0.1	-	1	+r
Geranium robertianum	-	+r	-	1	+r	0	0	1a	1	-
Sonchus oleraceus	-	-	-	1	+r	0	0	-	1	+r
Bryum caespiticium	1p	-	-	1	1p	0.1	0.4	-	1	-
Encalypta streptocarpa	-	-	+p	1	+p	<0.1	0.1	-	1	-
MNS - MCH - MCM		11	-	25	-	8		15-23-14		

In addition in 1: Polypodium vulgare 1b; Athyrium filix-femina +p; Schistidium apocarpum +p; in 2: Arenaria serpyllifolia, Senecio jacobea +p; Clematis vitalba, Lamium purpureum, Veronica chamaedrys, Galeopsis tetrahit, Cerastium fontanum +r; Didymodon rigidulus 2b; in 3: Campanula rapunculoides +r; Barbula unguiculata 2m; Lunularia cruciata 1a; Encalypta streptocarpa +p; in 4: Epilobium adnatum +p; Poa annua +r; in 5: Sedum acre, Artemisia vulgaris +r.

LEGEND

1. Maastricht (Limburg, Netherlands), rampart wall along Lang Grachtje near St. Pieterestraat.
2. Bad Wimpfen (Württemberg, West-Germany), supporting wall of ramparts near castle.
3. Salzburg (Austria), wall above Burggaststätte under staircase to Hoch Salzburg.
4. Stein (Limburg, Netherlands), wall of abbey.
5. Sélêstat (Bas-Rhin, France), rampart wall.

cheiri or *Corydalis lutea,* and usually they are found on fairly dry walls. *H. amplexicaule* was once recorded from an initial stage of *Corydalis lutea* vegetation (at Maestricht).

In the community with *Hieracium amplexicaule* species of dry, open sites are relatively common (representation 9% for both U and W values).

6.5 Vegetation with Corydalis lutea

Corydalis lutea is another species reputed to be an escape from cultivation in W.- and central Europe (e.g. PRAEGER, 1901: in Ireland "a garden escape in all cases", HEGI's Flora: "in W.- and central Europe established centuries ago, often seemingly indigenous"). According to Hegi's Flora, the species is presumably growing wild in Ticino and in the eastern Dolomites where it occurs on gravelly and rocky slopes.

LEFORT (1950)) is of the opinion that the species under discussion, which is nowadays found in mural vegetation in a few localities in the Grand-Duchy of Luxemburg, must have become introduced in recent times, because it is not mentioned in Tinant's *Flore Luxembourgeoise* of 1836. Conceivably *Corydalis lutea* has occasionally become naturalised after escaping from culture (like *Antirrhinum majus* and *Cheiranthus cheiri*), and even its occurrence in the United Kingdom may be exclusively subspontaneous, but that this holds true for all its localities in the European mainland is by no means certain. An argument against it being an introduction is once more the characteristic pattern of distribution following the greater rivers and their tributaries. In some areas the species is so common that it does not seem to be restricted in its distribution to the vicinity of rivers or streams (e.g. in the department Nord of France). It is of course possible that *Corydalis lutea* was indeed introduced as an ornamental many centuries ago and subsequently spread subspontaneously along river valleys. One must bear in mind, however, that nowadays (and presumably also in the past) plants are being taken from the walls to be grown in gardens instead of having been primarily introduced as garden plants. Whatever the case may be, the species behaves completely like a neophyte that has found a niche in a specific plant community distributed at least in the valleys of the rivers Rhine, Meuse and Loire, but also elsewhere (Netolice, valleys of the Moldau-Elbe, and according to SCHULZ & SUKOPP, 1965, also near Berlin). British stands with *Corydalis lutea* differ from those found on the mainland by a relatively high degree of presence of *Phyllitis scolopendrium, Polypodium vulgare, Ceterach officinarum* and *Centranthus ruber,* among other species. On the mainland there are two more or less clearly distinguishable categories of stands, the one more frequently noted on N.-facing and the other on S.-facing wall surfaces (Table 28 under c and d).

Table 28 Geographical distribution and exposure of stands of communities with Corydalis lutea and Asplenium ruta-muraria

a. inops
b. with Phyllitis scolopendrium
c. with Cystopteris fragilis
d. with Poa compressa

	exposure:	1	2	3	4	5	t	t%
geographical region:								
a	120	1w	1w	1w	2w	-	5w	42
a	131	2(1w)	-	1	-	-	3(1w)	25
a	132	-	-	-	2	-	2	17
a	200	-	-	-	-	2	2	17
a	t	3(2w)	1w	2(1w)	4(2w)	(2)	10(+2)	-
a	t%	30	10	20	40	-	-	100
b	111	-	5	3	3	-	11	100
b	t%	-	45	27	27	-	-	100
c	131	6(1w)	1	1w	-	-	8(2w)	61
c	132	-	1w	1	-	1	3(1w)	23
c	200	-	1	-	-	-	1	8
c	400	1w	-	-	-	-	1	8
c	t	7(2w)	3(1w)	2(2w)	-	1	12(+1)	-
c	t%	58	25	17	-	-	-	100
d	120	1	-	-	-	-	1	4
d	131	2	6(1w)	7	2(1w)	-	17	71
d	132	3	1	1	-	-	5	21
d	200	-	-	1	-	-	1	4
d	t	6	7	9	2	-	24	-
d	t%	25	29	38	8	-	-	100

Table 29 Communities of Corydalis lutea and Asplenium ruta-muraria

a. inops
b. with Phyllitis scolopendrium
c. with Cystopteris fragilis
d. with Poa compressa

	a.								b.						
Nr C	1	2	3	4	P	U W	MGT	MGP	5	6	7	P	U W	MGT	MGP
Date	12-7	23-7	18-8	16-4	*12st*				4-8	28-7	26-9	*11st*			
Year	61	66	60	66					64	64	61				
Nr R	282	524	276						446	298	526				
Ex	3	3s	1w	3s					3	4	2				
Geo R	131	400	120	200					111	111	111				
CH	30	80	10	30					40	30	5				
CM	-	-	3	-					-	2	3				
Corydalis lutea	3a	4a	2a	3a	*100*	2a-4a	27	27	3a	2a	1b	*100*	1b-3b	18	18
Asplenium ruta-muraria	+p	2m	1a	-	*83*	+r-2m	0.5	0.6	-	2m	1p	*55*	+p-2a	1	2
Linaria cymbalaria	2a	-	1b	-	*67*	1p-2a	3	4	2a	2b	1a	*73*	+r-2b	6	8
Tortula muralis	-	-	2m	-	*46*	+r-2m	0.7	2	-	-	+r	*73*	+r-2m	0.4	0.5
Taraxacum (div.) spec.	-	-	-	+p	*33*	+r-+p	0.1	0.1	+r	-	+r	*55*	+r-+p	<0.1	0.1
Epilobium montanum	-	-	-	1p	*17*	+p-1p	0.1	0.3	+p	-	-	*27*	+p	<0.1	0.1
Bryum capillare	-	-	+p	-	*25*	+p-2m	0.2	0.7	-	-	-	*27*	+p-2m	0.2	0.9
Sonchus oleraceus	+p	-	-	-	*17*	+r-+p	0.1	0.1	-	-	-	*9*	+p	<0.1	0.1
Ceratodon purpureus	-	-	2m	-	*25*	+p-2a	0.9	4	-	-	-	*17*	+p-2a	<0.1	0.1
Poa annua	-	-	-	2m	*25*	+r-2m	0.4	2	+p	-	-	*27*	+p	<0.1	0.1
Parietaria judaica	-	-	1a	-	*17*	+r-1a	0.1	0.8	-	-	-	*18*	+p-1a	0.1	0.8
Dryopteris filix-mas	-	-	+p	-	*25*	+r-+p	0.1	0.1	1a	-	-	*18*	+r-1a	0.1	0.8
Bryum caespiticium	-	-	-	-	*17*	+p-2m	0.2	1	-	-	-	-	-	-	-
Sagina procumbens	-	-	-	-	*8*	+p	<0.1	0.1	+p	-	-	*18*	+r-+p	<0.1	0.1
Asplenium trichomanes	-	-	-	-	-	-	-	-	-	+p	-	*36*	+p-2a	1	4
Homalothecium sericeum	-	-	-	-	-	-	-	-	-	+p	+a	*45*	+p-2b	2	4
Streblotrichum convolutum	-	-	-	-	-	-	-	-	-	-	-	*27*	+p-2b	2	7
Barbula vinealis	-	-	-	-	-	-	-	-	-	-	2m	*18*	2m	0.4	3
Geranium robertianum	-	-	-	-	-	-	-	-	+p	-	-	*18*	+r-+p	<0.1	0.1
Phyllitis scolopendrium	-	-	-	-	-	-	-	-	1p	1b	-	*64*	+r-1b	0.6	1
Centranthus ruber	-	-	-	-	-	-	-	-	-	2a	+p	*27*	+p-2a	0.7	3
Polypodium vulgare	-	-	-	-	-	-	-	-	-	-	1a	*27*	+p-1a	0.2	0.7
Ceterach officinarum	-	-	-	-	-	-	-	-	-	-	1p	*18*	+p-1p	<0.1	0.3
Holcus lanatus	-	-	-	-	-	-	-	-	-	-	+r	*18*	+r	0	0
Campanula rotundifolia	-	-	-	-	-	-	-	-	-	-	-	-	-	-	-
Bryoerytrophyllum recurvirostre	-	-	-	-	-	-	-	-	-	-	-	-	-	-	-
Sedum album	-	-	-	-	-	-	-	-	-	-	-	-	-	-	-
Encalypta streptocarpa	-	-	-	-	-	-	-	-	-	-	-	-	-	-	-
Barbula revoluta	-	-	-	-	-	-	-	-	-	-	-	-	-	-	-
Didymodon rigidulus	-	-	-	-	-	-	-	-	-	-	-	-	-	-	-
Cystopteris fragilis	-	-	-	-	-	-	-	-	-	-	-	-	-	-	-
Sedum dasyphyllum	-	-	-	-	-	-	-	-	-	-	-	-	-	-	-
Lamium album	-	-	-	-	-	-	-	-	-	-	-	-	-	-	-
Rhynchostegium murale	-	-	-	-	-	-	-	-	-	-	-	-	-	-	-
Leptobryum pyriforme	-	-	-	-	-	-	-	-	-	-	-	-	-	-	-
Poa compressa	-	-	-	-	-	-	-	-	-	-	-	-	-	-	-
Chelidonium majus	-	-	-	-	-	-	-	-	-	-	-	-	-	-	-
Poa angustifolia	-	-	-	-	-	-	-	-	-	-	-	-	-	-	-
MNS - MCH - MCM		*6*	-	*32*	-	*1 (3)*				*10*	-	*28*	-	*3 (7)*	

In addition in 2: Sambucus nigra 2b (2 specimen); in 4: Plantago major 1a; Stellaria media 1p; Polygonum persicaria, Geum urbanum, Poa nemoralis +r; in 5: Ficus carica +b; Dactylis glomerata, Soleirolia soleirolii +p; Erigeron mucronatus +r; in 7: Rubus spec. +r; in 8: Camptothecium lutescens 2b; in 10: Dactylis glomerata, Convolvulus arvensis +r; Gyroweisia tenuis 2a; in 11: Hieracium praecox 1b; Poa pseudocompressa +p; Epilobium collinum, Arenaria serpyllifolia, Leontodon hispidus, Schistidium apocarpum +r; in 12: Hieracium spec., Antirrhinum majus (cultivated form) +r; in 13: Hypnum cupressiforme 1p.

<u>LEGEND</u>

1. Etalle (prov. Luxembourg, Belgium), wall near the river Semois.
2. Metolice (South Bohemia, Czechoslovakia), wall of castle Kratochvile near tower.
3. Brughes (Belgium), canal wall opposite of Groene Rei near Paardenbrug.
4. Salzburg (Austria), wall foot underneath Festungskirche.
5. Falmouth (Cornwall, England), wall alongside Melvill Road nr 62.
6. Dartmouth (Devon, England), wall alongside road to Kingsbridge.
7. Kingston (Dorset, England), wall along road nr B 3069.
8. Rothenburg o. Tauber (Bavaria, Western Germany), rampart wall South of park near chapel.
9. Arbre (Namur, Belgium), rivulet wall alongside road to Rivière.
10. Mouthier (Doubs, France), wall alongside road nr N 67.
11. Luxembourg (town), wall near Um Bock.
12. Kanne (Limburg, Netherlands), wall of castle.
13. Montmort (Marne, France), wall alongside road nr N 51, opposite castle.

Distinction can be made between:
a. inops stands,
b. stands with *Phyllitis scolopendrium,*
c. stands with *Cystopteris fragilis,* and
d. stands with *Poa compressa.*

Stands of type (b) have only been recorded from S. England. Apart from the species mentioned above, the contribution by representatives of the *"Filici-Saginetum"* is fairly large: *Poa annua, Sagina procumbens* and *Dryopteris filix-mas.* Furthermore, *Parietaria judaica* appears on few occasions. These species also appear in stands of type (a) in inops vegetation. Perhaps such transitions towards the *"Filici-Saginetum"* should be considered as a separate entity, but the number of relevés is scanty (e.g. nos. 4 and 5 of the Table).

The majority of the inops stands is encountered on walls rising from a body of water. Although such stands are poor in species, *Corydalis lutea* may be very abundant and it is not at all impossible that this species tends to render the environment less suitable for the development of other species.

Stands occurring on the European mainland (groups c and d) are, among other things, distinguishable from (b) by the presence of *Campanula rotundifolia, Bryoerythrophyllum recurvirostre, Poa compressa* and *Chelidonium majus.* On better protected and relatively moist (and of course predominantly N.-facing) walls, *Bryoerythrophyllum recurvirostre, Cystopteris fragilis,* and sometimes (in the area bordering upon the submediterranean regions, in W. France and in the Lake Leman area of Switzerland, see no. 10) also *Sedum dasyphyllum* are found more commonly, whereas on drier walls *Poa compressa, P. angustifolia* and *Chelidonium majus* predominate. All these types of stands are connected by intermediate forms. In England *Chrysanthemum parthenium* was once represented in the stands. *Antirrhinum majus,* which exhibits optimal development in communities with *Cheiranthus,* was only recorded on few occassions. The affinity index (Poore's) between b and c is 0.53, between c and d 0.67, between b and d 0.59, and between b and (c+d) 0.62.

Stands with *Corydalis lutea* mostly develop from an initial community of *Linaria cymbalaria* and *Asplenium ruta-muraria.* The normal sequence is the familiar initial development of wall-rue followed by the appearance of *Linaria cymbalaria* and a progressive increase of the latter even after the first settlement of *Corydalis* whilst wall-rue is already on the decline. Ultimately *Linaria* is ousted out by *Corydalis* if the conditions are suitable. The percentage of coverage of *Corydalis* usually does not attain high values, percentages of over 40 per cent. being exceptional and almost always occurring near a body of water (e.g. no. 9). Nevertheless *Corydalis lutea* can survive rather long periods of drought. On dry walls the succession does not often progress beyond the early phase with a low representation of *Corydalis* plants and in very dry sites *Corydalis* is but rarely noted. Presumably this species can germinate more readily on a more alkaline substratum than e.g. *Cheiranthus* and thus can settle earlier among *Asplenium* and *Linaria cymbalaria.* In a stand with a coverage of the herbage layer of 70 per cent. of predominantly *Corydalis lutea* at Netolice (S. Bohemia, Czechoslovakia), containing e.g. *Asplenium ruta-muraria, Cystopteris fragilis* and *Leptobryum pyriforme,* a pH value of 8.4 was measured in the substratum, the highest value recorded in any other than the initial phase of vegetation development. However, according to Wiedman, cited in HEGI's Flora, *Corydalis lutea* grows on rock debris of "Urgestein" in the Bergamaskian Alps. On damp walls built of blocks of marl (as in Zuid-Limburg, Netherlands) the species grows very abundantly, but is was also found in Cornwall on jointed walls built of granite (in association with *Erigeron mucronatus* and *Soleirolia soleirolii,* see also no. 5, in which *Ficus carica* was present too!). *Corydalis lutea* is also somewhat nitrophilous. This is, among other things, evident from the fact that the first shrubby form to become established on walls with *Corydalis* is the black elder (*Sambucus nigra,* see no. 2), which also appears in lists of species recorded from walls with stands of *Corydalis lutea* in Henfstädt (Germany) by KAISER (1926), whose lists agree very satisfactorily with the relevé shown under (c) and (d) as far as the floristic composition is concerned.

Communities with *Corydalis lutea* are allied to both *Asplenium trichomanes* communities and *Parietaria judaica* communities, but the representation of *Parietaria* and of *Centranthus* in the stands is usually poor, both as far as presence and abundance are concerned. The fact that *Corydalis lutea* is often only represented in mural vegetation by a few specimens is probably the reason why TÜXEN (1937) considered it a faithful species of the *"Asplenium ruta-muraria-Asplenium trichomanes*-association". However, stands with *Corydalis* differ sufficiently from it, not only floristically but also in a structural and ecological respect, to deserve a separate place in the classification system. BÜKER (1939) published two relevés of mural vegetation with *Corydalis lutea* from Westphalia (W. Germany), one of which contained *Cystopteris fragilis* and the other one *Poa compressa.*

Here and there the stands come in contact with *Parietaria judaica* vegetation (e.g. in Luxemburg town), but the zone of mutual penetration is usually narrow. OBERDORFER (1954) gave two relevés from Graz (Austria) and Monti Berici (N. Italy), respectively, of *P. judaica* stands with a poor representation of *Corydalis lutea* ("+"), which stands he included in the *"Cheirantho-Parietarietum".* Communities with *Corydalis* can be fitted into Braun-Blanquet's system of classification as follows:

Alliance: *Cymbalario-Asplenion*

1. Association: Asplenio-Corydalidetum luteae
   - a. ? poetosum annuae
   - b. phyllitidetosum
   - c. cystopteridetosum
   - d. poetosum compressae.

The variants (c) and (d) could also be united; (a) shows transitions towards vegetation of trampled sites and towards vegetation of moist walls rich in nitrogen compounds; (b) towards eu-atlantic *Asplenium trichomanes* sociations, and (c) and (d) towards the *Sedo-Scleranthetea* and ruderal vegetation types.

In communities with *Geranium robertianum* the qualitative contribution of hemixerophytes is fairly high viz. 29% as against 15-19% in all other stands with *Corydalis lutea*. In the stands with *Poa compressa* the number of hygrophytes is low (U 6%, W 0.3% as against U 13-15% and W 0.2-1%, and, accordingly, also quantitatively unimportant). In all mural communities with *Corydalis lutea* myrmecochory is important (representation of myrmecochores U 19-30%, W 61-85%).

In a few localities in Bavaria (S. Germany), for instance at Creglingen, the species *Corydalis ochroleuca* was found in mural vegetation. The old records of its occurrence in the Low Countries undoubtedly referred to specimens escaped from cultivation. Horvatic (1963) recorded *Corydalis ochroleuca* from overshadowed rock crevices on limestone in the island of Pag (Yugoslavia). According to *Flora Europaea, C. ochroleuca* is a native of Italy and its ssp. *leiosperma* of the W. part of the Balkan peninsula.

6.6 Vegetation with Asplenium trichomanes and Tortula muralis

Asplenium trichomanes is generally considered to be a characteristic species of all chasmophytic communities (constituting the *Asplenietea rupestris*). It follows that there must be special reasons to discuss the stands of mural vegetation with this species separately instead of treating it as a characteristic element of all kinds of mural vegetation as is done, up to a point, with *A. ruta-muraria*. The latter species is not so commonly found in vegetation of rock crevices as *A. trichomanes*, possibly because it is more basophil, and it certainly is more resistant to appreciable fluctuations in the temperature. It is, furthermore, nitrotolerant and less skiadophilous than *A. trichomanes*. This combination of adaptations renders the wall-rue particularly suited to the mural habitat, and is most probably responsible for the wide-spread distribution of this fern as a follower of human culture.

The assignment of a special place to stands of mural vegetation with *A. trichomanes* is based on the following floristic, physiognomic and ecological criteria: 1. The occurrence of *A. trichomanes* often coincides with that of *Ceterach officinarum* and *Umbilicus rupestris*, in the eu-atlantic areas also with that of *A. adiantum-nigrum, Phyllitis scolopendrium, Polypodium vulgare* and *Bryum murorum*, and in the Mediterranean region with that of *Sedum dasyphyllum, Polypodium australe* and *Scorpiurium circinatum*. 2. Such stands are particularly rich in chasmophytic forms. 3. Exposures to the N. and the E. predominate and the environment is relatively poor in humus and in nitrogen compounds.

The stands can be roughly divided into four groups viz.

1. Stands with *Sedum dasyphyllum* of the Mediterranean and submediterranean areas;
2. Stands with *Asplenium adiantum-nigrum* and *Phyllitis* mainly of the eu-atlantic areas;
3. Stands with *Poa annua* and *Ceratodon purpureus* in the lowlands of W. Europe and (other) subatlantic areas; and
4. Stands with *Cystopteris fragilis* and with *Poa compressa* in the subatlantic and the montane areas.

In addition inops stands occur. Group 3 is in the first place negatively distinguished in respect of group 2 and more or less links up with group 4 by the precense of *Rhynchostegium murale.* A survey of relevés is shown in Table 30. *Asplenium trichomanes,* like several of the other species mentioned, does not exclusively occur in these mural communities, but also in various types of stands forming transitions towards other kinds of mural vegetation which can partly be classified as subassociations of those other vegetation type (compare 6.1.1, 6.1.2, 6.1.8, 6.1.11, 6.5 and 6.9.12). In Table 30 the notations of the presence and abundance of *Lepraria (cf.) aeruginosa* (which is of frequent occurrence in such stands) are omitted, because in the beginning they were not always recorded.

Asplenium trichomanes is found on both acid and basic rocks, but the specimens occurring in mural vegetation most probably all belong to the calciphilous subspecies *quadrivalens* D. E. Meyer. Generally speaking, the species does not act as a pioneer, but the pH of mural habitats of *A. trichomanes* is but rarely below 7.0. The fourteen pH values measured varied from 7.0 to 8.3 (mean value: 7.5). *A. ruta-muraria* was encountered on substrata with pH values lying between 5.7 and 8.7 (22 estimations, mean value: 7.4; compare Table 1).

6.6.1 Vegetation with Asplenium trichomanes and Sedum dasyphyllum

Communities of this kind were exclusively found in Mediterranean and submediterranean areas and in adjoining parts of area 132 (see Table 31).

Table 31 Geographical distribution and exposure of stands of the community of Asplenium trichomanes and Sedum dasyphyllum

exposure:	1	2	3	4	5	2w	3w	t	t%
geographical region:									
132	1	-	-	-	1	1	1	4	16
310	4	1	1	3	1	1	1	8	32
320	2	1	-	2	-	-	-	5	20
331 - 332	6	1	-	1	-	-	-	8	32
t	13	3	-	6	(1)	1	1	24(+1)	-
t%	54	13	-	25	-	4	4	-	100

The preference to N.-facing walls is manifest. The presence of *Parietaria judaica* is high but its abundance (and coverage) is always low. Transitions towards *Parietaria* communities (and, naturally, mostly to the community of *P. judaica* and *Sedum dasyphyllum,* see under 6.1.2)

are not rare. In about half of the relevés *Parietaria* is absent and the contribution of other species of the *"Parietarietum murale"* is always negligible. Nevertheless, one could combine the community under discussion with other *P. judaica* communities into a higher syntaxonomic unit. On the other hand, one could unite the community with the W.- and central European communities with *A. trichomanes.*

Species differentiating the stands in respect of other communities with *A. trichomanes* are, apart from *Sedum dasyphyllum* and *Parietaria judaica,* especially *Scorpiurium circinatum, Barbula acuta, Arenaria serpyllifolia, Saxifraga tridactylites, Rhynchostegiella tenella, Sedum rupestre, Poa bulbosa, Lamium amplexicaule* and *Lagoseris sancta.* The majority of these may occur in the *Parietarietum murale.* Furthermore, the community is differentiated by the almost complete absence of such species as *Epilobium montanum, Dryopteris filix-mas, Ceratodon purpureus* and *Bryum caespiticium.* The differentiation against the *Parietaria* communities is principally brought about by the relative abundance of mosses, the absence of *Phagnalon sordidum, Centranthus ruber* and *Antirrhinum majus,* and the insignificant contribution of agricultural weeds, in addition to the important contribution of *Ceterach officinarum, Sedum dasyphyllum, Umbilicus rupestris, Saxifraga tridactylites* and *Arenaria serpyllifolia.* The presence of *Asplenium ruta-muraria* is relatively low, a feature the community shares with Mediterranean *Parietaria judaica* communities. In Tuscany *Asplenium ruta-muraria* was only recorded once in a stand of the community, all other records being from the areas 310 and 132, which have a submediterranean or montane character. *Umbilicus rupestris* penetrates far into the eu-atlantic belt of W. Europe. According to DELVOSALLE (1968) this species more or less clearly follows the February isotherm of 5°C. Nitrogen compounds and phosphate of organic origin apparently favour the development of *Umbilicus.* In several localities the influence of bird life (guano) and of human activities was manifest by the very large and freely flowering specimens of *Umbilicus* (compare GILLHAM, 1956). At St. Sorlin (Ain, France) x*Asplenoceterach badense,* the hybrid of *Asplenium ruta-muraria* and *Ceterach officinarum,* was found. This hybrid was only described recently (MEYER, 1957). In the Cevennes (France) Chaenorrhinum *origanifolium* appears in the stands, and at Cavigliano (Ticino) *Dryopteris abbreviata,* a fern only recently discovered in Switzerland and very rare in that country.

Ceterach officinarum attains optimum development in this community. This fern, like *Umbilicus rupestris,* also occurs in *Asplenium trichomanes* vegetation in the eu-atlantic regions. It is interesting to note the aspect of stands with this fern in different climates. In the Mediterranean area *Ceterach* prefers N.-facing walls and hardly ever occurs on S.-facing ones, and yet it is often reputed to be a species adapted to dry habitats (e.g. in HEGI's Flora), but clearly this opinion is only partly justified. In any event this fern is not capable of surviving high temperatures, a temperature of 60°C already being fatal according to OPPENHEIMER & HALEVY (1962), and temperatures of about 60°C can easily be contained by S.-facing wall surfaces. Presumably *Ceterach* behaves like a "poikilohygric" organism. It is absent in sites with a constantly high relative humidity (in contrast to *Asplenium trichomanes*), but also avoids constantly dry sites. When the conditions turn unfavourable the species becomes dormant by an upward inrolling of the fronds, so that the ventral scales form a protective layer on the outside by providing an insulating air layer and by reflecting radiation. Inrolling of leaves during dry spells is also an adaptive feature of several species of *Asplenium* and *Polypodium*. In the Mediterranean region the summer season is usually the resting period for wall-dwelling specimens of *Ceterach*, but in central Europe it is inactive during the cold season. The species requires mild winters, because low temperatures have the same deleterious effect as excessive drought (compare PETTERSSON, 1962). In S. England *Ceterach* is found growing on both N.- and S.-facing walls, but the specimens on S.-facing wall surfaces have inrolled leaves and are usually dormant during the summer. It is striking that the species "retires" more and more to N.-exposures as it occurs farther away from the coast. In the submediterranean area this phenomenon can be observed in a direction from the mountains (e.g. the Cevennes) towards the coast. In either case there is a gradient of the average relative humidity. At altitudes exceeding 400 m *Asplenium trichomanes, Sedum dasyphyllum* and *Umbilicus rupestris* appear rather commonly on S.-facing walls.

Characteristic of the community under discussion is the presence of such winter annuals as *Saxifraga tridactylites, Arenaria serpyllifolia, Cerastium fontanum* and *Campanula erinus*. The first three species occur in W. Europe on S.-facing slopes in the dunes and in open sites, and in addition on tops of walls and on wall ledges. Conceivably the time and the duration of the suitable vegetation period in these sites approximately coincides with those of N.-facing walls in S. Europe.

6.2.2 Vegetation with Asplenium trichomanes and A. adiantum-nigrum

Stands of *A. trichomanes* with *A. adiantum-nigrum* and/or *Phyllitis scolopendrium* are typical of the eu-atlantic areas 111 and 112. Table 32 shows the variation of the exposure of the relevés with the area.

Table 32 Geographical distribution and exposure of stands of the community of Asplenium trichomanes and Asplenium adiantum-nigrum

exposure:	1	2	3	4	5	t	t%
geographical region:							
111	15	2	4	2	6	29	55
112	7	2	4	2	3	18	35
120	3	-	-	1	-	4	8
132	-	-	-	-	1	1	2
t	25	4	8	5	(10)	42(+10)	-
t%	50	10	19	12	-	-	101

Again a marked preference to N.-exposures is noticeable, but the stands on S.-facing walls are not very scarce, however. Seven of the stands (13%) were situated above a body of water, among which were the two relevés from area 120 on a S.- and on a W.-facing wall and the relevé from area 132. The stands in area 120 were relatively poor examples of this type, and likewise the stand in area 132 (at La Varenne near Nantes, i.e. not very far from area 112). In the areas 111 and 112 only one stand on a S.-facing wall was situated above a body of water, so that the overall picture does not change appreciably. On a few occasions the stands were growing on low walls though, or the ferns showed a lesser vitality than they usually exhibit on N.-facing walls. KENT (1960) drew attention to the fact that in Middlesex (England) *Asplenium trichomanes, Phyllitis scolopendrium* and *Polypodium vulgare* (all important species of the community) occur almost exclusively on N.- and on E.-facing walls and his data agree with the personal observations of the present author.

The common occurrence of *Ceterach* and of *Umbilicus* in the stands constitutes the most important point of similarity between this community and the Mediterranean *Asplenium trichomanes* communities. Characteristic, also in respect of *Asplenium* communities of more continental climate, is the presence of *Centranthus ruber, Epilobium lanceolatum* and *Bryum murorum* (in addition to *Asplenium adiantum-nigrum* and *Phyllitis scolopendrium*) and the frequent representation of *Polypodium vulgare, Taraxacum* spec., *Streblotrichum convolutum* and *Barbula unguiculata* (see Table 30).

In the most luxuriant stands all five aspleniid ferns indigenous in the area are found growing together: *A. adiantum-nigrum, A. ruta-muraria, A. trichomanes, Ceterach officinarum,* and *Phyllitis scolopendrium.* This combination was recorded in 7 (i.e. 13%) of the stands. *Ceterach* is hardly ever present without any one of the *Asplenium* species and nearly always does not show optimum development before several of these species are represented in the stand. In a later successional stage *Centranthus ruber* usually appears. The behaviour of *Asplenium adiantum-nigrum* and *A. ruta-muraria* in area 111 differs to some extent from that in area 112. The frequencies recorded in the combinations of species of *Asplenium* are shown in Table 33.

Table 33 Frequencies of combinations of Asplenium species in stands of the A. trichomanes and A. adiantum-nigrum community
Percentages of stands within the same geographical area are given in brackets

combinations	A. adiantum-nigrum + A. trichomanes	A. adiantum-nigrum + A. ruta-muraria	A. ruta-muraria + A. trichomanes	A. adiantum-nigrum + A. ruta-muraria + A. trichomanes	totals
geographical area					
111	-	3(*14*)	9(*41*)	10(*46*)	22
112 + 132	8(*50*)	1(*6*)	3(*19*)	4(*25*)	16
120	1	1	-	1	4

From this Table it becomes clear that not only *A. adiantum-nigrum* is relatively more common in mural vegetation in Brittany, but also the cohabitation of this species with *A. trichomanes* is more frequently found in less developed stands poor in species. *A. ruta-muraria,* on the other hand, is of rare occurrence in pioneer stands and early successional stages in Brittany in contradistinction to its behaviour in S. England. It seems as if in Brittany *A. adiantum-nigrum* replaces *A. ruta-muraria* as a pioneer of mural vegetation and forms a community with *Linaria cymbalaria* (but it usually becomes established before the latter species arrives). Such pioneer stands are strongly reminiscent of the communities of wall-rue with ivy-leaved toadflax found almost throughout W. Europe. Several authors report that *A. adiantum-nigrum* does not often occur on lime-containing substrata (see e.g. HEGI's Flora). It is indeed common in crevices of non-calcareous rocks, but in W. Europe it also grows on lime-containing rocks, and the mortar-joints of the walls on which it grows are of course always rich in lime. It is not very likely that the building stones themselves (usually granite in Brittany and Carboniferous limestone in Devon) exert an appreciable influence. Unfortunately the number of relevés from the granite walls in Cornwall (where the community does not seem to be common) is too small to justify a comparison. Of more frequent occurrence in this area are stands with *Sagina procumbens* and *Asplenium trichomanes* and here, too, *A. adiantum-nigrum* behaves in the same way as it does in Brittany. A further resemblance between the two areas is the occurrence of stands with *Erigeron mucronatus,* which are not found on walls in limestone regions elsewhere in Europe and in which *Asplenium adiantum-nigrum* plays an important part. The explanation of the singular behaviour of *A. adiantum-nigrum* in Brittany (and in Cornwall) can be sought in various directions. That a genetically different ecotype or physiological race is involved does not seem to be very probable. Possibly the composition of the mortar may be different from that of mortar mixtures used elsewhere and already from the very beginning not so rich in lime (e.g. by an admixture of non-calcareous components); however, this does not satisfactorily explain why the more basophil *A. ruta-muraria* only appears later in the succession when the pH, as may be expected, has become lower. It is, on the other hand, quite feasible that in England *A. adiantum-nigrum* appears later in the successional sequence on a more basic substratum. It is not quite certain if the unequal differentiation in different sites represents different phases of a succession series, but there are sufficient indications that this is indeed approximately the case. However, not all authors agree that *A. ruta-muraria* is restricted to lime-containing substrata. HEGI's Flora, LAWALREE (1950) and DURIN (1955), for instance, report its occurrence on acid rocks (one should always bear in mind that the primary decomposition products of non-calcareous rocks may be relatively rich in lime). MALCUIT (1928) calls the wall-rue a not exclusively but selectively calciphilous species.

An alternative explanation is a possible difference in climate. In the granitic areas mentioned the annual atmospheric precipitation is still higher than in e.g. Devonshire, and the relative humidity presumably stays at a high level more often and during longer uninterrupted periods. A third possibility is the predominance of *A. adiantum-nigrum* in natural habitats (rocky outcrops) in Brittany, i.e. a much higher dissemination capacity in respect of other elements of the community, but it is most unlikely that in Cornwall primarity is more important than selection by the environment.

In Table 30 some examples of stands are shown, which can be divided into two groups:

1. Stands without *Poa annua* and/or *Sagina procumbens* (columns b 1)
2. Stands with *Poa annua* and/or *Sagina procumbent* (columns b 2).

The second group forms a transition towards the *Filici-Saginetum*. These stands are, furthermore, differentiated by the presence of *Bryum argenteum* and *Epilobium lanceolatum* (which are almost entirely lacking in stands of group 1) and the absence of *Umbilicus rupestris*, *Bryum murorum* and *Catapodium rigidum*, and the low degree of presence of *Homalothecium sericeum*. However, the delimitation between these two groups is not very sharp and a clear-cut distinction is out of the question. Both groups of stands also exhibit the same distribution pattern. The ecological behaviour of species of *Barbula* and physiognomically similar forms in the various communities is interesting, a tendency being discernable for a progressive appearance (and subsequent dominance) of the following species as the amount of available nitrogen compounds becomes higher: *Didymodon rigidulus* - *Barbula vinealis* - *B. unguiculata* - *Streblotrichum convolutum* - *Ceratodon purpureus*. The latter species is especially common in the community of *Asplenium trichomanes* and *Poa annua* to be discussed presently. *Streblotrichum convolutum* grows together with *Ceratodon* and/or *Barbula unguiculata* more often than with *B. vinealis*. In principle all combinations may occur, but *Ceratodon* is only occasionally found in cohabitation with species of *Didymodon* or with *Barbula vinealis*. *Barbula vinealis* and *B. unguiculata* do not often appear in stands together, however, and other factors may play a role, e.g. the water content of the substrate. A corresponding sequence is shown by species of *Bryum*: *B. murorum* - *B. capillare* - *B. caespiticium* - *B. argenteum*; *B. bicolor* can possibly be inserted between *B. murorum* and *B. capillare*, but the number of records of this species is insufficient to draw more definite conclusions. Geographical factors are also involved: *B. murorum* is frequently encountered in Brittany and often in S. England, but only occasionally in the Mediterranean area and rarely in areas 120 and 131. (One might wonder if the same factors are responsible as in the case of *Asplenium trichomanes.*) The combination in one site of *B. capillare, B. caespiticium* and *B. argenteum* (usually also including *Ceratodon purpureus*) is very commonly found, and usually well-developed in the community of *Asplenium trichomanes* and *Poa annua,* but even better in the *"Filici-Saginetum"* (see 6.9.1 and 6.14). It is, incidentally, not certain as to how far these sequences, based primarily on floristic evidence, are at all indicative of the assimilable nitrogen content of the substratum. Conceivably other nutrients such as phosphates are involved and in some cases possibly the degree of atmospheric pollution plays a role e.g. in connection with tolerance of such substances as SO_2 and H_2S.

6.6.3 Vegetation with Asplenium trichomanes and Poa annua

Closely allied to the community discussed under "type 2" in the previous paragraph are communities with *Asplenium trichomanes* and *Poa annua,* which again are not sharply distinguishable from the former type of community. In a geographical sense the stands with *A. trichomanes* and *P. annua* are situated in more subatlantic regions (see Table 34).

Table 34 Geographical distribution and exposure of stands of the community of Asplenium trichomanes and Poa annua

exposure:	1	2	3	4	1w	2w	3w	4w	5w	t	t%
geographical region:											
111	-	-	2	-	-	-	-	-	-	2	13
120	2	-	-	-	2	1	-	-	-	5	31
131	5	-	1	1	1	-	-	-	1	9	56
t	7	-	3	1	3	1	-	-	(1)	15(+1)	-
t%	47	-	20	7	20	7	-	-	-	-	100

Here, too, a preference to exposures towards the N. is discernible, but in S. England also exposures towards the S. were recorded. The only "dry" S.-facing wall recorded in region 131 was a low one. In area 120 and in the lowlands of area 131, the stands more strictly seem to prefer N.-facing sites or the vicinity of water than they do in the eu-atlantic zone of W. Europe. The stands under discussion are primarily differentiated in a "negative" sense from the communities of *A. trichomanes* and *A. adiantum-nigrum* discussed in paragraph 6.6.2 by the absence of *Phyllitis scolopendrium, Ceterach officinarum, Centranthus ruber, Bryum murorum, Asplenium adiantum-nigrum* and *Linaria cymbalaria,* and the low degree of presence of *Sonchus oleraceus* and of *Homalothecium sericeum. Agrostis stolonifera, Rhynchostegium murale* and *Ceratodon purpureus,* on the other hand, exhibit a high degree of presence. The frequent occurrence of *Rhynchostegium murale* particularly points to a community of a not so extreme atlantic climate, as will also be clear from a comparison with the type of community to be discussed in the next paragraph; this moss has only been recorded on rare occasions (no. 8 in Table 30) in S. England where it is reputed to be rare (DIXON, 1924).

The stands with *A. trichomanes* and *Poa annua* are not only allied to (and linked by intermediate cases with) the communities with *A. trichomanes* discussed previously, but also, and more closely so, to the *"Filici-Saginetum"*. With this it also coincides better in a geographical respect. Transitions towards the still more continental communities of *A. trichomanes* with *Poa compressa* (see 6.6.4) have also been encountered but they are not so common. An example is shown in the Table in column 12.

The mean number of taxa recorded in the stands being only 9.6, the community is relatively poor in species.

6.6.4 Vegetation with Asplenium trichomanes and Poa compressa

Mural vegetation with *A. trichomanes* is not rare in those parts of Europe where the relative humidity is fairly high. TÜXEN (1937) was the first to recognise its stands as forming a separate phytosociological entity which he called an *"Asplenium ruta muraria-Asplenium trichomanes*-association". Its stands can be divided into two groups:

1. stands with *Cystopteris fragilis*
2. stands without *Cysteroptis fragilis.*

Table 35 shows the geographical distribution of the recorded stands.

Table 35 Geographical distribution and exposure of stands of the community of Asplenium trichomanes and Poa compressa

t. total
a. vegetation without Cystopteris fragilis
b. vegetation with Cystopteris fragilis

	exposure:	1	2	3	4	5	6	1w	2w	4w	6w	t	t%
geographical regions													
t	120	1	-	-	-	2	-	2	-	-	-	5	6
t	131	33	9	1	1	5	1	1	2	1	1	55	62
t	132	8	5	3	1	1	2	1	-	1	-	22	25
t	200	1	3	-		1			-	-	-	6	7
t	t	44	17	4	2	(9)	(3)	4	2	2	(1)	75(+13)	-
t	t%	59	23	5	3	-	-	5	3	3	-	-	100
a	120	1	-	-	-	2	-	1	-	-	-	4	6
a	131	23	8	1	1	4	1	-	1	-	-	39	62
a	132	8	4	3	1	-	2	1	-	1	-	20	32
a	200	-	-	-	-	-	-	-	-	-	-	-	-
a	t	32	12	4	2	(6)	(3)	2	1	1	-	54(+9)	-
a	t%	59	22	7	4	-	-	4	2	2	-	-	100
b	120	-	-	-	-	-	-	1	-	-	-	1	4
b	131	10	1	-	-	1	-	1	1	1	1	16	64
b	132	-	1	-	-	1	-	-	-	-	-	2	8
b	200	2	3	-	-	1	-	-	-	-	-	6	24
b	t	12	5	-	-	(3)	-	2	1	1	(1)	21(+4)	-
b	t%	57	24	-	-	-	-	10	5	5	-	-	100

Vegetation with *A. trichomanes* is common in area 131 and stands of it with *Cystopteris* are presumably fairly common in the W. part of area 200. A preference to N.-facing sites is again manifest and it is also quite clear that in W. Europe E.-facing walls, generally speaking, exhibit a greater ecological similarity to N.-facing than to S.-facing walls. The stands recorded from S.-facing sites were mostly growing on low walls or on walls built up against a mass of earth, which have a different ecology. Some examples are shown in Table 30 and these indicate that the stands with *Poa compressa* are differentiated from all other communities with *Asplenium* by the constant occurrence, or high degree of presence, of *Cystopteris fragilis, Gymnocarpium robertianum, Campanula rotundifolia, Leptobryum pyriforme* and *Didymodon rigidulus,* and from the partly sympatric community of *Asplenium* with *Poa annua* by the presence of e.g. *Polypodium vulgare, Chelidonium majus, Dactylis glomerata* and *Bryoerythrophyllum recurvirostre* and of the hypnid mosses *Homalothecium sericeum, Hypnum cupressiforme* and *Brachythecium rutabulum* (and, in a negative sense, by the absence, or relatively low presence of *Poa annua, Sagina procumbens, Agrostis stolonifera, Bryum argenteum* and *Ceratodon purpureus).* However, the majority of the positively differentiating taxa exhibits a relatively low degree of presence (which is even less than 30% in *all* cases). The structure of the community also differs appreciably from that of the stands with *Asplenium trichomanes* and *Poa annua* by the lower representation of therophytic forms and the greater contribution of hemicryptophytes.

The stands with *Cystopteris* are indistinctly differentiated from those without this fern by the presence (apart from *Cystopteris* itself) of *Gymnocarpium robertianum, Campanula rapunculoides, Epilobium collinum, Sonchus oleraceus, Encalypta streptocarpa, Mnium undulatum, Barbula vinealis, Didymodon rigidulus* and *Leptobryum pyriforme* and, conversely, by the absence of e.g. *Polypodium vulgare, Chelidonium majus, Mycelis muralis, Campanula rotundifolia, Encalypta streptocarpa* and *Brachythecium rutabulum.*

A number of taxa, although only attaining a presence of 10%, allow further differentiation from other communities with *Asplenium trichomanes.* Such forms as *Sedum album, Tortula intermedia, Festuca ovina, Fragaria vesca, Tortella tortuosa* and *Schistidium apocarpum* were recorded at least 5 times (i.e. in a representation of 6%). Perhaps stands with *Sedum album* and *Tortula intermedia* constitute an individual subentity which may be regarded as a link with *Sedo-Scleranthetea* communities with *Sedum* species, such as are found on wall tops and wall ledges. Arguments in favour of the assignment of a separate status to the stands with *Sedum album* are also supplied by 4 relevés containing *Phyllitis scolopendrium* recorded by LORENZONI (1961) in Italy (Valli di Natisone), in an area with an oceanic climate. Such stands can, in turn, be considered to represent a transition towards the communities of *A. trichomanes* and *A. adiantum-nigrum.*

Stands with *Cystopteris* are more frequently encountered in more sheltered or moister sites than stands without this species of fern, which, like the sometimes co-occurring *Gymnocarpium robertianum*, is never encountered in eu-atlantic and Mediterranean stands of mural communities (see also 6.9.13). Both species are adapted to short dry spells and presumably to considerable temperature fluctuations of short duration, conditions which most other European ferns do not survive. Of common occurrence is the combination of *Leptobryum pyriforme* and *Bryoerythrophyllum recurvirostre* which constitute a kind of stenocoenosis of their own resembling the moss communities of narrow steep inclines in the coastal dunes described as the *Bryoerythrophyllo-Tortuletum subulatae* (COESEL, 1962), in which *Ditrichum flexicaule* can be regarded as a "replacement" of the morphologically similar *Leptobryum pyriforme*. It is noteworthy that, especially at higher altitudes, *Leptobryum pyriforme* often appears to be replaced by a species with a corresponding habit viz. *Distichium capillaceum*; however, the two are occasionally sympatric.

IVIMEY-COOK & PROCTOR (1966) recorded two relevés of mural vegetation which they referred to the "*Ceterach officinarum-Asplenium trichomanes*-association". This was in conjunction with 6 relevés of chasmophytic vegetation, all from Burren (Ireland), in which the mural stands are differentiated by the presence of e.g. *Tortula muralis* and *Rhynchostegiella tenella* and the absence of the species *Festuca ovina* and *Sesleria coerulea*, both highly represented, however, in rock crevices. By the occurrence of e.g. *Geranium robertianum*, *Encalypta streptocarpa* and *Tortella tortuosa* these stands show a greater affinity to the central European stands with *Asplenium trichomanes* than to the eu-atlantic ones. It is not at all unlikely that the Irish stands represent a separate variant or type of which some singular constituents, such as the Mediterranean-Atlantic mosses *Tortella nitida*, *Trichostomum brachydontium* and *Anomodon viticulosus*, are the differentiating taxa.

In central France stands are sometimes encountered which are intermediate between *Asplenium trichomanes* communities and *Parietaria judaica* communities. Moreover, once a stand was recorded in which *Cystopteris fragilis* and *Sedum dasyphyllum* occurred:

Ornans (Doubs, France), wall along a staircase near the river Doubs. 65186, 11-5-1965. Exp. NNWs; herb layer 25%; moss layer 15%

Asplenium trichomanes	2b
„ ruta-muraria	+p
Linaria cymbalaria	2a
Tortula muralis	2a
Bryum capillare	+p
Homalothecium sericeum	1p
Cystopteris fragilis	1a
Leptobryum pyriforme	2m
Bryoerythrophyllum recurvirostre	+p
Barbula unguiculata	+p
Sedum dasyphyllum	+p
Parietaria judaica	1b
Bryum caespiticium	2m

The combination of *Sedum dasyphyllum*, *Parietaria* and *Cystopteris* is presumably extremely rare.

6.6.5 Asplenium trichomanes vegetation poor in species

Stands of *Asplenium trichomanes* vegetation poor in species have been found in all areas visited, but the lone occurrence of *A. trichomanes* is rare, in contrast to the occurrence of *A. ruta-muraria* or *Parietaria judaica* in pure (but sparse!) stands. The probable reason is that *A. trichomanes* does not often act as a pioneer and when occurring in pioneer vegetation in W.- or central Europe it is almost always associated with *A. ruta-muraria* and/or *Linaria cymbalaria.* In principle two kinds of stands containing few species can be distinguished viz. poor stands of extreme environments, and pioneer stages. The latter type of stand is rather uniform in a large area and mostly contains *A. ruta-muraria* and *Tortula muralis,* often also *Bryum capillare.* In the Mediterranean area *Parietaria judaica* may be present, and in Brittany and Cornwall *Asplenium adiantum-nigrum* usually replaces *A. ruta-muraria,* as we have seen. An example of a stand in a more extreme environment is found in relevé no. 21 (of Table 30). This is a stand in a rather dark well where relative humidity was constantly high. Stands with a poor representation of *A. trichomanes* are sometimes found on S.-facing walls. Relevé no. 19 is an example of a pioneer phase and no. 20 represents an "inops" form.

6.6.6 A survey of communities with Asplenium trichomanes and Tortula muralis

Stands of mural vegetation with *Asplenium trichomanes* can be classified within the floristic system as follows:

Alliance *Cymbalario-Asplenion*

	Association group of Asplenium trichomanes and Tortula muralis		
1.	Association	Sedo (dasyphylli)-Asplenietum trichomanis	Association Sedo (dasyphylli)-Asplenium trichomanis
2.	„	Asplenietum rutae-murariae-trichomanis	Association Asplenietum adianti-nigri-trichomanis
3.		asplenietosum adianti-nigri	ceterachetosum
4.		var. with Ceterach officinarum	poetosum annuae
5.		„ „ Poa annua	Sagino-Asplenietum trichomanis
6.		poetosum annuae	Encalypto-Asplenietum trichomanis
7.		poetosum compressa	? cystopteridetosum fragilis
8.		var. with Cystopteris fragilis	? sedetosum albi
9.		„ „ Sedum album	

Some of the coefficients of affinity (according to Poore) are as follows:

1	with	3	0.49
4	„	5	0.77
5	„	6	0.64
3	„	7	0.62
6	„	7	0.63

The allocation of the mural communities to the alliance of the *Cymbalario-Asplenion* is not altogether self-evident because (1) the stands of the community of type no. 1 could also be referred to the *Parietarion judaicae*; (2) community of type no. 3 (and possibly also no. 6) could be placed into a separate alliance of atlantic distribution. In a classification partly based on geographical distributions, separate Mediterranean, Atlantic and central European ("continental") alliances of mural vegetation could be recognised. As early as 1936 NORDHAGEN proposed the segregation from the *Potentillion caulescentis* of an individual oceanic alliance, or even an order with, as characteristic species, *Asplenium adiantum-nigrum, Phyllitis scolopendrium* and *Asplenium marinum*. In the present context the description of the order *"Potentilletalia caulescentis"* and the *"Potentillion caulescentis"* deserves some attention. It is quite clear that these concepts are, on the one side, founded on a very narrow basis, but, on the other hand, are very broad. MEIER & BRAUN-BLANQUET (1934) had, namely, enumerated as faithful species of the order and the alliance a number of taxa with a limited geographical distribution (endemic in e.g. the Cevennes or the Pyrenees). They described a number of associations of predominantly local significance, whilst attributing to the order and the alliance a very wide distribution, in which they were followed by a number of workers. The identity of these vegetational units is by no means clear, however (compare GAMS, 1954), and that is why OBERDORFER (1967) proposed to change their names into *Asplenietalia rutae-murariae* and *Asplenion rutae-murariae*. Possibly the original name of the alliance and even of the order can be maintained by assigning a more restricted distribution to them (e.g. "in the mountains of S.W. Europe"). The names suggested by Oberdorfer are not very appropriate either because the wall-rue also occurs in other orders and may presumably be regarded as a "characteristic species" of a superorder embracing all chasmophytic communities of basic rock formations (at least if one wishes to retain all communities of acid rocks in the same synsystematic class).

Communities of group 1 can also be regarded as a facies or a special variant of the *Parietarietum murale*, but they are always differentiated by the special environment, by a different structure, by a lower number of vascular plants and a higher number of bryophytes, and by the more frequent and more abundant occurrence of ferns, of *Sedum dasyphyllum* and of *Umbilicus rupestris*. The floristic distinction is more quantitative than qualitative, however. A number of relevés made in the Cevennes by BRAUN-BLANQUET (1915) are closely allied to those of communities of group 1 or could even be included in this type (e.g. the number 1 of his table of the wall-dwelling subassociation of the community of *Potentilla caulescens* and *Saxifraga cebennensis*). Rather closely related are also some relevés from Basse-Provence made by ARENES (1929), who gave lists of species of 3 stands of the so-called *"Filicetum murale"*.

Communities of type 3 are very similar to those described from Burren (Ireland) by IVIMEY-COOK & PROCTOR (1966) as the *"Ceterach officinarum-Asplenium trichomanes-association"*. They studied 8 stands, two of which grew on walls, and the remainder in rock crevices, the wall-dwelling ones being differentiated by the presence of *Tortula muralis* and *Rhynchostegiella tenella*. Also from Ireland, a *Ceterach-Cotyledon umbilicus*-association was described by BRAUN-BLANQUET & TÜXEN (1952), partly on account of data reported earlier by WEBB (1947). Stands of mural vegetation described by TÜXEN & OBERDORFER (1958) from N.W. Spain are likewise, at least partly, referable to this group 3 (their p. 16 and table under 'A').

Community group 6 links up with the *"Filici-Saginetum"*. The best known is group 7, first recognised by TÜXEN (1937) as we have seen, and later on again described and illustrated by means of relevés by e.g. KUHN (1937), GEHU

(1961) and Hübschmann (1967). Tüxen assigned the status of faithful species to *Corydalis lutea* on account of its occurrence in two(!) relevés; his view is not endorsed by the present author. The community is referred to in the phytosociological literature by the names of *Asplenietum rutae-murariae-trichomanis* (Knapp, 1948), *Tortulo-Asplenietum* (Oberdorfer, 1957), *Asplenietum trichomano-rutae-murariae* Tüxen 1937 (e.g., Oberdorfer, 1967), and *Asplenietum rutae-murariae* (Gehu, 1961).

These communities with *Asplenium trichomanes* show a particularly close relationship with chasmophytic vegetation of principally basic rocks in which *A. trichomanes* and *A. ruta-muraria* are often quite common. It is primarily the more exclusive wall-dwellers such as *Linaria cymbalaria* and *Tortula muralis* which give the mural stands their often characteristic appearance. Closely allied chasmophytic vegetation types have been described by e.g. Büker (1942), Ivimey-Cook & Proctor (1966) and Müller (1966). The latter recorded two stands from Hohentwiel (Baden-Württemberg, W. Germany), one of which from a mural site. He referred both to the *"Asplenio-Cystopteridetum"* (Kuhn 1937) Oberdorfer 1949. The relevés of Büker and Müller closely link up with communities of type 7.

Closely allied to these communities is the one of *Asplenium viride* and *Linaria cymbalaria* to be described in the following paragraph. The number of recorded relevés is too low to render the proposal of a *Cymbalario-Asplenietum viridis* more than a tentative one. This (provisional) community likewise has counterparts among chasmophytic stands lacking the more selectively wall-dwelling species.

In communities with *Asplenium trichomanes* many tetraploid species are represented viz. 28 to 49% for unweighted values and 39-62% for weighted values, the lower representation being found in stands with *Cystopteris fragilis* and the highest values in stands with *Poa annua* and in inops-vegetation. In the community of *Asplenium viride,* on the other hand, the contribution of the tetraploids is almost negligible (the values being U 4% and W 0.0%, respectively).

Communities with *Sedum dasyphyllum* deviate from all other kinds of mural vegetation in the high quantitative contribution of therophytic forms viz. 30% for unweighted values and 2% for weighted ones. The representation is 0-13% (U) and 0-2% (W), respectively, in all other communities with *Asplenium trichomanes*. The communities also differ in the appreciable amount of chamaephytes present (21% for unweighted and 30% for weighted values). Chamaephytes are also quantitatively important in communities with *Phyllitis* (11% and 29%, respectively, as against 3-11% and 4-20%, respectively, in all other stands with *A. trichomanes*). In communities with *Sedum dasyphyllum* the contribution by seasonal xerophytes is appreciable viz. 17-25% (against 0-3% and 0-1%, respectively, in all stands without *S. dasyphyllum*).

Hygrophytes are often encountered in stands with *Cystopteris fragilis* (representation U 27%, W 28%) and are quantitatively more important in stands with *Poa annua* (U 17-23%, W 5-8%) than in all other communities with *A. trichomanes* (representation U 9-15% and W 0.1-3%).

Anemochory is wide-spread among elements of all communities with *A. trichomanes*, and of particular importance for all taxa with a high dissemination capacity having diaspores as fine as dust (32-82% of the total, unweighted values, and 66-95%, weighted values, respectively). MEDWECKA-KORNAS (1950) reports figures as high as 60.0% and 77.8% for stands of *Asplenium ruta-muraria* and *A. trichomanes* in crevices of calcareous rock formations near Cracov (Poland). These values agree reasonably well with the figures obtained from my study of mural stands.

6.7 Vegetation with Asplenium viride

Stands with *Asplenium viride* were recorded from Austria and from the French Alps, all localities being situated in area 200. Both floristically and physiognomically the community resembles *A. trichomanes* communities of central Europe. Of the 6 relevés two were of N.-facing sites, 2 were of E.-facing and 2 of W.-facing walls. In all cases the walls were built of blocks of limestone.

Table 36 Community of Asplenium viride and Linaria cymbalaria

Nr C	1	2	3	P	U W	MGT	MGP
Date	15-7	15-7	23-7	6st			
year	66	66	63				
Nr R	453	454	268				
Ex	2	1	2				
CH	30	20	5				
CM	10	3	5				
Asplenium viride	2a	2a	1a	100	+p-2a	3	3
Linaria cymbalaria	2a	2a	-	83	2m-2a	6	7
Asplenium ruta-muraria	2a	1p	-	83	+r-2a	3	4
Encalypta streptocarpa	1b	1b	+p	83	+p-1b	2	2
Taraxacum (div.) spec.	+r	+p	+r	83	+r-+p	<0.1	<0.1
Tortula muralis	2m	+p	-	67	+p-2m	0.5	0.6
Geranium robertianum	-	-	-	50	+r-+p	<0.1	0.1
Festuca ovina	+r	-	+p	50	+r-+p	<0.1	0.1
Bryoerythrophyllum recurvirostre	2a	-	-	33	2a	3	8
Homalothecium sericeum	-	1b	-	33	1b	1	4
Amblystegium serpens	-	1p	-	33	1p-2a	1	4
Bryum caespiticium	2m	-	-	33	2m	1	3
Taxus baccata	-	+a	-	33	+a-1b	1	3
Barbula unguiculata	-	2m	-	33	2m	1	3
Oxalis acetosella	1b	-	-	33	+p-1b	0.6	2
Campanula rotundifolia	-	+r	-	33	+r-1a	0.3	0.8
Achillea millefolium	-	+p	-	33	+p	<0.1	0.1
Streblotrichum convolutum	-	-	+p	33	+p	<0.1	0.1
Epilobium montanum	-	-	-	33	+r-+p	<0.1	<0.1
Trisetum flavescens	-	+p	-	33	+r-+p	<0.1	<0.1
Anthriscus sylvestris	-	+r	-	33	+r	0	0
Ranunculus acris	-	+r	-	33	+r	0	0
Silene vulgaris	+r	+r	-	33	+r	0	0
MNS - MCH - MCM		14	-	17	-	7	

In addition in 2: Asplenium x meyeri, Collema spec. +p; Plantago media, Leontodon hispidus, Holcus lanatus, Galium spec. +r; Parthenocissus +b (transgressive); in 3: Distichium capillaceum 2m; Cystopteris fragilis 1b; Hieracium spec. +a; Bryum cf. bicolor 1p; Polystichum lonchitis +p; Poa alpina +p; Barbula vinealis +p; Thymus serpyllum +r.

LEGEND

1. Sankt Gertraudi (Tirol, Austria), wall of drive leading to a shead near Haus Matzen, alongside the high road nr 1.
2. Sankt Gertraudi (Tirol, Austria), wall alongside high road nr 1, near the gateway of the castle.
3. Lautaret (Hautes-Alpes, France), culvert underneath the road to Briancon.

The insufficient number of relevés does not justify a distinction between stands with *Cystopteris fragilis* and *Distichium capillaceum*, and the remainder of the stands (compare the similar case of stands of the community of *Sagina procumbens* and *Cystopteris fragilis*, see

6.9.13). Cohabitation of *A. viride* and *Cystopteris* is not at all unusual in rock crevices (compare e.g. OBERDORFER, 1938: description of *Cystopteris* communities in the northern Black Forest, W. Germany; OBERDORFER, 1949: *Asplenio-Cystopteridetum;* JACKSON, 1958, Lake District, England; LIPPERT, 1966: in the *Marchantia polymorpha-Cystopteris montana*-community with e.g. *Polystichum lonchitis* and in the *"Asplenio-Piceetum")*. The occurrence of *A. viride* with *Polystichum lonchitis* is also a common phenomenon in high montane regions; in the Hautes-Alpes (France) and in the Tyrol (Austria) the combination of these two species with *Cystopteris fragilis* was repeatedly noted in crevices in lime-stone and this community was provisionally indicated as *Polysticho-Asplenietum viridis* (SEGAL, 1963d). Two relevés of this community are shown in the following Table (the third relevé is an example of a stand referable to the *"Filici-Saginetum"*, compare 6.9.13).

Table 37 Community of Asplenium viride and Polystichum lonchitis in limestone crevices (nrs 1-2) and mural vegetation with Asplenium viride and Polystichum lonchitis (nr 3)

Nr C	1	2	3
Date	19-7	30-4	30-4
Year	63	63	63
Nr R	235	093	032
Ex	2	-	1
Height above sealevel (×100m)	22	19	16
CH	60	20	5
CM	70	20	8
Asplenium viride	4a	2b	1p
Polystichum lonchitis	+b	2a	+p
Cystopteris fragilis	-	+r	+p
Asplenium ruta-muraria	-	+p	+p
Asplenium trichomanes	-	+r	-
Cardamine resedifolia	2a	-	-
Saxifraga spec.	+p	-	-
Helianthemum alpestre	-	+r	-
Sagina procumbens	-	-	2m
Dryopteris filix-mas	-	-	+p
Poa annua	-	-	+p
Poa pratensis	-	-	+p
Veronica officinalis	-	-	+r
Thymus spec.	-	-	+r
Encalypta streptocarpa	2a	2a	2m
Tortella tortuosa	2a	2m	2m
Tortula ruralis var. norvegica	+p	+p	1p
Oxyrrhynchium swartzii	2b	+p	+p
Amblystegium serpens	2a	+p	+p
Ctenidium mollescum	2m	2a	-
Camptothecium lutescens	2b	-	+p
Mnium rostratum	+p		
Cladonia spec.	+p	-	-
Distichium capillaceum	-	2m	1p
Preissia quadrata	-	1a	+r
Schistidium apocarpum	-	-	2m
Bryum cf. bicolor	-	-	+p

LEGEND

1. Lautaret (Hautes-Alpes, France), rock crevices between Col du Lautaret and Col du Galibier.
2. Oberndorf (Tirol, Austria), rock crevices under the chapel over the Oberndorfer See.
3. Oberndorf (Tirol, Austria), garden wall of the guesthouse at the Oberndorfer See.

Indications of an ecological link between *Asplenium viride* and *Polystichum lonchitis* are also found in a paper by STERNER (1922) on stands in Oeland, Sweden, and in various field-trip reports in the *British Fern Gazette*.

Especially in the alpine regions, *A. viride* often occurs in more or less humid and usually overshadowed sites where lower average temperatures prevail and the vegetation period is shorter than in sites supporting stands with *A. trichomanes*. *A. viride* is also found in cohabitation with e.g. *Cystopteris regia* and *Viola biflora* on moist gravelly slopes and on rock debris (LIPPERT, 1966), and sometimes with *Polystichum lonchitis*. *A. viride* is reputed to be a calciphile (e.g. in HEGI's *Flora*), but according to JACKSON (1958) it occurs in the Lake District of England also on acid rocks. HORA (1947) estimated the range of pH tolerance of the species as (5.9-)6.6-7.8 in N. Wales, whereas the range of *A. trichomanes* lies between 5.6 and 7.0 in the same area. (At a mural site at Oberndorf, the Tyrol, Austria, I once recorded a pH value of 8.1 for a stand with *A. viride*, whereas my records concerning *A: trichomanes* all lie between pH 7.0 and 8.3.) In mural sites outside area 200 the species is rare, but it has been recorded from e.g. the department Seine-Inférieure of France (SENAY, 1952) and Brandenburg, E. Germany (KRAUSCH, 1955; SCHOLZ & SUKOPP, 1965).

In a mural stand recorded at Sankt Gertraudi (the Tyrol, Austria), the putative hybrid of *Asplenium viride* and *A. ruta-muraria* was found (three specimens!). This hybrid was first reported by MEYER (1958) from the Bavarian Alps. The Austrian plants had abortive spores, but this phenomenon may also occur in specimens of *A. ruta-muraria* according to Meyer. MELZER (1963) reported a find from the Mittagskogel (Burgenland, Austria). Both he and Meyer state that the hybrid is more like *A. ruta-muraria* than like *A. viride*, but in my material it was the other way around. Presumably their records are of the hybrid *A. ruta-muraria* ♀ x *A. viride* ♂, and mine of the reciprocal cross *A.viride* ♀ x *A. ruta-muraria* ♂. In the analogy of the terminology of hybrids between *A. trichomanes* and *A. septentrionale*, the latter hybrid represents an undescribed taxon for which I propose the name *A.* ×*heimansii* [1]).

A. viride is rather generally reputed to be a characteristic species of the order *Potentilletalia caulescentis* (MEIER & BRAUN-BLANQUET, 1934), but this order is not based on very solid evidence and the amplitude of *A. viride* was thus largely overrated. Stands with *A. viride* and *Linaria cymbalaria* resemble those of central European communities with *Asplenium trichomanes*, but the affinity index according to Poore, amounting to 0.46, justifies to my mind the recognition (mainly on floristic evidence) of separate communities with *A. viride* outside the *A. trichomanes* communities, in spite of a certain agreement in their structural features and ecology. Generally speaking the *A. viride* communities are somewhat richer in species, but the percentages of coverage are relatively low.

In Wales *A. viride* occurs in mural vegetation together with *A. trichomanes* (MENNEMA, 1968).

6.8 Vegetation with Polypodium

Vegetation in which species of *Polypodium* are dominant or codominant is physiognomically, and usually also ecologically, distinguish-

---

[1]) The taxon is named after my teacher Professor J. Heimans, the University of Amsterdam.

able from other mural communities, and often quite readily so. The thick rhizomes of this fern group can develop in joints and cracks that are not too narrow, subsequently to grow over the wall surface sending out fine root-hairs into small crevices. The wall-ferns often grow in the uppermost joint just below the covering slab. The special character of these sites, in an ecological sense and otherwise, has been pointed out in paragraph 3.1.4. Sometimes *Polypodium* is found as a kind of "horizon" below the stands developed on top of a wall or on wall ledges, provided a considerable layer of litter and humus has accumulated, and thus may form a zonation by being situated between the apical stand and the mural community proper. The former often contains species of the *Sedo-Scleranthetea,* or sometimes communities with *Cheiranthus cheiri* or *Centranthus ruber.* The latter supports communities in which frequently *Asplenium trichomanes* is conspicuous. In a purely floristic sense the stands with *Polypodium* are hardly separable from other mural communities. They are usually related to communities with *Asplenium trichomanes* (see 6.6) and, *Polypodium* frequently being represented in *Asplenium* communities, the stands with *Polypodium* can be allotted a place among the *Asplenium* communities.

Communities with *Polypodium* are, geographically speaking, closely linked with geographical variants of *Asplenium* communities. In the eu-atlantic areas 111 and 112 both *P. vulgare* and *P. interjectum* are represented, the latter usually being associated with the first. In a few sites in Devon (Dartmouth) *P. australe* was also recorded, but never in quantity and always in association with *P. vulgare.* In sites where *Polypodium* is plentiful, the other species of the *Asplenium* communities are often more or less suppressed and thus scantily represented or absent, and in this respect the stands with *Polypodium* are "negatively" characterised. *Ceterach officinarum* and *Asplenium trichomanes* are usually entirely lacking; *Phyllitis scolopendrium* is somewhat better represented. In addition there is a sprinkling of species of the *"Filici-Saginetum",* as is also the case in area 120. Sometimes typical species of both alliances are lacking and the stands are simply sociations of *Polypodium.* In area 131, stands rich in *Polypodium* associated with *Asplenium trichomanes* and *Cystopteris fragilis* are found (see 6.6.4), and, as in area 132, sometimes stands with *Cheiranthus* and *Poa compressa* (see 6.1.8.2). In the submediterranean and Mediterranean areas *Polypodium australe* occurs, frequently in cohabitation with *P. vulgare,* in mural vegetation resembling the *Polypodietum serrati,* a community of steep rocky inclines described by BRAUN-BLANQUET (1931). Such stands can, however, also be interpreted as a form of the community of *Asplenium trichomanes* and *Sedum dasyphyllum* rich in *Polypodium* (compare 6.6.1). However, it may also be a community more closely allied to impoverished *Parietaria judaica* communities. In Table 38 a few examples are shown.

Table 38 Communities with Polypodium species

Nr C	1	2	3	4	5	6	7	8	9	10	11	12	13
Date	31-7	31-7	17-8	28-9	26-3	31-3	12-5	12-5	2-11	31-7	1-8	23-4	23-4
Year	64	64	60	65	61	61	65	65	65	64	64	62	62
Nr R	353	338	272	1283	004	007	192	190	1318	350	368	200	199
Geo	111	111	120	132	131	131	131	131	131	111	111	310	310
Ex	3	3	2w	4	1	2	1	1	2	1	3	4	4
CH	10	50	79	70	25	20	70	30	20	50	25	45	20
CM	s	1	70	20	30	30	100	40	1	20	30	40	0
Species number	3	5	14	5	9	12	13	13	8	9	16	14	6
Polypodium vulgare	2a	2b	3a	3a	2b	2b	4b	2b	2b	+p	1b	1a	1a
Polypodium interjectum	-	-	-	-	-	-	-	-	-	2a	2a	-	-
Polypodium australe	-	-	-	-	-	-	-	-	-	-	-	2a	2a
Tortula muralis	2p	+p	2b	-	2m	1a	2m	-	1p	2a	1p	+r	-
Asplenium trichomanes	2m	-	-	3a	1p	1a	2a	-	-	-	+p	-	-
Linaria cymbalaria	-	3a	3a	1p	-	-	+p	-	-	3b	2a	-	-
Streblotrichum convolutum	-	1p	1p	-	-	-	-	-	-	+p	2m	-	-
Bryum capillare	-	1p	-	-	1p	+p	-	-	-	2a	-	-	-
Rhynchostegium confertum	-	-	3b	-	2b	-	-	-	-	-	-	-	-
Bryum argenteum	-	-	2m	-	-	+p	-	-	1p	-	-	-	-
Bryum caespiticium	-	-	1p	-	-	1p	-	-	+p	1p	-	-	-
Brachythecium rutabulum	-	-	1p	-	2a	-	1p	-	-	-	-	-	-
Dryopteris filix-mas	-	-	1p	-	-	-	-	-	-	-	-	-	-
Dryopteris carthusiana	-	-	+p	-	-	-	-	-	-	-	-	-	-
Poa annua	-	-	+p	-	-	-	-	-	-	-	-	-	-
Mycelis muralis	-	-	1a	-	-	-	-	-	-	-	-	-	-
Erigeron canadensis	-	-	+p	-	-	-	-	-	-	-	-	-	-
Asplenium ruta-muraria	-	-	-	1p	1a	+r	-	-	-	-	2a	-	-
Homalothecium sericeum	-	-	-	2b	-	+b	4b	3a	-	-	+p	2a	-
Hypnum cupressiforme var. tectorum	-	-	-	-	2a	+p	2b	-	-	-	-	1b	-
Chelidonium majus	-	-	-	-	+r	+r	-	-	-	-	-	+r	-
Ceratodon purpureus	-	-	-	-	-	2a	-	-	+p	2a	3a	-	-
Camptothecium lutescens	-	-	-	-	-	1a	-	-	-	-	-	-	-
Cystopteris fragilis	-	-	-	-	-	-	+p	2a	-	-	-	-	-
Mnium undulatum	-	-	-	-	-	-	2m	1p	-	-	-	-	-
Barbula unguiculata	-	-	-	-	-	-	2m	-	-	-	-	-	-
Bryoeythrophyllum recurvirostre	-	-	-	-	-	-	+p	-	-	-	-	-	-
Saxifraga rosacea ssp. sponhemica	-	-	-	-	-	-	+p	-	-	-	-	-	-
Cardaminopsis arenosa	-	-	-	-	-	-	+p	-	-	-	-	-	-
Geranium robertianum	-	-	-	-	-	-	-	+p	-	-	-	-	-
Taraxacum spec.	-	-	-	-	-	-	-	+p	-	-	-	-	-
Encalypta streptocarpa	-	-	-	-	-	-	-	2m	-	-	-	-	-
Didymodon rigidulus	-	-	-	-	-	-	-	2m	-	-	-	-	-
Epilobium montanum	-	-	-	-	-	-	-	+r	-	+p	+r	-	-
Cheiranthus cheiri	-	-	-	-	-	-	-	-	+p	-	-	-	-
Linaria vulgaris	-	-	-	-	-	-	-	-	+p	-	-	-	-
Phyllitis scolopendrium	-	-	-	-	-	-	-	-	-	-	+p	-	-
Epilobium lanceolatum	-	-	-	-	-	-	-	-	-	-	+p	-	-
Arenaria serpyllifolia	-	-	-	-	-	-	-	-	-	-	+r	-	-
Dactylis glomerata	-	-	-	-	-	-	-	-	-	-	+r	-	-
Sonchus oleraceus	-	-	-	-	-	-	-	-	-	-	+r	-	-
Sagina procumbens	-	-	-	-	-	-	-	-	-	-	+r	-	-
Sedum dasyphyllum	-	-	-	-	-	-	-	-	-	-	-	2a	1b
Umbilicus rupestris	-	-	-	-	-	-	-	-	-	-	-	2b	+r
Ceterach officinarum	-	-	-	-	-	-	-	-	-	-	-	+p	-
Scorpiurium circinatum	-	-	-	-	-	-	-	-	-	-	-	3a	-
Barbula acuta	-	-	-	-	-	-	-	-	-	-	-	2m	-
Pleurochaete squarrosa	-	-	-	-	-	-	-	-	-	-	-	+p	-
Grimmia pulvinata	-	-	-	-	-	-	-	-	-	-	-	+p	-
Lagoseris sancta	-	-	-	-	-	-	-	-	-	-	-	+r	-
Parietaria judaica	-	-	-	-	-	-	-	-	-	-	-	-	1b
Antirrhinum majus	-	-	-	-	-	-	-	-	-	-	-	-	1b
Hyoscyamus albus	-	-	-	-	-	-	-	-	-	-	-	-	()
MNS - MCH - MCM				10	-	38	-	30		-		-	

LEGEND

1. Exeter (England), wall in front of chapel opposite police-station in Heavitree Road.
2. Exeter, wall behind Richmond Street.
3. Brughes (Belgium), Verwersdijk near Kandelaarstraat.
4. Vézelay (Yonne, France), wall alongside the Rue des Ursulines near Le Pontot.
5. Üffeln (Niedersachsen, Western Germany), wall alongside high road.
6. Bramsche (Niedersachsen, Western Germany), wall in centre of village.
7. Bouillon (Belgium), wall below the castle facing the river Semois.
8. Bouillon, wall alongside the road to Arlon, 0.2 km S. of the bridge.
9. St. Truiden (Limburg, Belgium), wall alongside Mijnstraat, opposite nr 17.
10. Exeter, wall facing river Exe between Commercial Road and Edmund Street.
11. Ottery St. Mary (Devon, England), wall alongside Ridgeway, opposite nr 18.
12. Mourèze (Hérault, France), wall opposite of churchyard.
13. Mourèze, wall of house, in shadow.

Mural vegetation with a coverage of *Polypodium vulgare* exceeding 25%, and stands with representation of *Asplenium trichomanes,* are not quite so unusual as the table might suggest, because the table is not intended to be a good representation of the relations between various stands with *Polypodium.* The appreciable variation in the number of species per site does not plead in favour of a separate status of the stands in a phytosociological sense.

A study of the three species of *Polypodium* has also been made in other environments (see MEINDERS-GROENEVELD & SEGAL, 1967), e.g. of *P. vulgare* and *P. interjectum* in dunes, hedgerows and pollard willows in W.- and central Europe, and of *P. australe* on steep rocky cliffs in S. Europe. It appeared that the first two species may be present in a number of altogether different communities and may (co-)dominate in the stands. Often the stands occurring in different ecological environments share few other species or none at all.

Stands growing on big boulders and on slopes of rock debris have been described as a *Polytrichum-Polypodium*-association by PREIS (1937) and as the *"groupement à Polypodium vulgare"* by LEBRUN et al. (1949). Again, such stands have only a local significance. JURKO & PECIAR (1963) are of the opinion that in the Carpathian Mountains a class *Polypodietea* can be distinguished, but elsewhere in Europe there are no grounds for such a distinction on the basis of floristic criteria. These workers distinguished two orders viz. the *Ctenidio-Polypodietalia* and the *Hypno-Polypodietalia,* each containing only one alliance and a single association, and occurring on lime-rich and on acid rock formations, respectively. The associations may have a local significance. Earlier, the order *Anomodonto-Polypodietalia* Bolos & Vives (1957) had been recognised in Spain (see MARTÍNEZ, 1960), with the subordinate alliances *Bartramio-Polypodion serrati* Bolos & Vives 1957 and *Homalothecio-Polypodion serrati* Braun-Blanquet (1931)1947. The latter was supposed to be a new name for Braun-Blanquet's *Polypodion serrati* and included all stands of base-rich rocks. Relevés taken in S. France at sites on rock formations poor in lime do not show a great resemblance to relevés recorded in Spain. Such relevés, recorded by the present author at Ramatuelle, contain but few species of the *Asplenietea rupestris.* Elsewhere *Polypodium australe* was growing abundantly in scrub vegetation and also on old tree stumps. Weighing all the evidence, I believe that the recognition of an alliance *"Polypodion serrati"*, either in a very broad or in a rather narrow circumscription, is unacceptable on floristic grounds, although such a syntaxonomic entity could be meaningful from a structural and an ecological point of view. The alleged characteristic species *Homalothecium sericeum* and *Porella platyphylla,* and presumably also *Anomodon viticulosus,* have been found in other communities, including Mediterranean ones.

6.9 Vegetation with Sagina procumbens and Poa annua

The close relationships between stands of mural vegetation and stands of vegetation of paved surfaces (streets, etc.) are especially evident in stands that contain species commonly occurring in vegetation of trampled habitats, in particular *Sagina procumbens, Poa annua, Bryum argenteum* and *Ceratodon purpureus.* The mural stands can be ranked in one of two categories viz. (1) a large group of stands containing differentiating species not normally encountered in vegetation of trampled sites (such as the typical wall-dwellers *Tortula muralis* and *Linaria cymbalaria,* and several species of ferns), and (2) a smaller assembly of stands without such differentiating species, more closely allied to stands of trampled ground, and containing some singular differentiating taxa not so usually found in mural vegetation (such as *Plantago major* and *Polygonum heterophyllum*). Between the two categories, which shall be referred to by the names of *Filici-Saginetum* and *Sagino-Bryetum argentei,* respectively, there are some transitional cases, and the two names are not intended as an *a priori* recognition of the two categories as separate associations (although this could be done), but mainly for pragmatic reasons. These names can be replaced by such phrases as: "stands of mural vegetation with *Sagina procumbens, Poa annua,* and some typical wall-dwellers and ferns". The communities are upon the whole hygrophilous and for that reason often found on walls facing canals, rivers and streams. Furthermore they show a pronounced preference for high concentrations of nitrogen (and presumably also of phosphate of usually organic origin) in the substratum. The enrichment with such compounds is brought about by rain water trickling down or by aerial deposits of particles (especially in towns and along busy roads), soot and ash probably also contributing.

Stands of mural vegetation with *Sagina procumbens* and *Poa annua* have not been noted in Mediterranean regions where they are replaced by *Parietaria judaica* communities.

6.9.1 The Filici-Saginetum

For the *Filici-Saginetum* 217 relevés were assembled which fall into four groups. Of these relevés, 148 (i.e. 68%) were made in area 120, 39 (18%) in areas 111 (the majority) and 112, 20 (9%) in area 131, and the remainder in 132 (1 stand), 200 (6 stands) and 400 (7 stands). The *Filici-Saginetum* is the most common and wide-spread type of mural vegetation in the W. European lowland, but it is also most probably the only more specific mural community of the more continental climatic regions of Europe. In S. England the community is quite common, albeit (as yet?) not of such a frequent occurrence as stands with *Asplenium* but without *Sagina* and *Poa annua.* In Brittany and in Normandy the *Filici-Saginetum* is decidedly scarce.

The name *Filici-Saginetum* expresses the cohabitation of *Sagina procumbens* with several species of ferns including, apart from species of *Asplenium,* various leptosporangiate taxa of which only *Cystopteris fragilis, Polypodium vulgare, Phyllitis scolopendrium* and *Dryopteris filix-mas* commonly appear in other kinds of mural vegetation, principally in communities with *Asplenium trichomanes* (see 6.6). In the *Filici-Saginetum, Athyrium filix-femina* is an additional species and for each of its subordinate syntaxonomic units differential fern species can be indicated. Of all the leptosporangiate species native in W. Europe, only two have never been found in mural vegetation viz. the calcifuges *Thelypteris limbosperma* and *Blechnum spicant*. Even of species preferring acid soils (such as *Dryopteris cristata* and *Osmunda regalis*) wall-dwelling individuals have been recorded: *Dryopteris cristata* from Amsterdam, Rotterdam and Utrecht, *Osmunda* from Buckfastleigh, (Devon, England). In addition, some introduced ferns occur especially in the *Filici-Saginetum* viz. *Polystichum falcatum* and *Pteris cretica* (Segal, 1962b).

Phyllitis scolopendrium is one of the most characteristic taxa of the *Filici-Saginetum,* becoming scarcer from W. to E. in Europe and in central Europe being mainly restricted to N.- and E.-facing slopes in ravine forests (known by such names as *Ulmo-Aceretum* Issler 1924) and to the walls of water wells (compare Kotlaba, 1962). In the ravine forests its development is optimal, but at mural sites its fronds seldom attain a length of over 30 cm. In the steep ravine forests usually stenocoenoses can be discerned and conceivably *Phyllitis* is an element of one of these stenocoenoses consisting of rocky sites with some rock debris and humus inside such forests (see the discussion by Segal in Mayer, 1966). In the Bavarian and Swabian Alps, at greater elevations approaching 1800 m above sea-level, a community of *Phyllitis scolopendrium, Cystopteris fragilis* and *Gymnocarpium robertianum* is found on slopes of limestone debris whose stands are rather similar to stenocoenoses of the so-called *Phyllitido-Aceretum* Moor 1952 of lower altitudes. On slopes covered with calcareous gravel and rock debris, *Phyllitis* is also found growing together with *Asplenium ruta-muraria* and *A. viride* (Hegi). In Luxemburg, on slopes alongside the Syre near Manternach, the species under discussion is represented in stenocoenoses occurring in cracks and crevices between boulders in mixed stands with *Asplenium trichomanes, Polypodium vulgare* and *Cystopteris*. Optimum development of *Phyllitis* was observed in this locality in the most sheltered places at the foot of rocks and boulders with a faintly ruderal character (evident from the occurrence of *Urtica dioica*). In S. England *Phyllitis* was observed in similar situations, but also frequently in hedgerows alongside roads, in a community with *Polystichum aculeatum* and some nitrophiles such as *Urtica dioica* and *Eupatorium cannabinum*. The poorer development of *Phyllitis* in stands of mural vegetation as compared to its forest habitats could be caused by an excessive quantity of light. Many authors call it a shade-loving species (e.g. Lawalree, 1950). Its opulent occurrence on slopes of rock debris in central Europe does not support this idea, however. Conceivably the species requires, apart from moisture, shade for the germination of its spores and the development of its gametophyte, conditions it may find in the joints and cracks of a

Table 3 a. Communities of Parietaria judaica and Bromus madritensis
a1. with Mercurialis annua
a2. with Umbilicus rupestris
a2-1. and Centranthus ruber
a2-2. without Centranthus ruber
b. Community of Parietaria judaica and Asplenium trichomanes
c. Community of Asplenium trichomanes and Sedum dasyphyllum

	a1.							a2-1.							a2-2					
Nr C	1	2	3	P	U W	MGT	MGP	4	5	6	P	U W	MGT	MGP	7	8	9	P	U W	MGT
Date	3-4	20-4	21-4	56st				20-4	20-4	7-4	16st				31-3	25-4	29-4	32st		
Year	62	62	62					62	62	62					62	65	65			
Nr R	031	137	158					141	135	052					015	064	095			
Ex	2	3	3					1	4	2					3	1	4			
Geo R	332	331	331					331	331	310					332	320	320			
CH	30	20	60					30	15	5					20	25	35			
CM	s	s	-					s	5	s					15	4	20			
Parietaria judaica	2b	2a	2b	98	+r-4b	20	21	1a	2a	1b	81	+p-3b	12	14	2b	2b	2b	100	+p-3a	15
Sonchus oleraceus	+p	+p	2a	41	+r-2a	0.2	0.5	+r	+p	+r	81	+r-1p	0.1	0.1	+r	+r	-	56	+r-1a	0.1
Tortula muralis	+p	+r	-	27	+r-3a	1	4	+r	2m	+r	63	+r-2m	0.5	0.8	2a	2m	1p	65	+r-2a	2
Linaria cymbalaria	-	-	-	38	+p-2b	2	6	-	-	-	31	1b-2b	3	9	-	-	2a	40	+r-2b	2
Barbula acuta	+r	+p	-	21	+r-2b	0.7	3	-	2m	+r	31	+r-2m	0.5	2	2m	2m	-	31	+p-2b	1
Ficus carica	-	-	+p	23	+r-1p	<0.1	<0.1	-	-	-	31	+r-+a	0.2	0.6	-	-	-	19	+r-+a	<0.1
Bromus madritensis	+p	+r	-	38	+r-1a	0.1	0.2	-	-	-	31	+r-+p	<0.1	<0.1	.a	+r	-	59	+r-1a	0.2
Centranthus ruber	-	1a	-	36	+r-4b	3	8	2b	1a	+p	100	+r-2b	5	5	-	-	-	-	-	-
Mercurialis annua	+p	1a	+p	46	+r-2b	0.4	0.8	-	-	-	19	+r-1b	0.2	1	-	-	-	16	+r-1p	<0.1
Lamium amplexicaule	1p	-	-	30	+r-1p	<0.1	0.1	-	-	-	13	+r	0	0	+p	-	-	34	+r-1p	<0.1
Veronica cymbalaria	1p	+r	-	21	+r-1a	0.1	0.3	-	-	-	25	+r-+p	<0.1	0.1	-	-	+p	28	+r-1p	<0.1
Phagnalon sordidum	1b	1b	2b	29	+r-2b	1	5	1a	-	-	38	+r-2b	2	7	-	-	-	6	+r-1b	0.1
Fumaria capreolata	-	-	+p	21	+r-1a	0.1	0.3	-	-	-	19	+r	0	0	-	-	+p	19	+r-+p	<0.1
Mentha rotundifolia	-	-	1p	16	+r-1b	0.2	1	-	+r	-	25	+r-1p	<0.1	0.1	-	-	-	22	+r-1p	<0.1
Calamintha nepeta	+r	+p	-	16	+r-1b	0.1	0.7	-	-	-	-	-	-	-	1p	-	-	19	+p-1a	0.1
Rubus (div.) spec.	-	-	-	14	+r-+a	<0.1	0.2	+r	-	-	13	+r-+a	0.1	0.8	-	+r	-	25	+r-+b	0.2
Fumaria officinalis	+p	-	-	16	+r-+p	<0.1	0.1	-	-	-	19	+r-+p	<0.1	<0.1	1p	-	-	16	+r-1b	0.1
Lagoseris sancta	+r	-	-	16	+r	0	0	-	-	+r	19	+r-+p	<0.1	<0.1	-	-	-	12	+r-+p	<0.1
Stellaria media	1p	-	-	14	+r-1p	3	0.2	-	+r	-	19	+r-1p	<0.1	0.2	+r	-	-	12	+r-1p	<0.1
Bromus sterilis	+r	+p	-	23	+r-2m	0.1	0.4	-	-	-	13	+p-1p	<0.1	0.3	-	-	-	9	+p	<0.1
Matthiola incana	-	-	1a	11	+r-2a	0.3	2	-	-	-	-	-	-	-	-	-	-	-	-	-
Veronica hederifolia	+p	-	-	20	+r-1a	<0.1	0.2	-	-	-	6	+r	0.1	0	+r	-	-	16	+r-1p	<0.1
Hedera helix	-	-	+r	13	+r-+p	<0.1	<0.1	-	-	-	19	+r-+a	0.1	0.5	-	-	-	6	+r	0
Oxalis corniculata	-	-	-	14	+r-2a	0.3	2	-	-	+r	44	+r-+p	<0.1	<0.1	+p	-	-	31	+r-2a	0.3
Dactylis glomerata	-	-	1p	13	+r-1p	<0.1	0.2	+p	+p	-	44	+r-1b	0.2	0.6	1p	-	-	22	+r-1p	0.1
Veronica polita	-	-	-	13	+r-2m	0.1	0.5	-	-	-	25	+r-1p	<0.1	0.1	+p	-	-	28	+r-1a	0.1
Geranium rotundifolium	+r	-	+r	13	+r	0	0	-	-	+r	13	+r-+p	<0.1	0.1	+r	-	-	34	+r-1p	<0.1
Catapodium rigidum	-	-	-	7	+r-+p	<0.1	0.1	+r	+p	+p	50	+r-+p	<0.1	0.1	-	+p	-	34	+r-1b	0.2
Sedum sediforme	-	+r	-	5	+r	0	0	-	+r	-	25	+r-+p	<0.1	<0.1	+p	-	-	9	+r-1p	<0.1
Geranium molle	-	-	-	2	+r	0	0	-	-	-	25	+r-2a	0.7	3	-	-	-	3	+r	0
Reichardia picroides	-	-	-	5	+r-1p	<0.1	0.1	-	-	-	25	+r-+p	<0.1	0.1	-	-	-	6	+r	0
Picris hieracioides	-	-	-	5	+r-+p	<0.1	<0.1	-	-	-	19	+r-1p	<0.1	0.2	-	-	-	-	-	-
Petrorhaga prolifera	-	-	-	-	-	-	-	-	-	-	13	+p	<0.1	0.1	-	-	-	-	-	-
Antirrhinum latifolium	-	-	-	2	+p	<0.1	0.1	1a	1p	-	19	+p-1a	<0.1	0.7	-	+p	-	6	+p-1b	0.1
Avena barbata	-	+p	-	7	+p-1a	<0.1	0.5	-	-	-	19	+r-+p	<0.1	<0.1	-	-	-	3	+r	0
Arabidopsis thaliana	-	-	-	-	-	-	-	-	-	+p	19	+r-+p	<0.1	0.1	-	-	-	3	1p	<0.1
Antirrhinum majus	-	-	-	7	+r-+p	<0.1	<0.1	-	-	+p	19	+p-1p	<0.1	0.2	-	-	-	3	2b	0.6
Cardamine hirsuta	-	-	-	-	-	-	-	-	+r	-	19	+r-1p	<0.1	0.1	-	-	1p	12	+r-1p	<0.1
Bryum capillare	-	-	-	-	-	-	-	-	-	+r	19	+r-1p	<0.1	0.1	-	-	-	9	1p-2m	0.1
Poa pratensis	-	-	-	7	+r-+p	<0.1	<0.1	-	-	+p	13	+r-+p	<0.1	0.1	-	-	-	-	-	-
Senecio viscosus	-	-	-	-	-	-	-	-	+r	-	13	+r-+p	<0.1	0.1	+p	-	-	9	+p	<0.1
Minuartia mutabilis	-	-	-	-	-	-	-	-	-	-	13	+r-+p	<0.1	0.1	-	-	-	3	+r	0
Erigeron mucronatus	-	-	-	4	1b	0.1	4	-	-	-	13	+r-1a	0.1	0.8	-	-	-	9	+p	<0.1
Clematis flammula	-	-	-	5	+r	0	0	-	+r	-	13	+r	0	0	-	-	-	6	+p	<0.1
Urtica dioica	-	-	-	-	-	-	-	-	-	+r	13	+r	0	0	-	-	-	6	+r	0
Thymus vulgaris	-	-	-	5	2a	0.4	8	-	+r	-	13	+r	0	0	-	-	-	3	+p	<0.1
Sonchus asper	-	-	-	2	+p	<0.1	0.1	-	-	-	13	+r	0	0	+r	-	-	3	+r	0
Poa annua	-	-	-	9	+r-2a	0.2	2	-	-	-	13	+s	0	0	-	+p	-	3	+p	<0.1
Cerastium glomeratum	-	-	-	-	-	-	-	-	-	-	19	+p-1a	0.1	0.7	+p	-	-	22	+r-1p	<0.1
Crepis vesicaria ssp. taraxacifolia	-	-	-	5	+r	0	0	-	-	-	13	+r	0	0	-	-	-	12	+r-1p	<0.1
Lepidium graminifolium	-	-	-	5	+r	0	0	-	-	+r	13	+r-+p	<0.1	0.1	1p	-	-	16	+p-1p	<0.1
Saxifraga tridactylites	-	-	-	-	-	-	-	-	+r	-	19	+r-2m	0.2	1	-	-	2m	6	2m	0.2
Inula viscosa	-	-	-	-	-	-	-	+r	-	-	19	+r-+p	<0.1	0.1	-	-	-	6	+r-1b	0.1
Homalothecium sericeum	-	-	-	-	-	-	-	+p	-	-	19	+p-3b	4	21	-	-	-	9	1b-3a	1
Taraxacum (div.) spec.	-	-	-	5	+r	0	0	+r	+r	-	13	+r	0	0	-	-	-	16	+r-+p	<0.1
Hypnum cupressiforme	-	-	-	-	-	-	-	-	-	-	13	+p-2a	0.5	4	-	-	+p	9	+p-2m	0.1
Barbula vinealis	-	-	-	-	-	-	-	-	-	-	13	1p-2b	1	9	-	-	2a	16	2m-3a	2
Polypodium vulgare	-	-	-	-	-	-	-	-	-	1p	13	+r-1p	<0.1	0.2	-	-	-	6	+p	<0.1
Bryum caespiticium	-	-	-	4	+p-2m	<0.1	1	-	+p	-	13	+p-1p	<0.1	0.3	-	-	-	3	1p	<0.1
Arenaria serpyllifolia	+p	-	-	7	+r-+p	<0.1	0.1	-	-	+r	13	+r-+p	<0.1	0.1	1p	+p	-	28	+r-1p	<0.1
Campanula erinus	-	+r	-	9	+r-+p	<0.1	<0.1	-	-	-	25	+r-1p	0.1	0.2	-	-	+p	47	+r-2a	0.8
Senecio vulgaris	+r	-	-	9	+r	0	0	-	-	-	19	+r-1a	0.1	0.5	+p	-	-	26	+r-+p	<0.1
Umbilicus rupestris	-	-	-	5	1p-2b	0.4	7	-	-	2m	69	+r-2a	1	2	+p	+p	-	65	+r-3a	3
Sedum dasyphyllum	-	-	-	-	-	-	-	+p	+r	+r	63	+r-2b	2	4	-	-	-	53	+r-2a	0.7
Ceterach officinarum	-	-	2a	4	+r-2a	0.2	4	1b	-	+p	44	+r-1b	0.4	0.9	1p	+r	1a	59	+r-2b	1
Asplenium trichomanes	-	-	-	-	-	-	-	-	-	+p	19	+p-1p	<0.1	0.3	+p	1p	+p	28	+r-1b	0.2
Scorpiurium circinatum	-	-	-	2	2a	0.1	8	-	-	-	13	1a-2a	0.6	5	-	-	2b	19	+p-2b	2
Poa bulbosa	-	-	-	7	+r-+p	<0.1	<0.1	-	+p	-	13	+p-1p	<0.1	0.3	-	-	1p	16	1p-1a	0.1
Streblotrichum convolutum	-	-	-	4	1p-2m	0.1	0.2	-	-	-	6	+r	0	0	-	-	-	25	+r-2a	0.5
Euphorbia peplus	-	-	1p	9	+r-1p	<0.1	0.2	-	-	-	6	+r	0	0	+r	-	-	22	+r-+p	<0.1
Euphorbia helioscopia	-	+r	-	9	+r-+p	<0.1	<0.1	-	-	-	-	-	-	-	+p	-	-	12	+r-+p	<0.1
Cerastium semidecandrum	-	-	-	-	-	-	-	-	-	-	-	-	-	-	-	1p	+p	16	+p-1p	<0.1
Erigeron canadensis	-	-	-	7	+r-1a	<0.1	0.4	-	-	-	-	-	-	-	1p	-	-	12	+r-1p	<0.1
Pleurochaete squarrosa	-	-	-	2	2m	<0.1	3	-	-	-	-	-	-	-	-	1p	-	12	+r-2m	0.1
Sedum album	-	-	-	-	-	-	-	-	-	-	6	+p	<0.1	0.1	-	-	-	6	+r-1p	<0.1
Barbula revoluta	-	-	-	-	-	-	-	-	-	-	-	-	-	-	+p	-	2m	6	+p-2m	0.2
Schistidium apocarpum	-	-	-	-	-	-	-	-	-	-	-	-	-	-	-	-	-	6	+r-2m	0.1
Polypodium australe	-	-	-	-	-	-	-	-	-	-	-	-	-	-	-	-	-	3	2b	0.6
Asplenium ruta-muraria	-	-	-	-	-	-	-	-	-	-	-	-	-	-	-	-	-	3	+r	0
Brachythecium rutabulum	-	-	-	-	-	-	-	-	-	-	-	-	-	-	-	-	-	-	-	-
Chelidonium maius	-	-	-	2	1a	<0.1	2	-	-	-	-	-	-	-	-	-	-	-	-	-
Rhynchostegiella tenella	-	-	-	-	-	-	-	-	-	-	-	-	-	-	-	-	-	-	-	-
MHS - MCH - MCM		11	-	31	-	2 (13)			20	-	38	-	7	(16)		19	-	28	-	11 (18)

Table 30 Communities of Asplenium trichomanes and Tortula muralis

a. with Sedum dasyphyllum
b. with Asplenium adiantum-nigrum
b1 and Ceterach officinarum
b2 and Sagina procumbens
c. with Poa annua and Ceratodon purpureus
d. with Poa compressa
d1 and Geranium robertianum
d2 and Cystopteris fragilis
e. inops

	a.							b1.						
Nr C	1	2	3	P 25st	U W	MGT	MGP	4	5	6	P 39st	U W	MGT	MGP
Date	14-4	25-4	11-4					9-8	2-8	4-10				
Year	62	62	65					62	62	61				
Nr R	115	214	023					568	535	568				
Ex	1	1	2s					4	1	1				
Geo R	332	310	132					112	112	111				
CH	12	60	25					25	10	45				
CM	7	20	10					3	30	4				
Asplenium trichomanes	2a	4a	2a	82	+p-4a	10	10	2m	1a	2b	82	+p-2b	7	9
Tortula muralis	+p	+r	1p	68	+r-2a	1	2	2m	1p	2m	95	+r-2a	2	2
Asplenium ruta-muraria	-	-	2a	32	+p-2a	1	4	-	1a	1a	72	+r-2b	2	2
Linaria cymbalaria	-	1b	-	40	+r-2a	1	3	2b	2a	2b	69	+r-3b	6	9
Homalothecium sericeum	-	-	1a	42	+p-2b	3	6	+r	-	-	54	+r-4a	6	11
Bryum capillare	-	-	-	24	+p-2m	0.5	2	-	+p	+p	26	+r-1p	0.1	0.3
Streblotrichum convolutum	-	-	2m	16	+p-2m	0.2	1	+p	-	+p	51	+p-2b	2	4
Ceterach officinarum	1b	1b	1b	88	+p-3b	5	5	-	+r	+r	44	+r-2b	2	4
Hypnum cupressiforme	-	-	-	28	+p-5a	6	20	-	-	-	8	+p-2a	0.4	5
Barbula vinealis	-	-	2m	16	+p-2m	0.3	2	-	-	-	23	+r-2b	0.7	3
Chelidonium majus	-	-	-	16	+r-1a	0.1	0.4	-	-	-	-	-	-	-
Polypodium vulgare	-	-	-	16	+p-3a	0.3	2	-	+p	1p	62	+p-2a	0.8	1
Hedera helix	-	-	-	16	+r-+b	0.1	0.4	-	-	-	21	+r-+a	<0.1	0.2
Sedum rupestre	-	-	-	80	+p-2a	0.4	2	-	-	-	-	-	-	-
Saxifraga tridactylites	1p	-	-	20	+p-2m	0.2	1	-	-	-	3	+p	<0.1	0.1
Arenaria serpyllifolia	+p	-	-	16	+r-2m	0.1	0.6	-	-	-	6	+p-1p	<0.1	0.3
Umbilicus rupestris	+p	-	-	32	+r-4b	4	11	+p	-	-	10	+r-+p	<0.1	<0.1
Sedum dasyphyllum	1p	1a	1a	80	+r-2a	2	2	-	-	-	-	-	-	-
Parietaria judaica	1b	+r	-	48	+r-1b	0.3	0.7	-	-	-	-	-	-	-
Scorpiurium circinatum	1b	-	-	40	+p-3a	4	11	-	-	-	3	2a	0.2	8
Barbula acuta	2m	-	-	38	+r-2b	2	6	-	-	-	3	1p	<0.1	0.4
Catapodium rigidum	-	-	-	12	+p-1p	<0.1	0.2	-	-	-	18	+p-1a	0.1	0.4
Poa bulbosa	+r	-	-	16	+r-1a	0.1	0.5	-	-	-	-	-	-	-
Lagoseris sancta	+r	-	-	16	+r	0	0	-	-	-	-	-	-	-
Rhynchostegiella tenella	-	2b	1p	12	+p-2b	1	3	-	-	-	-	-	-	-
Campanula erinus	+r	-	-	12	+r-1p	<0.1	0.3	-	-	-	-	-	-	-
Geranium lucidum	+r	-	-	12	+r	0	0	-	-	-	-	-	-	-
Lamium amplexicaule	-	-	-	12	+r	0	0	-	-	-	-	-	-	-
Sonchus oleraceus	-	-	-	8	+r	0	0	-	-	+p	28	+r-+p	<0.1	<0.1
Barbula unguiculata	-	-	-	8	2m	0.2	3	-	2m	2m	26	+r-2a	0.9	4
Bryoerytrophyllum recurvirostre	-	-	-	8	+p-2m	0.1	1	-	-	-	16	+p-2m	0.3	2
Bryum caespiticium	-	-	-	8	+r-1p	<0.1	0.2	-	-	-	13	+r-2m	0.1	1
Ceratodon purpureus	-	-	-	4	+p	<0.1	<0.1	-	-	-	13	+p-2m	0.1	1
Dactylis glomerata	-	-	-	4	+r	0	0	-	-	-	16	+r-+p	<0.1	0.1
Phyllitis scolopendrium	-	-	-	-	-	-	-	+p	1a	+p	56	+p-2b	2	3
Asplenium adiantum-nigrum	-	-	-	-	-	-	-	1b	+r	1a	54	+r-2a	0.9	2
Taraxacum (div.) spec.	-	-	-	-	-	-	-	()	+r	+p	41	+r-+p	<0.1	<0.1
Epilobium montanum	-	-	-	-	-	-	-	+r	-	+r	38	+r-+p	<0.1	0.1
Bryum murorum	-	-	-	-	-	-	-	-	-	-	26	+p-2a	0.3	1
Dryopteris filix-mas	-	-	-	-	-	-	-	-	-	+r	13	+r-+p	<0.1	<0.1
Centranthus ruber	-	-	-	-	-	-	-	-	-	+a	21	+r-+a	<0.1	0.2
Poa pratensis	-	-	-	-	-	-	-	-	-	+r	15	+r-1p	<0.1	0.1
Buddleja davidii	-	-	-	-	-	-	-	-	-	-	10	+r-+a	<0.1	0.4
Antirrhinum majus	-	-	-	-	-	-	-	-	-	-	10	+r-2a	0.2	2
Sagina procumbens	-	-	-	8	1p-2m	0.1	1	-	-	-	-	-	-	-
Poa annua	-	-	-	-	-	-	-	-	-	-	3	+p	<0.1	0.1
Epilobium parviflorum	-	-	-	-	-	-	-	+p	-	-	10	+r-+p	<0.1	<0.1
Bryum argenteum	-	-	-	-	-	-	-	-	-	-	3	+p	<0.1	0.1
Epilobium lanceolatum	-	-	-	-	-	-	-	-	-	-	6	+r-+p	<0.1	0.1
Holcus lanatus	-	-	-	4	+r	0	0	-	-	-	8	+r-+p	<0.1	<0.1
Athyrium filix-femina	-	-	-	-	-	-	-	-	-	-	-	-	-	-
Agrostis stolonifera	-	-	-	-	-	-	-	-	-	-	3	+p	<0.1	0.1
Rhynchostegium murale	-	-	-	4	+p	<0.1	0.1	-	-	-	3	+p	<0.1	0.1
Encalypta streptocarpa	-	+p	-	8	+p-1a	0.1	0.8	-	-	-	-	-	-	-
Poa compressa	-	-	-	4	+p	<0.1	0.1	-	-	-	8	+p	<0.1	0.1
Geranium robertianum	-	-	-	4	+p	<0.1	0.1	+r	-	-	16	+r-1p	<0.1	0.1
Brachythecium rutabulum	-	-	-	-	-	-	-	-	-	-	-	-	-	-
Sedum album	-	-	-	8	+p-2a	0.3	4	-	-	-	-	-	-	-
Poa angustifolia	-	-	-	-	-	-	-	-	+r	-	3	+r	0	0
Campanula rotundifolia	-	-	-	-	-	-	-	-	-	-	-	-	-	-
Cystopteris fragilis	-	-	-	-	-	-	-	-	-	-	-	-	-	-
Mnium undulatum			8m	4	3m	0.1	3	-	-	-	-	-	-	-
Leptobryum pyriforme	-	-	-	-	-	-	-	-	-	-	3	2m	0.1	3
Didymodon rigidulus	-	-	-	8	+p-2m	0.1	1	-	-	-	3	2m	0.1	3
Galium mollugo	-	-	-	4	+r.	0	0	-	-	-	3	+r	0	0
Hypericum perforatum	-	-	-	4	+r	0	0	-	-	-	-	-	-	-
Epilobium collinum	-	-	-	-	-	-	-	-	-	-	-	-	-	-
Gymnocarpium robertianum	-	-	-	-	-	-	-	-	-	-	-	-	-	-
Campanula rapunculoides	-	-	-	-	-	-	-	-	-	-	-	-	-	-
MBS - MCH - MCM		13	-	24	-	21 (24)			13	-	20	-	16 (18)	

	b2.							b.				c.		
Nr C	7	8	9	P 13st	U W	MGT	MGP	P 52st	U W	MGT	MGP	10	11	12
Date	9-8	2-8	4-10									25-5	15-8	2-5
Year	62	64	61									65	60	61
Nr R	570	393	585									205	253	050
Ex	4	2a	2									3	1	1
Geo R	112	111	120									131	120	131
CH	10	30	70									30	20	30
CM	40	12	1									10	25	20
Asplenium trichomanes	2m	2m	2m	92	1p-3a	7	7	86	+p-3a	7	8	2m	2m	2a
Tortula muralis	+r	+p	1p	100	+r-2a	2	2	98	+r-2a	2	2	2a	2b	2a
Asplenium ruta-muraria	-	1p	3b	62	+r-3b	4	7	69	+r-3b	2	3	3a	2a	1b
Linaria cymbalaria	1a	2b	-	62	+p-2b	5	8	67	+r-3b	6	9	-	-	2a
Homalothecium sericeum	3a	-	-	33	2a-3a	4	16	46	+r-4a	5	12	-	-	-
Bryum capillare	-	-	-	38	+p-2a	0.7	2	29	+r-2a	0.2	0.8	2m	+p	+p
Streblotrichum convolutum	-	-	-	46	+p-2a	1	3	50	+p-2b	2	4	-	-	-
Ceterach officinarum	1b	+p	-	38	+r-1b	0.4	1	42	+r-2b	1	3	-	-	-
Hypnum cupressiforme	-	-	-	8	+p	0.1	0.1	8	+p-2a	0.3	4	-	-	-
Barbula vinealis	-	-	-	15	1p-2m	0.2	1	21	+r-2b	0.6	3	-	-	-
Chelidonium majus	-	-	-	-	-	-	-	-	-	-	-	-	-	-
Polypodium vulgare	1b	1b	-	62	+r-1b	0.8	1	62	+r-2a	0.8	1	-	2a	-
Hedera helix	-	+r	-	15	+r-+p	<0.1	0.1	19	+r-+a	<0.1	0.2	-	-	-
Sedum rupestre	-	-	-	-	-	-	-	-	-	-	-	-	-	-
Saxifraga tridactylites	-	-	-	-	-	-	-	2	+p	<0.1	0.1	-	-	-
Arenaria serpyllifolia	-	-	-	8	+p	<0.1	0.1	6	+p-1p	<0.1	0.2	-	-	-
Umbilicus rupestris	-	-	-	-	-	-	-	8	+r-+p	<0.1	<0.1	-	-	-
Sedum dasyphyllum	-	-	-	-	-	-	-	-	-	-	-	-	-	-
Parietaria judaica	-	-	-	15	+r	0	0	4	+r	0	0	-	-	-
Scorpiurium circinatum	-	2m	-	8	2m	0.2	3	4	2m-2a	0.2	5	-	-	-
Barbula acuta	-	-	-	-	-	-	-	2	1p	<0.1	0.4	-	-	-
Catapodium rigidum	-	-	-	-	-	-	-	13	+p-1a	0.1	0.4	-	-	-
Poa bulbosa	-	-	-	-	-	-	-	-	-	-	-	-	-	-
Lagoseris sancta	-	-	-	-	-	-	-	-	-	-	-	-	-	-
Rhynchostegiella tenella	-	-	-	-	-	-	-	-	-	-	-	-	-	-
Campanula erinus	-	-	-	-	-	-	-	-	-	-	-	-	-	-
Geranium lucidum	-	-	-	-	-	-	-	-	-	-	-	-	-	-
Lamium amplexicaule	-	-	-	-	-	-	-	-	-	-	-	-	-	-
Sonchus oleraceus	-	+r	-	33	+r-+p	<0.1	<0.1	27	+r-+p	<0.1	<0.1	+r	-	-
Barbula unguiculata	-	2m	-	31	1p-2a	1	3	27	+r-2a	1	4	2m	-	-
Bryoerytrophyllum recurvirostre	1a	2m	-	23	1a-2m	0.5	2	17	+p-2m	0.4	2	-	-	-
Bryum caespiticium	-	+p	+p	15	+p	<0.1	0.1	13	+r-2m	0.1	0.9	1p	2m	+p
Ceratodon purpureus	+p	-	1p	15	+p-1p	<0.1	0.3	13	+p-2m	0.1	0.9	-	2m	2m
Dactylis glomerata	-	-	-	8	1a	0.1	2	13	+r-1a	<0.1	0.3	-	-	-
Phyllitis scolopendrium	-	+r	-	64	+r-1b	1	1	56	+r-2b	1	3	-	-	-
Asplenium adiantum-nigrum	1a	1p	+p	77	+r-3a	4	5	60	+r-3a	2	3	-	-	-
Taraxacum (div.) spec.	-	+r	+r	31	+r-+p	<0.1	<0.1	38	+r-+p	<0.1	<0.1	-	-	-
Epilobium montanum	-	1p	-	23	+r-1p	<0.1	0.1	35	+r-1p	<0.1	0.1	-	-	-
Bryum murorum	-	-	-	-	-	-	-	19	+p-2a	0.2	1	-	-	-
Dryopteris filix-mas	-	1a	2b	15	1a-2b	2	10	13	+r-2b	0.4	3	-	-	-
Centranthus ruber	-	-	-	15	1a-1b	0.4	3	19	+r-1b	0.1	0.7	-	-	-
Poa pratensis	-	-	-	8	+p	<0.1	0.1	13	+r-1p	<0.1	0.1	-	-	-
Buddleja davidii	-	-	-	8	+r	0	0	10	+r-+a	<0.1	0.3	-	-	-
Antirrhinum majus	-	-	-	-	-	-	-	6	+r-2a	0.2	2	-	-	-
Sagina procumbens	1p	1p	2m	54	+r-2m	0.4	0.8	13	+r-+p	0.1	0.8	1p	-	-
Poa annua	-	+p	1a	77	+r-1a	0.2	0.2	21	+r-1a	<0.1	0.2	-	+r	+r
Epilobium parviflorum	-	-	-	23	+p-1p	0.1	0.3	13	+p-1p	<0.1	0.1	-	-	-
Bryum argenteum	-	+p	-	23	+p-1p	<0.1	0.2	28	+p-1p	<0.1	0.1	+p	2a	-
Epilobium lanceolatum	+r	+p	-	15	+r-+p	<0.1	0.1	8	+r-+p	<0.1	0.1	-	-	-
Holcus lanatus	-	-	-	15	1a	0.2	2	10	+r-1a	0.1	0.6	-	-	-
Athyrium filix-femina	-	-	-	15	+p-2m	0.2	1	4	+p-2m	0.1	1	-	-	-
Agrostis stolonifera	-	-	-	16	+r-+p	<0.1	0.1	6	+r-+p	<0.1	0.1	-	-	-
Rhynchostegium murale	-	+p	-	8	+p	<0.1	0.1	3	+p	<0.1	0.1	-	-	-
Encalypta streptocarpa	-	-	-	-	-	-	-	-	-	-	-	-	-	-
Poa compressa	-	-	-	-	-	-	-	6	+p	<0.1	0.1	-	-	-
Geranium robertianum	-	-	-	-	-	-	-	13	+r-1p	<0.1	0.1	-	-	-
Brachythecium rutabulum	-	-	-	8	+p	<0.1	0.1	3	+p	<0.1	0.1	-	-	2a
Sedum album	-	-	-	-	-	-	-	-	-	-	-	-	-	-
Poa angustifolia	-	-	-	-	-	-	-	3	+r	0	0	-	-	-
Campanula rotundifolia	-	-	-	-	-	-	-	-	-	-	-	-	-	+r
Cystopteris fragilis	-	-	-	-	-	-	-	-	-	-	-	-	-	-
Mnium undulatum	-	-	-	-	-	-	-	-	-	-	-	-	-	-
Leptobryum pyriforme	-	-	-	-	-	-	-	3	2m	<0.1	3	-	-	
Didymodon rigidulus	-	-	-	8	2m	0.1	3	4	2m	0.1	3	-	-	-
Galium mollugo	-	-	-	-	-	-	-	3	+r	0	0	-	-	-
Hypericum perforatum	-	-	-	-	-	-	-	-	-	-	-	-	-	-
Epilobium collinum	-	-	-	-	-	-	-	-	-	-	-	-	-	-
Gymnocarpium robertianum	-	-	-	-	-	-	-	-	-	-	-	-	-	-
Campanula rapunculoides	-	-	-	-	-	-	-	-	-	-	-	-	-	-
MBS - MCH - MCM		13	-	26	-	11 (12)		13 - 21 - 16 (18)					10	-

In addition in 1: Hornungia petraea 1p; Bromus spec. +r; Orthotrichum diaphanum +p; in 2: Stellaria media +r; in 3: x Asplenoceterach badense +r; Encalypta vulgaris 2m; in 5: Camptothecium lutescens 2b; Didymodon tophaceus 1p; Tortula intermedia +a; in 8: Oxalis corniculata +r; Plasteurhynchium striatulum 2m; Metzgeria furcata +p; in 9: Thelypteris palustris +r; Asplenium x perardii +r; Urtica dioica (+r); in 12: Oxyrrhynchium swartzii 2a; Neckera complanata 2m; Dicranella subulata var. curvata 2m; Preissia quadrata 1b; Tortula ruralis +a; Mnium stellare +p; Encalypta vulgaris +r; in 13: Glechoma hederacea +r; Encalypta vulgaris var. mutica 1a; in 14: Amblystegium serpens +p; Origanum vulgare 1b; Fragaria vesca +p; Sedum reflexum +r; in 15: Sedum album 2m; Polypodium interjectum 1b; Sonchus asper +p; Lactuca serriola +r; in 16: Stellaria media, Senecio vulgaris, Galium aparine +r; Helictotrichon pubescens (+r); Cladonia pyxidata 1b; in 17: Corylus avellana +r; in 18: Polystichum lonchitis, Senecio vulgaris +r; in 21: Hygroamblystegium tenax 2m.

LEGEND

1. Notre-Dame-de-Londres (Hérault, France), wall of abbey.
2. Ste. Enimie (Lozère, France), wall in a narrow street.
3. St. Sorlin (Ain, France), wall near communal laundering place.
4. Brasparts (Finistère, FRance), wall alongside churchyard.
5. Le-Mont-St.-Michel (Manche, France), wall between gardens above the Eglise paroissale.
6. Buckfastleigh (Devon, England), wall in Boss Road.
7. Pleyben (Finistère, France), wall around church.
8. Topsham (Devon, England), wall alongside gardens of Ashford Street nrs 6-14.
9. Enkhuizen (Noord-Holland, Netherlands), sluice wall near the Dromedaris.
10. Libeek (municipality St. Geertruid, Limburg, Netherlands), garden wall of farm of Houbiers.
11. Ghent (Belgium), under bridge over the river Leye near Gravensteen castle.
12. Luxembourg (town), wall between Rue St. Ulrich and railway bridge.
13. Pavillonchamps (Liège, Belgium), wall of castle garden.
14. Neckarsteinach (Baden-Württemberg, Western Germany), wall alongside Gemünderstrasse facing orchard.
15. Champagney (Côte-d'Or, France), wall alongside road to Pesmes.
16. Habay-la-Neuve (Luxembourg, Belgium), bridge wall.
17. Sprendorf (Rheinland-Pfalz, Western Germany), wall opposite Hotel Viktoria.
18. Solbad Hall (Tirol, Austria), wall facing Milserstrasse opposite nr 9.
19. Karlstejn (Tschechoslovakia), wall of castle near well.
20. St. Saphorin (Vaud, Switzerland), wall above staircase.
21. Boxtel (Noord-Brabant, Netherlands), inside wall of round well of Stapelen castle.

wall. Its frequent occurrence in areas 111 and 112 does not necessarily imply that a high annual rainfall is required for its development, because this fern is capable of absorbing water from the air through its leaves, even if the air is not saturated with water vapour (POTTS & PENFOUND, 1948). A nearly constant high relative humidity is of course a favourable environmental factor, possibly also for the development of *Asplenium adiantum-nigrum*.

In the more continental subordinate syntaxon with *Gymnocarpium robertianum, Phyllitis* is of rare occurrence, and, as far as mural vegetation is concerned, cohabitation of these two ferns, and of *Phyllitis* and *Cystopteris* (the latter being a differentiating species of that subordinate taxon), is restricted to a few sites in W. Europe. LEBRUN et al. (1949) distinguished a *"groupement à Phyllitis scolopendrium et Cystopteris fragilis"* in W.- and N. Belgium, which community is found in water wells and is considered to be a fragment of other forms of mural vegetation.

A more penetrating study of the stands with *Gymnocarpium robertianum* and with *Cystopteris fragilis* in central and E. Europe is required before one can decide whether these two groups of stands deserve the status of separate phytosociological syntaxa.

6.9.1.1 Vegetation with Sagina procumbens and Aspidiaceae

The distribution of the stands with *Aspidiaceae* is shown in Table 40, which also includes inops-expressions and transitions towards stands with *Sagina procumbens* and *Asplenium trichomanes*. The term "inops" is not appropiate for *all* the stands, because the communities have their own differentiating taxa even though their degree of presence may be low. Moreover, the community is geographically fairly well separated from the other subordinate syntaxa of the vegetation type under discussion.

Table 40 Geographical distribution and exposure of stands of the community of Sagina procumbens and Aspidiaceae

exposure:	1	2	3	4	5	6	1w	2w	3w	4w	t	t%
geographical region:												
111	7	3	2	1	2	1	-	-	-	1	17	18
112	-	1	-	-	-	-	-	-	-	-	1	1
120	1	6	3	6	2	1	20	12	4	17	72	76
131	3	-	-	1	1	-	-	-	-	-	5	5
t	11	10	5	8	(5)	(2)	20	12	4	18	88(+7)	-
t%	12	11	6	9	-	-	23	14	5	20	-	100

Of these relevés, 76% come from area 120 (which have been studied rather thoroughly), and 18% from area 111. The stands are often transitional towards communities with *Asplenium trichomanes* (compare 6.6.3), especially in 111. Of the stands recorded in area 120, 53 (i.e. 74%) were situated on walls facing a body of water. One should of course bear in mind that in this area steep banks of masonry, which include the quay-sides and the water fronts of canals in many Dutch and Flemish urban areas, are quite common. In spite of the comparatively large numbers of stands not facing a mass of water, one must conclude that the community is hygrophilous, because mural sites other than brick-work water fronts include parts of walls alongside leaking rainpipes or drains, low walls with a considerable inflow of capillary water, walls periodically soaked when streets are cleaned by spraying, walls of greenhouses and gardens receiving sprayed water, and walls built up against bodies of earth and raised earth walls (gardens, vineyards, dikes, cemeteries, etc.). The same applies to the seemingly "dry" walls of area 131, usually occurring in localities not so far away from area 120. In S.W. England, where the higher atmospheric humidity and precipitation plays a considerable role, there are indeed "dry" walls supporting stands of the community, but such cases often include the walls of narrow basement stories between the lowest stories or cellars and the streets, where desiccation is not very important, and wet cleaning of the wall surfaces takes place. Such basement walls have been found at Exeter, but also elsewhere (Edinburgh, Scotland).

Phyllitis scolopendrium is rather common in mural vegetation in S. England. In area 120 it is of rare occurrence, but if occurring on walls, it nearly always appears in the community under discussion, as it does in area 131 where the species was only recorded on rare occasions. In area 111 transitions towards the community of *Asplenium trichomanes* and *A. adiantum-nigrum* are found, but the link is, in this area, chiefly formed by the community of *Sagina procumbens* and *Asplenium trichomanes*. There are also transitions towards the Atlantic *Parietaria judaica* community, to all other types of mural vegetation found in the area (the resemblance to vegetation with *Soleirolia* being particularly noteworthy, see 6.10), and, in not very steep sites, to the non-mural *Sagino-Bryetum argentei* and *Plantagini-Lolietum*.

The community is only rarely developed on S.-facing walls and even the number of S.-facing sites standing over a mass of water is lower than that of similar ones with different exposures. The most favourable exposure is apparently to the N., particularly in area 131.

As an element of mural vegetation, *Pteridium aquilinum* is almost entirely restricted to the community discussed in this paragraph, in many cases being represented by its f. *umbrosum* (Borb.) J. Schmidt. It remains to be seen whether this form has any taxonomic significance the more so because wall-dwelling specimens hardly ever attain a length of more than a few decimeters and, accordingly, remain sterile as a rule. Although the bracken does not attain optimum development on walls, its frequent occurrence does not justify its reputation as a calcifuge, suggested in many publications (e.g. LAWALREE, 1950). In point of fact the species is regularly encountered on lime-rich soils in W. Europe and elsewhere, as was pointed out in 1933 by LITARDIERE, who estimated the pH of the rhizosphere of bracken plants at Versailles and in Corsica and measured pH 8.3 and 8.4, respectively. Bracken prefers walls rich in humus. The natural habitat of this tall fern is disturbed areas in forests, whereas in eu-atlantic regions particularly it is found in some other biotopes (especially heath moors). It is often found on deforested patches in woods in association with *Chamaenerion angustifolium* and, to my mind, its stands should be ranked in Braun-Blanquet's system among the *Epilobietea angustifolii* rather than among the *Quercetea robori-petraeae* to which it is usually referred (see e.g. WESTHOFF, 1966). Bracken is not restricted in its distribution to forests on poor soil types, and may also be found in vegetation of recently deforested patches in ravine forests on calcareous soils with *Fraxinus excelsior* in N. France (compare DURIN, 1955). The habitats are nearly always "secondary" environments, the disturbances not necessarily always being of an anthropogenic nature. Conceivably, the original natural habitat was constituted by the small open places caused by a fallen tree in a forest. Occasionally the bracken occurs in great numbers after an invasion by birds has disturbed a natural succession series (see WEBB & GLANVILLE, 1962) e.g. by the enrichment with nitrogen compounds. Both the structure and the ecology of stands with *Pteridium aquilinum* point to a separate vegetational syntaxon, the *Pteridietum aquilini.*

In England, the neophyte *Buddleja davidii* was represented in seven relevés of the *Filici-Saginetum.* The species was first introduced around 1890 as a garden plant. It has been growing in abundance on rubble heaps on sites bombed during World-War II in the London area (see KENT, 1960) and was also recorded in mural vegetation in Brittany and Normandy. It is rarely encountered as an escape in the Netherlands, where it does not behave like a neophyte. This alien seems to thrive on substrata rich in lime and in nitrogen. Its occurrence in stands of the *Filici-Saginetum* seems to indicate a certain advanced successional stage, especially found on strongly decomposed walls or on wall tops, but it is not clear to me to what vegetation type this may ultimately lead. Presumably *B. davidii* can form scrub vegetation with the equally nitrophilous *Sambucus nigra,* a species often encountered in similar mural habitats in the W. European lowlands. *Buddleja davidii* is not rare in the region of Lake Como where it occurs along stream beds (SUTTER, 1962) and also behaves like a neophyte. A third species which is almost entirely restricted in its occurrence in the community to mural stands in eu-atlantic climates is *Catapodium rigidum.* These species, though rather exclusive, do not have a sufficiently high degree of presence to serve as differentiating taxa for a separate eu-atlantic syntaxonomic unit and the same applies to *Asplenium adiantum-nigrum* when appearing in the community under discussion (and possibly indicating transitional cases towards the community with *A. adiantum-nigrum* and *A. trichomanes*). A species that deserves special attention in area 120 is *Polystichum falcatum,* an escape from cultivation repeatedly appearing in mural vegetation in Dutch towns, always in stands of the *Filici-Saginetum* and in many instances occurring with *Phyllitis scolopendrium.* This is most striking because the latter is by no means common in the Netherlands. In a single case the site consisted of glass-covered air-shafts of the cellar of an old building (Mariënhof, Amersfoort) with a constantly high relative humidity; all other relevés are from brickwork walls of canals and a stone dam through a ditch. *Phyllitis scolopendrium* and *Polystichum falcatum* have a similar leaf texture which may be indicative of more or less identical environmental prerequisites for their occurrence. In Holland, and also in Flanders (e.g. at Courtray), they are usually dependent on the immediate vicinity of some open water. *Polystichum falcatum* was represented in 9 relevés (i.e. in 9% of the stands). It has also been found as an escape outside the Low Countries viz. in France at Lingostière (Alpes-Maritimes) in 1937, on a wall at Moissac (Tarn-et-Garonne) in 1940, and later on a wall at Nice, in Switzerland at Brissago (Ticino), also on a wall (DHIEN, 1964); and in England (Cheltenham, Gloucestershire) "it is naturalised in some quantity, growing together with *Phyllitis scolopendrium*" (Milne-Redhead in WALLACE, 1968). In Ohio (U.S.A.) the species has also been found growing on a wall (GOSLIN, 1958).

The natural area of distribution of *Polystichum falcatum* includes S.E. Asia and Japan, the Sandwich Islands, E. Africa (Kilimanjaro) and S. Africa. It remains to be seen whether it will ultimately become a neophyte in W. Europe, because it is not very frost-resistent and a truly spontaneous increase in number at a mural site has only happened on few occasions (Amsterdam, Monnikendam); in all other cases one or two specimens were present which maintained themselves for a number of years but sometimes suddenly disappeared.

In the community the following additional differentiating species appear: *Dryopteris carthusiana, D. dilatata* (9 stands) and *Epilobium parviflorum.*

Special mention must be made of a variant with *Spergularia rupicola* and *Sagina maritima,* which was, for instance, observed on several walls at Penzance (Cornwall). This variant is not restricted to quay-side walls in sea-ports, but it is exclusively found within a narrow coastal strip of a few kilometers wide. In three relevés from Devon and Cornwall the following taxa were consistently represented: *Spergularia rupicola, Asplenium adiantum-nigrum, Phyllitis scolopendrium, Dryopteris filix-mas, Polypodium vulgare, Pteridium aquilinum, Poa annua, Sagina procumbens* and *Sonchus oleraceus. Buddleja davidii, Sagina maritima, Epilobium montanum, Chamaenerion angustifolium, Streblotrichum con-*

volutum and *Barbula unguiculata* were recorded twice, and *Asplenium trichomanes, Athyrium filix-femina, Soleirolia soleirolii, Antirrhinum majus* and *Tortula muralis* were recorded once. In several English and French Channel ports the community was fragmentarily developed on quay-side walls.

6.9.1.2 Vegetation with Sagina procumbens and Asplenium trichomanes

Stands of the community of *Sagina procumbens* and *Asplenium trichomanes* are indubitably best developed in the eu-atlantic belts of Europe. Table 39 shows a summary of the relevés and Table 41 the distribution and exposures.

Table 41 Geographical distribution and exposure of stands of the community of Sagina procumbens and Asplenium trichomanes

exposure:	1	2	3	4	5	1w	2w	3w	5w	t	t%
geographical region:											
111-112	5	2	3	3	1	1	-	1	-	16	52
120	3	2	-	2	-	2	1	-	1	11	35
131-132	-	1	-	2	1	-	-	-	-	4	13
t	8	5	3	7	(2)	3	1	1	(1)	28(+3)	-
t%	28	18	11	25	-	11	4	4	-	-	100

The most favourable exposition is to the north and the most unfavourable to the south. This preference is much clearer than in the previously discussed community. South-facing stands were, moreover, exclusively found in the eu-atlantic areas. In one case (no. 6) the site was a wall alongside a gutter on which grew *Adiantum capillus-veneris, Hygroamblystegium tenax* and *Rhynchostegium murale.* This moss is not common in the S. of England. *Buddleja davidii* was noted three times in stands in S. England. The best differentiating species are (apart from *Asplenium trichomanes*): *A. adiantum-nigrum, Epilobium adnatum, E. lanceolatum* and *Bryum murorum.* The abundance of *Phyllitis* and of *Polypodium vulgare* is striking. The majority of these species is rare or entirely lacking in areas 120, 131 and 132, where an inops form is developed or a variant in which *Poa compressa, Chelidonium majus, Erigeron canadensis* and *Bryoerythrophyllum recurvirostre* appear (mainly in 131). The variant forms transitions towards the community of *Sagina procumbens* and *Gymnocarpium robertianum,* the community with *Asplenium trichomanes* towards allied communities with *Asplenium trichomanes* and of course towards other mural communities with *Sagina procumbens.*

The average number of species per stand is 15. This number is higher than in the community of *Sagina procumbens* with *Aspidiaceae,* but the total coverage is usually somewhat lower than it is in the latter.

6.9.1.3 Vegetation of Sagina procumbens with Gymnocarpium robertianum and/or Cystopteris fragilis

Stands of vegetation of *Sagina procumbens* with *Gymnocarpium robertianum* and/or *Cystopteris fragilis* are not known from the eu-atlantic zones of Europe and they are presumably of more common occurrence as one moves farther to the east. Distribution and exposure are shown in Table 42.

Table 42 Geographical distribution and exposure of stands of the communities of Sagina procumbens and Gymnocarpium robertianum and/or Cystopteris fragilis

exposure:	1	2	3	4	5	1w	2w	3w	4w	t	t%
geographical region:											
120	-	-	-	-	1	-	2	1	2	6	21
131	3	-	-	4	-	1	1	-	1	10	36
200	2	1	-	1	1	1	-	-	-	6	21
400	2	-	-	2	-	2	-	-	-	6	21
t	7	1	-	7	(2)	4	3	1	3	26(+2)	-
t%	27	4	-	27	-	15	11	4	11	-	99

Areas 200 and 400 have not been very exhaustively studied and the community may be quite common there. Again the poor representation at S.-facing sites and the prevalence of N.-facing ones is evident. W.-facing stands predominate over E.-facing ones. It is remarkable that it is exactly in the most Atlantic area 120 that the combination is especially found on walls near water, whereas one would expect that taxa with a more continental distribution are not so much dependent on a constantly high relative humidity in W. Europe. It must be pointed out in this connection that the "dry" walls of central and E. Europe are usually situated in places where relative humidity is "unduly" high for those regions. This can be deduced from the fact that stands of the community, sometimes even fine examples, were found growing on walls rising up from a river in Karlovy Vary (Czechoslovakia) and in other places in that country and in Austria (which stands could, unfortunately, not be studied in detail). Dr. H. Sukopp kindly placed a relevé of such a stand from W. Berlin at my disposal.

Gymnocarpium robertianum is especially known from slopes of limestone debris and from open places in escarpment forests on rich soils. This fern exhibits some evident morphological adaptations to drought, viz. a glandular pubescence and mechanical tissue. According to HEGI's Flora, the species can grow on dry, even S.-exposed, slopes. However, in other publications (see e.g. KUHN, 1937) it is reported to be sciophilous, and LIPPERT (1966) mentions its occurrence on N.-facing slopes of rock debris and gravel or, in rock crevices, also on S.-facing slopes provided its root-system remains in the shade. Cohabitation with *Cystopteris* has been noted by a number of workers (e.g. TÜXEN, 1937); in the relevés of mural vegetation the combination was recorded eight times (i.e. in 28% of the stands). Now and then *G. robertianum* was found in association with *G. dryopteris* and twice their hybrid ($\times$ *G. hybridum*) was recorded (in Amsterdam and at Lautaret, Hautes-Alpes, France).

The few finds of *Bryum murorum* in area 120 (and these always in association with *Phyllitis scolopendrium*) indicate transitional cases towards the subordinate syntaxon described in the preceding paragraph. Transitions towards the community of *Sagina procumbens* and *Lycopus europaeus* have also been noted e.g. on walls facing the river Moldau in Prague.

Differentiating species of the subordinate syntaxon are (apart from *Gymnocarpium robertianum* and *Cystopteris fragilis*): *Poa compressa, Leptobryum pyriforme, Bryoerythrophyllum recurvirostre, Funaria hygrometrica* and *Encalypta streptocarpa.* A variant with *Distichium capillaceum* is especially found in high-montane regions: in the French Alps (Lautaret), in the Tyrol (Gschnitz), in N. Bohemia (Chribská, no. 9).

The community as a whole resembles the community of *Asplenium trichomanes* and *Cystopteris fragilis* mainly in its floristic composition (compare the relevé under no. 7). It is striking that the stenocoenosis of *Leptobryum pyriforme* and *Bryoerythrophyllum* is sometimes supplemented by *Tortula marginata* (compare 6.14).

Special mention must be made of a relevé from Merano (Trentino, Italy):

Nr. 66082, 29-4-1966, Merano, wall alongside the Via Roma opposite no. 66; exposition NW; inclination 86°; CH 12%; CM 2%.

Asplenium ruta-muraria	2m	Arenaria serpyllifolia	+r
,, trichomanes	2m	Poa pratensis	+r
Cystopteris fragilis	2m	Sonchus oleraceus	+r
Sagina procumbens	2m	Pimpinella magna	r
Taraxacum (3 spec.)	1b	Betula verrucosa	r
Sedum dasyphyllum	1p	Larix decidua	(r)
Poa annua	1p	Eurhynchium pulchellum	1a
Dryopteris filix-mas	+p		
Plantago major	+p	Amblystegium serpens	+p
Oxalis cernua	+p	Barbula vinealis	+p

This stand is intermediate between the community of *Sagina procumbens* with *Cystopteris,* and the one of *Asplenium trichomanes* with *Sedum dasyphyllum.* Possibly stands of this type (with *Sagina procumbens* and *Sedum dasyphyllum*) occur more often in area 310.

A transition towards the community of *Asplenium viride* and *Linaria cymbalaria* has been discussed in 6.7 (see Table 36, no. 3).

6.9.1.4 Vegetation with Sagina procumbens and Lycopus europaeus

Stands with *Sagina procumbens* and *Lycopus europaeus* are only found alongside water and they are quite common on walls of canals in towns in area 120 (compare Table 43).

Table 43 Geographical distribution and exposure of stands of the community of Sagina procumbens and Lycopus europaeus

exposure:	1w	2w	3w	4w	5w	t	t%
geographical region:							
111	1	2	-	-	-	3	5
120	20	11	14	12	1	58	91
131	-	1	1	-	-	2	3
400	1	-	-	-	-	1	2
t	22	14	15	12	(1)	63(+1)	-
t%	35	22	24	19	-	-	100

The preference for N.-facing walls is not very pronounced and in area 120 the community is also common on S.-facing walls, which is not at all surprising, considering that the stands are usually developed at the water-level or just above it, where the dash of waves and capillary suction maintain a moist substratum. Often a zonation is present on canal sites, with a zone of the *Lycopus* community below a stand of the *Sagina procumbens-Aspidiaceae* community. The zone with *Lycopus* is usually only a few decimetres in height. The sites are characterised by a very steep gradient in the moisture content of the substratum, wave-dash and fluctuation of the water-level sometimes complicating the situation. In this environment with considerable changes across small distances, elements of the *Bidention* and of the *Agropyro-Rumicion crispi* are at home, and the character of a community of a disturbed habitat is enhanced by the frequent admixture of especially species of the *Filipendulion* and the *Magnocaricion* which usually appear in secondary environments. In principle all species of these alliances (which are not only floristically but also physiognomically and ecologically well characterised) can serve as differentiating species against other types of mural vegetation, but in particular *Lycopus europaeus, Rorippa sylvestris, Epilobium roseum, Ranunculus sceleratus, Rumex crispus, Scutellaria galericulata, Galium palustre, Alnus glutinosa, Agrostis stolonifera, Holcus lanatus* and *Angelica sylvestris. Lycopus europaeus* (represented in 83% of the stands) is a species appearing in *Bidention* communities as well as *Magnocaricion* and *Filipendulion* communities. *Rorippa sylvestris, Ranunculus sceleratus* and *Epilobium roseum* are *Bidention* species (natural habitat: a substrate rich in nitrate subjected to appreciable changes in the moisture content). *Rumex crispus* is a characteristic element of the *Agropyro-Rumicion crispi* (natural habitat: sites subjected to considerable and usually aperiodical environmental changes). *Scutellaria galericulata* and *Galium palustre* are *Magnocaricion* species (natural habitat: secondary environments on wet ground), and *Angelica sylvestris,* finally, is a constituent of the *Filipendulion* (natural habitat: secondary environments on soils with a changing moisture content). The majority of these species appear in other alliances also: *Galium palustre,* for instance, in the *Bidention*, and the species of damp hay-fields *Agrostis stolonifera* and *Holcus lanatus* in the *Filipendulion. Alnus glutinosa* seldom attains a height exceeding a couple of feet, but in a few places on not so steeply inclined banks stands were observed which greatly resemble a fragmentary, young alder carr and presumably represent the potential succession. The almost complete absence of species of willow is striking; it is not clear whether diaspore size is prohibitive or if the environment is not suitable for *Salix. Betula pubescens* was repeatedly observed in the stands. Dozens of other species of the alliances mentioned have been recorded in the community, including *Thelypteris palustris.*

In S. England (Salisbury) *Adiantum capillus-veneris* was once recorded. A type of stand in which *Angelica archangelica* and *Parietaria judaica* appear, deserves special mention. This was encountered several times in the tidal river estuaries of W. Holland (Dordrecht, Kralingse Veer) just above the high-water line.

The steep ecological gradient renders the selection of a homogeneous experimental surface virtually impossible. Theoretically it would be a single line parallel to the water-level and the relevés must be considered in this light. In reality the various species have a place of their own somewhere along the gradient, *Rorippa sylvestris* and *Ranunculus sceleratus* growing closest to the water-level, *Matricaria maritima* ssp. *inodora, Rumex obtusifolius* and *R. crispus* usually a little higher up, *Lycopus* and *Scutellaria* appearing still higher as a rule, and only at a higher level again such species as *Filipendula ulmaria, Valeriana officinalis* and *Agrostis stolonifera* being found. In a great many sites there is so much "telescoping" of the zonation that distinct horizontal patterns can hardly be distinguished.

The community is connected with that of *Sagina procumbens* and *Aspidiaceae* by intermediate stands, but transitions towards other communities with *Sagina procumbens* are much rarer, only those towards the combination of *S. procumbens* with *Cystopteris* being found a little more often. On a few occasions transitions were noted towards the W.-European *Parietaria judaica* communities, chiefly in Flanders.

6.9.2 The Sagino-Bryetum argentei

Mural vegetation with *Sagina procumbens* and *Poa annua* but without typical wall-dwelling taxa can be interpreted as an impoverished expression of the *Filici-Saginetum*, or, alternatively, as special cases of the *Sagino-Bryetum argentei,* which is essentially a community of trampled habitats. As a rule the stands contain species of moist habitats belonging to the groups enumerated in the preceding paragraph. The floristic composition points to a syntaxon referable to the *Sagino-Bryetum argentei*. A few examples are shown in Table 44.

Table 44 Sagino-Bryetum argentei on brick walls

Nr C	1	2	3	4	5
Date	14-9	9-7	22-10	12-10	5-6
Year	64	66	63	63	62
Nr R	671	213	519	492	308
Ex	2	2	4	4	2
CH	10	20	25	15	15
CM	20	1	5	<1	17
Poa annua	2a	2m	2b	1p	2m
Sagina procumbens	2m	2a	2a	2m	2m
Bryum argenteum	+p	2m	1p	1p	-
Ceratodon purpureus	1p	-	2m	2p	-
Bryum capillare	+p	-	2m	2p	-
Marchantia polymorpha	2b	-	-	-	-
Lycopus europaeus	-	2a	1b	2a	1b
Bryum caespiticium	-	2m	+p	-	-
Plantago major	-	-	+r	1b	+r
Taraxacum (div.) spec.	-	-	+p	+p	+r
Rumex crispus	-	-	1p	1b	-
Epilobium roseum	-	-	1p	+p	-
Sonchus oleraceus	-	-	1p	+p	-
Agrostis stolonifera	-	-	1a	-	-
Tussilago farfara	-	-	-	+a	+r
Rumex hydrolapathum	-	-	-	-	1b
Peucedanum palustre	-	-	-	-	1a
Alnus glutinosa j	-	-	-	-	1p

In addition in 2: Glechoma hederacea +r; in 3: Funaria hygrometrica +p; in 4: Matricaria matricarioides, Epilobium montanum, Senecio vulgaris, Rumex conglomeratus, Sonchus asper +p; Sium erectum +r; in 5: Poa pratensis, Angelica sylvestris, Festuca rubra, Cicuta virosa, Carex otrubae, Erigeron canadensis, Clematis vitalba j +p; Stachys palustris, Plantago lanceolata, Elytrigia repens, Dactylis glomerata, Poa trivialis, Betula pubescens +r.

LEGEND

1. Stein (Limburg, Netherlands), damp wall basis of farm in Kerkstraat
2. Leiden (Zuid-Holland, Netherlands), canal wall of Lage Rijndijk below nr 9a.
3. Schiedam (Zuid-Holland, Netherlands), canal wall of Lange Haven in front of nr 100.
4. Amsterdam (Netherlands), sloping, jointed basalt bank of Ruysdaelkade in front of nr 37. Inclination 30°.
5. Apeldoorn (Gelderland, Netherlands), canal wall of Apeldoorn canal near the E 8 through-way.

6.9.3 A survey of mural vegetation with Sagina procumbens and Poa annua

All communities discussed under 6.9 can, in conjunction with mural vegetation with *Soleirolia*, be classified in the floristic system as follows:

A. Alliance: *Cymbalario-Asplenion*
   Association group of Sagina procumbens and Tortula muralis:
   1. Association: Linario-Soleirolietum
   2. ,, Filici-Saginetum
   3. aspidietosum
   4. lycopctosum
   5. asplenietosum
   6. gymnocarpietosum

B. Alliance: *Polygonion avicularis*
   1. Association: Sagino-Bryetum argentei
   2. lycopetosum

The relationships between these syntaxa and communities of trampled sites is discussed in detail in 8.3. The coefficients of affinity according to Poore are as follows (alliance A):

(3) in respect of (4): 0.64
(3) ,, ,, ,, (5): 0.76
(3) ,, ,, ,, (6): 0.71
(4) ,, ,, ,, (5): 0.58
(4) ,, ,, ,, (6): 0.54
(5) ,, ,, ,, (6): 0.66

Communities (3) and (4) have been referred to the *Sagino-Bryetum argentei*, subassociation with *Ceratodon purpureus*, fern-rich variant, subvariant with *Asplenium*, and subvariant with *Ranunculus sceleratus*, respectively, by VAN KONINGSDAAL & REYNDERS (1956), the poorest stands being taken as the typical subvariant by these authors.

In the communities with *Sagina procumbens* the contribution of the hygrophytes is always rather substantial (scores: U 33-44%, W 31-59%). In stands with *Lycopus europaeus* the representation of the telmatophytes is appreciable (score: U 11%, W 25%; in all other communities with *S. procumbens* these figures are 0.0-0.4% and 0.0%, respectively) and the number of hemixerophytes is low (score: U 4%, W 5%; in all other communities: U 10-17% and W 14-31%). In stands with *Gymnocarpium robertianum* the geophytes are relatively numerous (their score being: U 10%, W 22%, in all other stands these figures are: U 2-3% and W 1-3%, respectively), the therophytes are poorly represented (2% as against 6-10% in the other stands) and the number of chamaephytes is much lower (1% as against 18-24% in all other stands), so that the structure of the stands with *G. robertianum* is rather aberrant.

In the *Filici-Saginetum* the score is relatively low for all normally chasmophytic, rock-, or wall-dwelling forms (viz. 8-20% U and 16-31% W), especially in stands with *Lycopus europaeus* (8% U, 16% W). These figures are always much higher in all other types of mural vegetation and especially the quantitative contribution may be as high as 50% and over. On the other hand, especially in stands with *Aspidiaceae*, the representation of forest- and scrub elements is relatively more important (scores: U 12%, W 11%, as against U 6-10% and W 2-4% in all other stands of the *Filici-Saginetum*). The proportion of riparian species and helophytes is, naturally, high in all stands with *Lycopus europaeus* (the score being: U 11%, W 18%).

6.10 Vegetation with Soleirolia soleirolii

Soleirolia soleirolii (syn. *Helxine soleirolii*) is a native of Corsica, Sardinia, Capri and the Balearics, which is often cultivated as a greenhouse- or pot plant and has run wild in a number of localities in England and France. It became firmly established in S. England and in Brittany, where it behaves like a neophyte. Like the other European *Urticaceae*, *Soleirolia* is nitrophilous. Little is known of the ecology of its natural habitat in its original area other than the statements of BRIQUET (1910) and CONTANDRIOPOULOS (1957) that its grows on rocks and on walls in damp places below 600 m alt. I recorded the species in Corsica in 1956 in the immediate vicinity of human settlements on humid rocky cliffs and on humid walls (e.g. near Luri and Sisco), in association with species also found in cohabitation with it in W. Europe such as *Poa annua, Sagina procumbens* and *Tortula muralis*. At that time no relevés were made.

In Devon and in Cornwall I recorded *Soleirolia* on walls in Totnes, Topsham, Charleston, Paignton, St. Blazey, Falmouth, Penryn, Mable, Helston, Mullion and Goldsithney. I also found it on a wall in Edinburgh (Scotland) and at Dinan (Brittany). In the Channel Islands it is said to be common on walls (WALLACE, 1949). Stands with *Soleirolia* clearly exhibit a relation with the *Filici-Saginetum* and, to a less extent, with the eu-atlantic wall-dwelling communities with *Asplenium trichomanes* and *Phyllitis scolopendrium,* but the absence of *Ceterach officinarum* is striking and the contribution of *Asplenium trichomanes* and *A. ruta-muraria* is rather small. Several hypnid mosses, such as *Brachythecium rutabulum, B. salebrosum* and *Rhynchostegium confertum* often attain a high degree of coverage. In contrast to the *Filici-Saginetum,* the contribution of *Bryum argenteum* and *B. caespiticium* is not so important. These differences give the stands with *Soleirolia* a special aspect which is strengthened by the high degrees of coverage attained by *Soleirolia,* which in 60% of the recorded cases exceeds 40% and sometimes approaches 100%. The average number of species of the stands is 13 and only in three relevés the number of species was below 9 or over 17. This constancy of composition is also suggestive of a separate syntaxonomic unit. Table 45 gives an idea of the community. In the calculations, three relevés were taken into account with a low degree of coverage of *Soleirolia* ($< 5\%$) and presumably representing initial stages of a succession. If these three are not taken into account in calculations concerning *Soleirolia,* the mean coverage percentage of this species becomes as high as 47% whilst that of the other species hardly changes. Of the 17 relevés the following exposures were noted:

N to NE	10	times	or	59%
ENE to E	2	,,	,,	12%
Overshadowed	2	,,	,,	12%
S to W	3	,,	,,	18%

In the majority of the cases the stands occur on low walls or on the lower parts of walls, often immediately above the base, where, also in S.-facing stands, the humidity of the substratum may be relatively high. *Soleirolia* seems to tolerate an inflow of topsoil and humus particles better than *Parietaria judaica* and the latter only appeared twice in the stands with *Soleirolia.* On the other hand the moisture requirements of *Soleirolia* are higher. *Soleirolia* also grows on pavements and roads, usually in the immediate vicinity of its mural stands (see 8.2.6) and, singularly, often associated with species not usually found growing in cracks of horizontal paved surfaces such as *Linaria cymbalaria* and *Epilobium montanum.* In Mullion (Cornwall) *Soleirolia* was found growing in a lawn, together with e.g. *Sagina procumbens.* In the prevailing oceanic climate these Dicots competed succesfully with the grasses.

Another taxon conspicuous in the list of species of the community is *Epilobium nerterioides,* originally coming from New Zealand but now naturalised in England, where it appeared in relevés made at Falmouth and Penryn and seems to be wide-spread on walls in the Lizard peninsula. According to DAVEY (1961) this species occurs in various communities but especially "as a coloniser of open habitats, on moist and well-drained substrata of stony or gritty texture". It was often seen in association with *Sagina procumbens,* also in the Netherlands where it occurs subspontaneously in the Amstelveen area (near Amsterdam) on peaty soil. Professor A. D. J. Meeuse (Amsterdam) collected *Epilobium nerterioides* on a wall without *Soleirolia* at Berwick-upon-Tweed in N. England (Border).

Table 45 Community of Soleirolia soleirolii and Linaria cymbalaria

Nr C	1	2	3	P 17st	U W	MGT	MGP
Date	4-8	4-8	28-8				
Year	64	64	64				
Nr R	437	442	302b				
Ex	2	1	1				
CH	70	70	60				
CM	10	0	15				
Soleirolia soleirolii	4b	4a	4a	100	1a-4b	39	39
Linaria cymbalaria	1a	+p	1b	82	1a-2b	4	5
Taraxacum (div.) spec.	+p	1a	+p	82	+r-1a	0.4	0.5
Poa annua	1p	2a	1p	76	+p-2a	2	2
Epilobium montanum	+p	+p	+p	65	+p-1p	<0.1	0.2
Tortula muralis	+p	-	+p	47	+p-2m	0.5	1
Sonchus oleraceus	-	+p	+r	47	+r-1p	<0.1	0.1
Sagina procumbens	-	-	-	35	+r-2a	0.7	2
Phyllitis scolopendrium	1b	-	-	35	+p-1b	0.3	1
Asplenium adiantum-nigrum	2a	-	-	30	+p-2a	1	3
Dryopteris filix-mas	-	+p	-	30	+p-2a	0.7	3
Polypodium vulgare	+r	-	-	30	+r-2a	0.5	2
Brachythecium rutabulum	2a	-	-	24	+p-2a	0.8	3
Homalothecium sericeum	-	-	1b	24	+r-2a	0.7	3
Barbula vinealis	-	-	-	24	+p-2m	0.4	2
Ceratodon purpureus	-	-	2m	24	+p-2m	0.4	2
Bryum capillare	-	-	2m	24	+p-2m	0.2	1
Convolvulus arvensis	-	+a	+p	24	+p-+a	0.2	0.8
Hypnum cupressiforme	-	-	+p	24	+p	<0.1	0.1
Brachythecium salebrosum	-	-	2a	18	2a-3a	3	19
Centranthus ruber	-	-	-	18	+p-+b	0.3	2
Asplenium ruta-muraria	-	-	+p	18	+p-2m	0.2	1
Athyrium filix-femina	-	-	-	18	+p	<0.1	0.1
Epilobium roseum	-	-	+p	18	+p	<0.1	0.1
Poa pratensis	-	+p	-	18	+p	<0.1	<0.1
Buddleja davidii	-	-	-	18	+r-+p	<0.1	0.1
Bryum murorum	+p	-	-	18	+r-+p	<0.1	0.1
Rhynchostegium confertum	-	-	-	12	2m-2a	0.6	5
Barbula convoluta	-	-	-	12	2m	0.3	3
Erigeron mucronatus	+r	-	-	12	+r-1b	0.2	2
Plantago major	-	1a	-	12	1a	0.2	2
Bryoerythrophyllum recurvirostre	2m	-	-	12	2p-2m	0.2	2
Barbula unguiculata	2m	-	1p	12	1p-2m	0.2	2
Amblystegium serpens	2m	-	-	12	1p-2m	0.2	2
Asplenium trichomanes	-	-	2m	12	1p-2m	0.2	1
Linaria purpurea	-	2a	-	12	+p-2a	<0.1	0.8
Epilobium parviflorum	-	1p	-	12	1p	<0.1	0.4
Bryum caespiticium	+p	-	-	12	+p-1p	<0.1	0.3
Epilobium nerterioides	-	+p	-	12	+p-1p	<0.1	0.3
Bryum argenteum	-	-	-	12	+p	<0.1	0.1
Umbilicus rupestris	-	-	-	12	+r-+p	<0.1	<0.1
Parietaria judaica	-	-	+p	12	+r-+p	<0.1	<0.1
Senecio vulgaris	-	-	-	12	+r	<0.1	<0.1
Lunularia cruciata	-	2a	-	1×	-	-	-
MNS - MCH - MCM		13		-	44	-	10

LEGEND

1. St. Blazey (Cornwall, England), garden wall alongside the road to St. Austell.
2. Falmouth (Cornwall, England), wall alongside Melvill Road near Sea View Road.
3. Charleston (Devon, England), wall alongside road to Frogmore.

6.11 Vegetation with Chrysanthemum parthenium

In a number of stands of mural vegetation in the areas 111, 120 and 131 *Chrysanthemum parthenium* occurs, usually in a combination with species reminiscent of the *Filici-Saginetum* but in which *Sagina procumbens* is lacking and the contribution of *Poa annua* and *Bryum caespiticium* is relatively unimportant. Most of the stands are not unequivocally referable to one of the syntaxa previously discussed and show a sufficient mutual relationship that they can be segregated as a separate community.

Table 46 Geographical distribution and exposure of stands of the community of Chrysanthemum parthenium and Linaria cymbalaria

exposure:	1	2	3	4	5	1w	2w	4w	t	t%
geographical region:										
111	2	-	2	-	-	-	-	-	4	36
120	-	-	-	1	1	1	-	1	4	36
131	-	-	1	-	-	-	1	1	3	27
t	2	-	3	1	(1)	1	1	1	10(+1)	-
t%	20	-	30	10	-	10	10	20	-	100

There is no clear preference to a certain exposure, but the community was always encountered in more or less sheltered and not very dry sites which were, upon the whole, rich in nitrogen-containing soil particles. Presumably the nitrophilous character is to be ascribed to the fact that in 4 out of the 11 cases the *Chrysanthemum parthenium* community was growing on not quite perpendicular walls (angle of inclination $< 85°$) and may prefer an inclination smaller than 90°.

The composition is shown in Table 47. It changes to some extent from W. to E., in area 111 e.g. *Phyllitis scolopendrium, Centranthus ruber,* and *Buddleja davidii* being represented, and in area 131 *Chelidonium majus, Poa angustifolia* and *Geranium robertianum.* However, the small number of relevés renders the recognition of subordinate units decidedly premature.

Apart from *Buddleja davidii, Sambucus racemosa, S. ebulus, Fraxinus excelsior* and *Acer pseudoplatanus* were recorded, but each of them only once. Their presence could point to a development in the direction of scrub vegetation related to ravine forests, but with a sprinkling of nitrophilous species.

Chrysanthemum parthenium is a garden escape, presumably native in S.E. Europe and Asia Minor, now naturalised in W. and central Europe and especially encountered at somewhat ruderal sites and on arable land. It can maintain itself in mural vegetation for many years on end in sites in Holland and W. Germany (and presumably also in England).

The average number of species per stand is 10, the highest values (18 and 14) each occurring once and the lowest number recorded being 7. The fluctuations are not excessive and this may be taken as a possible indication of the autonomous status of the community in respect of other types.

Table 47 Community of Chrysanthemum parthenium and Linaria cymbalaria

Nr C	1	2	3	P	U W	MGT	MGP
Date	1-7	30-10	1-8	*11st*			
Year	62	61	64				
Nr R	412	610	378				
Ex	3	4	2				
Inclination	89	90	85				
Geo R	131	120	111				
CH	12	15	40				
CM	s	5	30				
Chrysanthemum parthenium	1b	1a	2a	*100*	+r-2b	7	7
Linaria cymbalaria	+p	+r	2b	*82*	+r-5a	16	19
Asplenium ruta-muraria	2m	1b	-	*64*	+r-2a	2	3
Ceratodon purpureus	-	+p	2b	*55*	+p-2b	3	5
Bryum argenteum	+r	+p	+p	*45*	+r-2m	0.5	1
Sonchus oleraceus	-	+p	+p	*45*	+p-1a	0.3	0.7
Chelidonium majus	+r	-	-	*36*	+r-2b	2	5
Taraxacum (div.)spec.	+p	+p	+p	*36*	+p	<0.1	0.1
Tortula muralis	-	2m	-	*27*	+p-2a	1	4
Bryum capillare	-	-	2m	*27*	+p-2m	0.5	2
Poa angustifolia	1a	-	-	*27*	1p-2m	0.5	2
Poa pratensis	-	-	1p	*27*	+r-1p	<0.1	0.3
Poa annua	-	-	1p	*27*	+r-1p	<0.1	0.2
Geranium robertianum	-	-	-	*18*	+p-2b	2	9
Phyllitis scolopendrium	-	-	-	*18*	+p-2a	0.7	1
Dryopteris filix-mas	-	-	-	*18*	+p-1b	0.1	0.8
Hieracium lachenalii	-	-	-	*18*	+p	<0.1	0.1
MNS - MCH - MCM		*10*	-	*30*	-	*7*	

In addition in 1: Antirrhinum majus 1a; Poa pseudocompressa +p; Ficus carica, Capsella bursa-pastoris, Sambucus racemosa +r; in 2: Festuca ovina 2m; Tortula intermedia 1p; Poa compressa, Erigeron canadensis, Glechoma hederacea, Galium mollugo ssp. erectum +p; Dactylis glomerata, Panicum miliaceum, Artemisia vulgaris, Plantago lanceolata +r; in 3: Centranthus ruber 2a; Rubus ulmifolius +b; Epilobium roseum, Acer pseudoplatanus +a; Barbula vinealis 2a.

LEGEND

1. Aschaffenburg (Bayern, Western Germany), wall along side the river Main near the castle.
2. 's-Hertogenbosch (Noord-Brabant, Netherlands), city wall (Westwal) in front of nrs 40-44.
3. Exeter (Devon, England), Exe river wall alongside New Bridge Street.

In stands with *Chrysanthemum parthenium* anthropochory plays an important part, the representation of man-dispersed forms being 20% (U) and 21% (W), respectively. The contribution of ruderal taxa is exceptionally high viz. 24% (U) and 23% (W), respectively. However, species of forest and scrub vegetation also attain a fairly high score (U 9%, W 2%).

6.12 Other cormophytic vegetation on walls

In Chapter 7.4 the initial stands of *Linaria cymbalaria* and *Asplenium ruta-muraria* will be discussed and it will be pointed out that they may serve as the initial phases of a number of more or less diverse communities (as will also be clear from various passages in this chapter). The number of relevés of other communities is too low to justify more definite conclusions. The following are noteworthy:

1. A community with *Asplenium billotii*, recorded from Brittany and presumably optimally developed in the French Basque Country and in N.W. Spain (see Jovet, 1941, and Allorge, 1941).
2. A community with *Asplenium marinum*, also observed in Brittany. Chasmophytic stands with this species were studied to some extent in Brittany and in S. England and it appears that they are referable to the *Asplenietea rupestris*. The allocation of the community in question in the *Potentillion caulescentis* (Braun-Blanquet & Tüxen, 1952) can not be endorsed. There are at least two vicariating communities, the one on calcareous rocks (containing e.g. *Asplenium ruta-muraria, Phyllitis scolopendrium* and *Fissidens cristatus*) and the other on granitic substrata (with e.g. *Asplenium*

billotii, *A. adiantum-nigrum* and *Sedum anglicum*). Close to the sea *A. marinum* often occurs without accompanying species. Communities with *A. marinum* can be included in an association assembly with the above-mentioned communities in the centre of which the second could possibly be allocated to a higher chasmophytic syntaxon of acid rock formations. The very special place of these communities perhaps justifies a separate alliance to incorporate them (*Asplenion marini*). In Brittany (Fort de la Latte) *A. marinum* was found in mural vegetation with *Parietaria judaica* and *Daucus gummifer*, a combination also mentioned by ROUX & LAHONDERE, (1960) from Camaret-sur-Mer.

3. A community with *Asplenium septentrionale* and *A.×alternifolium* was found e.g. in Luxemburg (Moulin Bourscheid).

6.13 The classification of mural vegetation

Table 48 shows a matrix in which all species are included that were represented in at least 3% of the relevés. For all combinations a 2 x 2 divergency table was set up. The mathematical evaluation was carried out by means of an "Electrologica EX 8" computer of the "Mathematisch Centrum", Amsterdam.

The stronger the association, the smaller the probability (P). The following categories were distinguished:

P	<	0.01	(positive association X, negative association O)
P		0.01—0.05	(,, ,, x, ,, ,, o)
P	>	0.05—0.10	(,, ,, +, ,, ,, —)
P	⩾	0.10	(a blank)

The species were entered in the original matrix alphabetically. From this matrix a second version was made by placing the species with the lowest incidence of co-occurrence as far apart as possible and all other species in respect of these positions. This arrangement may not be taken as indicative of a close phytosociological relationship of species appearing in close proximity in the matrix, because one can not arrange the elements in a meaningful way along anything but a single co-ordinate axis. One must also consider whether the pooling of all the material from such a large geographical area is permissable. Generally speaking, one may conclude that the species with a low matrix number predominantly appear in stands with *Parietaria judaica* and those with a high number in stands with *Poa annua* and *Sagina procumbens*, etc. From the second matrix one can deduce that *Umbilicus rupestris* and *Sedum dasyphyllum* exhibit a fairly strong association with *Parietaria* communities. This result we may interpret as another justification of the classification of Mediterranean communities with *Asplenium trichomanes* in a higher syntaxon together with the *Parietaria* communities on floristic grounds, if the relatively small biomass of the species characteristic of stands with *Parietaria* and especially the different structure did not plead against such an arrangement.

In the matrix *Asplenium ruta-muraria* shows a negative association with *Parietaria judaica*. This would presumably not be the case if a matrix had been set up in which the W. and central European communities were included. The absence of the wall-rue in Mediterranean *Parietaria* communities is manifest and of such communities a relatively large number of relevés were made.

In the centre of the matrix a special group of species appears which occurs in various communities in central Europe, e.g. in stands with *Parietaria judaica,* with *Corydalis lutea,* or with *Asplenium trichomanes.* The group includes e.g. *Poa compressa, P. angustifolia, Geranium robertianum, Chelidonium majus, Hieracium lachenalii* and *Mycelis muralis* (a species not entered in the matrix).

Cormophytic mural communities can be classified in the class *Asplenietea rupestris* and the majority can be combined in the order *Tortulo-Cymbalarietalia* for which the following species are characteristic: *Asplenium ruta-muraria, A. trichomanes, Ceterach officinarum, Umbilicus rupestris, Sedum dasyphyllum, Hieracium amplexicaule, Linaria cymbalaria, Antirrhinum majus, Poa compressa, Tortula muralis* and *Barbula vinealis.* They are partly characteristic taxa of the class, and partly species also appearing in others orders. Two alliances can be distinguished viz.

1. The *Parietarion judaicae* with the principal characteristic species *Parietaria judaica, P. lusitanica, Centranthus ruber, Erigeron mucronatus, Phagnalon sordidum, Veronica cymbalaria, Capparis spinosa, Mercurialis annua* ssp. *huetii, Oryzopsis miliacea, Antirrhinum latifolium* and *Cheiranthus cheiri.* Many other differentiating species can be recognised in the matrix.
2. The *Cymbalario-Asplenion* with the following principal characteristic and differentiating species: *Asplenium* spec. div. (particularly *A. adiantum-nigrum*), and numerous other ferns such as *A. viride, Phyllitis scolopendrium, Polypodium* spec. div., *Dryopteris filix-mas,* and *Cystopteris fragilis,* and also *Corydalis lutea, Chrysanthemum parthenium, Soleirolia soleirolii, Epilobium montanum, E. lanceolatum, Sagina procumbens, Bryum murorum, Barbula unguiculata* and *Rhynchostegium murale.* In this alliance two groups of communities can be distinguished viz.
   a. Stands with predominantly *Asplenium trichomanes* and *Ceterach officinarum,* the community with *Corydalis lutea* also belonging here; if deemed necessary to be called the suballiance *Tortulo-Asplenion,* and
   b. Stands with *Sagina procumbens* and *Poa annua,* including stands with *Chrysanthemum parthenium* and with *Soleirolia,* which could be referred to as the suballiance *Tortulo-Saginion.* The communities of this group usually occur in sites richer in nitrogen and moister than those supporting stands of communities of (a).

6.14 Bryophyte communities

The study was predominantly concerned with communities of vascular plants, but all mosses present were always recorded. Especially from parts of the areas 120 and 131 about 100 relevés were obtained of stands of bryophytes which included only a few vascular plants or none at all. Another

150 relevés were made but not processed in the computer, and about 100 were available from students participating in the project. Somewhat surprisingly the data are inadequate for the delimitation of a number of typical mural communities. It is hoped that more evidence can be obtained to complete the study at some later stage. A number of communities could be recognised because sufficient data were already in hand; the results will not be documented *in extenso,* but only be given in the form of a summary.

The impression was gained that in many instances the bryophyte layer of a mural community with vascular plants represents an impoverished expression of bryophyte stands without vascular plants. Another peculiarity I noticed is that the bryophyte layers of stands which are closely allied from a floristic point of view and belong to the same community are not necessarily all of the same floristic composition. The phenomenon is most frequently observed in open stands e.g. in those with *Asplenium trichomanes.* In such cases there can hardly be any functional bonds between the vascular plants and the mosses present in the same stand. In communities with *Parietaria* a bryophyte layer is often poorly developed.

In the area studied the most wide-spread community, encountered in sites where the substratum contains enough nitrogen, is vegetation of *Tortula muralis, Bryum caespiticium, B. argenteum, B. capillare* and *Ceratodon purpureus.* This species combination may occur all by itself, but also in association with other species and may be considered characteristic of all communities containing such nitrogen-converting species as *Sagina procumbens* and *Poa annua* (in particular, of the *Tortulo-Saginion,* see 6.13).

Characteristic of a hard rocky substratum is the combination of *Grimmia pulvinata, Orthotrichum anomalum, O. diaphanum, Schistidium apocarpum* and *Tortula ruralis,* occurring on e.g. concrete and basalt walls. On roofs and roof-tiles of asbestos cement also *Tortula virescens* is sometimes present.

Barbula unguiculata is regularly seen on moderately damp walls, often accompanied by *Streblotrichum convolutum.* On marl walls and on other lime-containing, rapidly decomposing walls a community with *Barbula revoluta* develops in which usually also the two last-mentioned species appear. On walls built up against vineyards in Switzerland, *Barbula revoluta* was noted in combination with *Didymodon rigidulus* and *Grimmia pulvinata* var. *africana.* At drier mural sites *Barbula fallax* and *B. hornschuchiana* may be present, especially on horizontal ledges. On very dry walls of soft lime-stone, species of *Aloina* may be expected to occur. In places subjected to considerable changes in the moisture content of the substratum, especially in the more continental zone of Europe and in S. Europe, *Leptobryum pyriforme* and *Funaria hygro-*

metrica are found. On calcareous masonry the *Leptobryum* species constitutes a characteristic community in association with *Tortula marginata* and *Gyroweisia tenuis, Amblystegium varium* often appearing in later successional stages. In central and S. Europe, on very damp limestone the following combination may be found: *Eucladium verticillatum, Didymodon tophaceus, Gymnostomum aeruginosum* and *Pellia endiviaefolia*; in central Europe the combination of *Encalypta streptocarpa* with *Tortella tortuosa* is not rare, often found with *Ctenidium molluscum* and *Fissidens cristatus* in later successional stages.

Communities containing pleurocarpic mosses are mostly successional stages of communities with ultimately acrocarpic forms and are often found in places with "more favourable" water relations. Common and widespread are *Homalothecium sericeum* and *Hypnum cupressiforme* var. *tectorum*.

On walls built of calcareous substances, a community with *Oxyrrhynchium swartzii* and *O. schleicheri,* often associated with *Brachythecium salebrosum* and sometimes with *Fissidens taxifolius,* develops in damp situations, and on drier walls of this kind, especially in sites rich in nitrogen, a *Rhynchostegium murale* community usually develops. In places very rich in assimilable nitrogen compounds trivial species such as *Brachythecium rutabulum* are found. On damp and usually also more or less overshadowed walls, especially towards their base, communities with *Brachythecium velutinum* or *Rhynchostegium confertum* are often encountered. In very damp situations, particularly near the edge of a body of water, a community with *Leptodictyum riparium* is fairly often seen, but on walls of lime-stone in such places stands of *Conocephalum conicum* are usually developed. This species may cover large surface areas of the wall and is mostly associated with *Mnium punctatum, Oxyrrhynchium swartzii* and *Brachythecium salebrosum.* In S. Europe mural stands with *Scorpiurium circinatum* are of fairly common occurrence, and also communities with *Porella platyphylla, Neckera complanata* and *Homalothecium sericeum*; on moist walls and wall bases a community with *Plasteurhynchium meridionale* is found. In mural vegetation in E. Europe the combination of *Leucodon sciurioides* and *Orthotrichum lyellii* was repeatedly recorded. In stands with pleurocarpic mosses on calcareous substrata, *Amblystegium serpens* (particularly its var. *saxicola*) is a common element. Apart from *Conocephalum,* also other thallose liverworts are sometimes dominant on damp walls and wall bases, particularly *Marchantia polymorpha* and *Lunularia cruciata.* The latter is firmly naturalised in W. and central Europe.

Descriptions of communities more or less clearly resembling those enumerated in the aforegoing summary can be found in e.g. AMANN (1922), STODIEK (1937), KUHN (1937), WALDHEIM (1944), VON KRUSENSTJERNA (1945), SMARDA (1947), BARKMAN (1947b), HÜBSCHMANN

Table 49 Matrix of Bryophytes found on walls
See text under 6.14.

1	2	3	4	5	6	7	8	9	10	11	12	13	14	15	16	17	18	19	20	21	22	23	24	25	
	X	X	x	X	X		X	X	X		X	X			-	+									1 Bryum caespiticium
		X	x	X	X	X	x			X		+		x											2 Leptobryum pyriforme
			X		X	+	X	X	X				X												3 Amblystegium serpens
					+	X	X	X	X	+		x	X		+			+				-			4 Rhynchostegium murale
						X	X	x		x		X	x	X	X					+	+			o	5 Barbula unguiculata
						x		X	X		X	+			o			+	+						6 Funaria hygrometrica
								X	o	X	-	X	X	+	X										7 Bryoerythrophyllum recurvirostre
								X	X	x		X	X		X			X							8 Brachythecium rutabulum
									X		X	X	X	X	x	x	X	X	X			O			9 Bryum capillare
										o	X	X	X	o	O	X	X	x	x			O	O	O	10 Ceratodon purpureus
											o	X			X			x			x				11 Encalypta streptocarpa
												X			O	X	X	X	X	o		-	O		12 Bryum argenteum
													X	X	X	+	X	+	X	X	X	X	X	X	13 Tortula muralis
														X	X	X	X	X	X		+		X	X	14 Hypnum cupressiforme
															X			x		X	+			X	15 Streblotrichum convolutum
																	X	X	x	X	X		X	X	16 Homalothecium sericeum
																	X	X	X			x	+		17 Orthotrichum diaphanum
																		X	X		X				18 Grimmia pulvinata
																			X				x		19 Schistidium apocarpum
																				x					20 Orthotrichum anomalum
																								X	21 Bryum murorum
																								X	22 Tortula intermedia
																							X		23 Barbula acuta
																								X	24 Scorpiurium circinatum
																									25 Barbula vinealis

(1950, 1967), Poelt (1954), Nick-Navratil (1960), Pankow & Fischer (1965).

Table 49 shows a matrix designed according to the same principles as the one discussed in 6.13, but only those bryophytes are taken into account that had been recorded in at least 2% of the stands of cormophytic mural communities.

7 Succession of mural vegetation

The establishment of a vegetation cover on a wall is intimately associated with the advancement of the wall's decomposition (see 3.2.2) and this holds *mutatis mutandis* for the progress of succession, the final stage of the succession series being determined by the ultimate degree of extremity of the environment. In the case of an autogenic process of succession this means that the ultimate successional stage is dependent on the capacity of the plant cover to change the habitat. However, as is the rule, the succession series is not exclusively autogenic or allogenic and the decomposition is a function of both the plant growth and the environmental (especially the climatological) factors.

In practice we frequently encounter vegetation maintaining and perpetuating itself in the course of time (subclimax, *Dauergesellschaft*). In such cases the progress of decomposition proceeds so slowly that only after a very long time span (usually of several centuries) has it sufficiently advanced to permit a progression of the succession series. The time scale can be more or less deduced from vegetation on walls dating from the Roman era. The oldest walls that were never repaired or renovated are usually strongly decayed and the masonry has assumed the character of a ruin. Such an advanced decomposition is associated with a number of changes favouring the development of vegetation: a large surface for germination of diaspores; more room for root development; a greater amount of easily assimilable substances, both from decomposition products and from biotic metabolic processes; an accumulation of extraneous matter; a finer texture and better water economies; a lesser inclination; changed biotic influences; an increased settlement of organisms of various kinds. Generally speaking, the accessibility has increased and it may, moreover, have been improved by direct or indirect (cattle !) anthropogenic influences. On very decayed walls typical chasmophytic vegetation is almost non-existant and the obligatory epiliths are much reduced in numbers or have completely disappeared. In a dry (micro)climate, provided strong anthropogenic influences are lacking, the plant cover initially develops in the direction of vegetation characteristic of dry, open sites with e.g. species of *Sedum* such as *S. album* and *S. acre* (towards *Sedo-Scleranthetea*). Under human influence the development tends to lead rather to ruderal vegetation (towards *Artemisie-*

tea). In the long run shrubs can develop in all cases. In very humid sites in the W. European coastal plain e.g. on walls of canals, *Alnus glutinosa* is of common occurence on walls on which typical, riparian forms had previously taken root (see 6.9.14). In sites rich in nitrogen compounds, generally under anthropogenic or ornithogenic influence, there is usually a characteristic development of the shrubs of *Sambucus nigra* and in southern England also of the neophytic *Buddleja davidii*. In relatively undisturbed places sometimes rather dense scrub vegetation of species of the *Prunetalia* develops, vegetation of *Lycium halimifolium* being particularly conspicuous. In southern Europe vegetation develops which links up with types known as maquis and garigues. In more oceanic climatic zones the development is often rather singular, the establishment of stands of *Fraxinus excelsior* and *Acer pseudoplatanus* being strongly suggestive of a development towards a ravine forest (*Schluchtwald*). Occasionally, e.g. on the old castle walls at Caen (W. France), virtual scrub vegetation of these species is present with trees up to 4.5 m tall and an undergrowth of *Clematis vitalba* and *Hedera helix*. *Hedera* vegetation is more often transitional between chasmophytic and scrub vegetation. The sequences succeeding the phases of epilithic and chasmophytic vegetation are not discussed here, because they were dealt with among the various vegetation types.

In this chapter, special attention is paid to the development of the epilithic and chasmophytic stages, and of vegetation on relatively little decomposed, vertical walls which can be semi-permanent.

The most extreme circumstances do not permit a development of living organisms. Macroscopic organisms are not found on walls in very dry sites or on walls subjected to considerable fluctuations in temperature. The plant growth is usually, or perhaps always, initiated by the establishment of bacteria, fungi and/or blue-green algae. The study of these stages is a much neglected subject. The subsequent succession leads, sometimes via lichen vegetation, to vegetation of bryophytes which either represents the ultimate phase or shows a progressive development to vegetation of vascular plants. However, the latter does not rarely develop without a previous stage of bryophyte or lichen vegetation. The development of thallophytic vegetation is only summarily treated here and requires a special study. All stages mentioned can become terminal stages of vegetation development. Normally the development tends to proceed in the direction of a mesophytic vegetation, irrespective of the relative degree of humidity or drought prevailing during previous successional stages. A change of this kind, associated with a less extreme environmental condition, permits progressive succession. The time span in which the type of vegetation changes is, accordingly, dependent on this change in the environment.

Occasionally the succession is a cyclic one, e.g when moss cushions attain such a size and weight that the gravitational force gains the upper hand over the forces which make the moss cling to the wall. Such a 'cyclic' succession is, however, a step in a more secular succession (as is the rule) in that the vegetation cover has contributed towards alterations in the habitat and tends to do so more effectively in every subsequent cycle, for instance by promoting the decomposition of the masonry and, indirectly, by the development of biomass during a longer period of time than before. In 'natural' sites, cyclic succession of epilithic (and epiphytic) vegetation can be brought about by erosion, the sensitivity to erosion increasing as the development of moss vegetation proceeds (see e.g. PATON, 1956, who described a case with a cycle period of 4½ years on a sandstone substratum in Britain, and SJØGREN, 1964).

7.1 Fungi and algae

There are only a few scattered reports on the occurrence of fungi and algae on walls.

As far as the *Algae* are concerned, the following forms have been recorded in humid sites, receiving only atmospheric humidity: species of the genera *Oedogonium, Trentepohlia (T. aurea), Hormidium (H. flaccidum), Chlorococcum, Cladophora (C. glomerata), Gloeocapsa* and *Gloeocystis,* and in places rich in nitrogen compounds, e.g. on flagstones of pavements, *Prasiola crispa* and *Porphyridium cruentum* (compare RISBETH, 1948 and GAMS, 1958; also personal records). Of the *Fungi* there is a report on the occurrence of *Phoma hibernica* and *P. pigmentivora* (NICOT, 1951), of *Psalliota arvensis* (WOODELL & ROSSITER, 1959) and a list of 18 species recorded by W. J. REYNDERS (1956) on the walls of canals in Amsterdam, some of which have since been found in similar sites in other towns. Of particular interest are *Rhodophyllus strigosissimus* and *R. byssisedes* which, in Europe, are not very common. *Armillariella mellea* was found on walls in several towns in the form of toadstools, but its hyphae are probably always connected with neighbouring tree-roots. In nearly all cases the fungi occur in the deeper holes in and between the bricks and presumably form small individual coenoses not or seldom interlocked with those of the other plant forms found on the same wall. Their sensitivity to atmospheric pollution seems to be considerably less than that of lichens.

7.1.1 Vegetation of Protococcus

On walls which are very damp during long periods, but are much drier during short or occasionally longer intervals, *Protococcaceae* often develop, especially with *Apatococcus* playing an important role (compare BARKMAN; 1958). According to SCHORLER (1914), *Stichococcus* usually grows together with *Pleurococcus* (on sandstone). The taxonomic evaluation of the members of these difficult groups is disregarded here and the generic name *Protococcus* is used in a very broad sense (and thus includes e.g. *Pleurococcus*). These forms do not seem to have any sexual reproduction, dispersal taking place by division, the plants being carried along by water and by air currents. The growth habit of *Protococcus* colonies is reminiscent of that of pulverulent lichens.

Protococcus vegetation is not exclusively epilithic (as on walls), but can also be epiphytic; it is not known with certainty to what extent the same, or different, species or the same mass relations are involved. A growth of *Protococcaceae* occupies its own place and can accordingly be classified in separate eco-systematic units (*Protococcion,* etc.). Such growths form a dull green, pulverulent and water-repellent layer which may cover large areas of a wall including its mortar-joints. The structure of such vegetation is extremely simple. *Protococcus* is supposed to be nitrophilous and strongly toxitolerant (compare BARKMAN, 1958). In a number of towns e.g. on walls of Amsterdam canals, it occurs in places where lichens, probably due to aerial pollution, are absent for miles around (see under 3.1.6). In such sites the nitrogen content can indeed be high. In the Netherlands the preference of *Protococcus* is for sites with a marked fluctuation in the degree of moisture, where W. and S.W. winds prevail. This is clearly illustrated by its occurrence on walls facing W. or S.W. where the atmospheric precipitation is highest but the desiccation by insolation can be considerable. *Protococcus* is, furthermore, often found on walls near leaking taps and drain pipes, faulty gutters, and other places which periodically get very wet. The type of substratum is irrelevant, the taxon apparently having a wide range of tolerance regarding chemical and physical factors. It does not occur in permanently moist or permanently dry habitats. *Protococcus* vegetation is capable of maintaining itself indefinitely and in this case it acts not only as a pioneer stage but also as a terminal succession phase.

7.2 Lichen vegetation

Crustaceous lichens often act as pioneers on solid substrata, a considerable part of the species living epilithically or epiphytically, but many species (or perhaps even whole genera) are obligatory epiliths or obligatory epiphytes and they frequently exhibit a marked preference for a certain type of substratum. Among the epilithic species, for instance, there are a good few exclusively found on limestone or only occurring on granitic rocks. Altogether over 200 species of lichens were recorded from walls during the present investigation, but the naming of the material has not yet been completed and vegetation of lichens is not treated here *in extenso.*

Among the material collected on walls the following genera are well represented: *Caloplaca, Lecanora* (e.g. *L. lentigera), Verrucaria, Xanthoria* (e.g. *X. parietina* and *X. aureola*), *Physcia* (e.g. *P. orbicularis* and *P. adscendens*), *Acarospora, Candellariella* (e.g. *C. aurella*), *Buellia* (especially *B. canescens*), *Roccella, Dermatocarpon, Pertusaria* and *Cladonia* (e.g. *C. fimbriata* and *C. coniocraea).* Of rarer occurrence are, e.g. *Ramalina* (e.g. *R. duriaei* and *R. intermedia,* the latter was recorded for the first time in the Netherlands, SEGAL, 1968 b), and *Peltigera,* especially *P. canina* var. *rufescens* (= *P. rufescens*). On marl a characteristic combination was encountered with e.g. *Toninia coeruleo-nigricans.* On walls near the coast I recorded, among other species, *Verrucaria maura* and *Caloplaca marina,* the latter species often in a zone above the former. Above both there was often a zone of *Xanthoria parietina.*

Many of the more typical wall-dwellers, such as a number of species of *Caloplaca* (e.g. *C. murorum), Physcia* and *Xanthoria,* are nitrophilous. Lichens are, generally speaking, drought-resistant, but they require at least a periodic moistening or a high relative humidity. Lichens withstand considerable fluctuations in temperature in a dry state when no assimilation takes place and the reserve substances stored in the thallus are being used up. In unfavourable conditions e.g. when the light intensity is low, pulverulent lichenous forms are frequently encountered. Such forms may represent early developmental stages of various groups of species which do not develop any further and are often referred to *Lepraria.* The very common *L. aeruginosa,* which shows a preference for fern vegetation (quite often it is associated with *Asplenium trichomanes*) and usually occurs in places where there is an appreciable accumulation of extraneous soil particles and of humus, is in fact only named by the colour.

A succession series can initiate with such pulverulent (and 'imperfect') forms, but the more usual sequence is
crustaceous lichens ⟶ foliaceous lichens ⟶ fruticose lichens.

The succession from crustaceous to foliaceous lichens was described by Linnaeus as early as 1762 (compare PLITT, 1927), and the whole sequence was noticed by various workers (e.g. especially in epiphytic vegetation, DUDGEON, 1923; HILITZER, 1925; OCHSNER, 1928; BARKMAN, 1958). Each of the successional stages can represent a terminal stage. Vegetation of foliaceous lichens can be followed by vegetation of acrocarpic mosses. In these sequences the following tendencies can be recognised: A change-over from a strong attachment to the substratum to a much weaker one; from a simply constructed thallus to more branched types which raise themselves up above the substratum; from relatively slow-growing forms to relatively fast-growing ones. Within a category of growth forms the same trends can be observed; among the foliaceous lichens for instance, from forms with ascendent margins to forms with adnate margins (compare PLITT, 1927).

A common group of lichens, (predominantly foliaceous types), is the one which KUSAN (1933) referred to the *Physcion caesiae* (compare MOTYKA, 1927: *Physcietum caesiae*), and which includes both nitrophilous or nitrotolerant and fairly toxitolerant species such as *Physcia caesia, P. dubia, P. adscendens, P. orbicularis, Caloplaca murorum* and *Lecanora muralis*. In vegetation of fruticose lichens species of *Ramalina,* but especially certain species of *Cladonia,* play an important role. A rather common combination is that of *Cladonia fimbriata, C. coniocraea* and *C. pyxidata* var. *chlorophaea.*

One gets the impression that in this sequence these species of *Cladonia* occur in drier habitats, that *C. coniocraea* is the most nitrophilous of the three, and that *C. pyxidata* prefers walls on which a deposit of humus or soil has accumulated. These species are capable of overgrowing acrocarpic mosses. Stands of *Cladonia* are usually much better developed on horizontal walls than on vertical ones, but on vertical wall surfaces the development of crustaceous and foliaceous lichens prevails. Horizontal parts of walls are also preferred by mosses.

7.3 Succession of bryophytes

Bryophyte vegetation exhibits the same trends as lichen vegetation, acrocarpic mosses often preceding pleurocarpic forms (compare HILITZER, 1925; BIRSE & GIMINGHAM, 1954; BARKMAN, 1958). *Gyroweisia tenuis* and *Barbula revoluta,* both of small stature, appear as pioneers on soft or much decomposed types of stone, but of more common occurrence are the nitrophilous and toxitolerant species *Tortula muralis* and *Bryum argenteum* (the latter particularly in towns). These species are very active humus producers and collectors of extraneous matter. With their protonemata they are firmly attached to the substratum and during periods of temporary drought the attachment becomes even firmer, because the filaments of their protonemata get firmly stuck on to the substratum. *Barbula revoluta,* in addition, seems to be capable of accelerating the decomposition process of softer types of stone by penetrating superficially into such substrata when they are wet. In order to make a decent gathering of this moss (and of *Gyroweisia tenuis*) one must often take along a piece of the stone or brick.

As a rule the establishment of mosses is not preceded by the development of lichen vegetation, the mosses usually getting a first foothold in the mortar joints, not rarely along the edges of the stones or bricks and especially on the upper edge of the joints where the rate of decomposition is usually the fastest and the desiccation the least owing to the weak protection of the somewhat protruding stone (or brick). Both *Tortula muralis* and *Bryum argenteum* can sometimes establish themselves on a clean horizontal wall surface in less than two years. This could be repeatedly established in the Netherlands. However, it always takes much longer, even if the conditions are most favourable, before a stand has developed with a coverage exceeding 20%. Only if the inflow of extraneous dust particles is excessive can this process be accelerated. The coverage of a stand of exclusively acrocarpic species is seldom much higher than 20%, for that matter. *Tortula muralis* is the most ubiquitous pioneer in rather diverse sites. In the subsequent phases, more particularly in habitats rich in nitrogen compounds, *Ceratodon purpureus* and species of *Bryum* are often encountered, together with *B. argenteum, B. caespiticium, B. capillare,* and, especially in Brittany, *B. murorum.* Further, species of *Barbula* are of some importance, especially *B. vinealis,* and *Streblotrichum convolutum,* whilst in the Mediterranean area *B. acuta* is also important. On harder substrata, such as concrete, *Grimmia pulvinata, Orthotrichum anomalum* and other species of *Orthotrichum* and *Tortula muralis* are well represented. In these habitats vegetation of acrocarpic mosses usually represents the terminal stage on vertical wall surfaces, but on horizontal surfaces moss cushions can act as the germination bed of higher plants, especially of dwarf therophytes and *Sedum* species. In all other cases an increasing number of pleurocarpic forms appear, especially *Homalothecium sericeum, Amblystegium serpens, Brachythecium velutinum, B. rutabulum, Rhynchostegium murale* and *R. confertum* which also create favourable conditions for the germination of diaspores of higher plants.

7.4 Succession of cormophytes

Clements (1928) described a number of stages of a succession which are adopted here in a slightly amended form. During the development of an ecosystem the following sequence takes place (compare Segal, 1969):

1. **initiation** i.e. the complex of causes changing a habitat in such a way that an initial stage of a succession (primary or secondary) can develop;
2. **settlement** during which selection by the environment plays a mayor part;
3. **interference** i.e. the interrelations of individuals and taxa including competition, allelopathy, parasitism, saprophytism, symbiosis and other forms of interaction and interdependence (e.g. the less intimate ones of lianas, epiphytes and arboricole animals with trees);
4. **consolidation** during which phase vegetation develops, regulation begins and equilibria become established, and, finally,
5. **stabilisation** during which phase the equilibria between the ecological activities of the various organisms and between the habitat and the organisms contained therein are stabilised.

For each individual species the sequence includes the following phases:

1. **migration** i.e. the arrival of its diaspores;
2. **ecesis** i.e. germination, establishment and subsequent colonisation;
3. **aggregation** i.e. the vegetative and generative propagation leading to the formation of spatial patterns;
4. **competition** i.e. the competition with other plants;
5. **reaction** i.e. the reaction of the plant to its environment; for light, water, nutrients, etc.; and
6. **stabilisation** i.e. the consolidation of the equilibrium between plant and environment which leads at least to the integration of the individual in its own stenocoenosis, which has a certain internal stability.

7.4.1 Initiation and settlement

Walls are built by man and the development of mural vegetation only takes place if the conditions have become favourable. These conditions have been discussed in detail in Chapter 3. The migration is associated with historical factors and with the accessibility of the site and, hence, with the dissemination capacity of each individual species. In this connection one may raise the question of how applicable to higher organism is the so-called *'Law of Beyerinck'*, as formulated by BAAS-BECKING (1934): *"Everything is everywhere, but the environment is selective"*. It is of course evident that not 'everything' is 'everywhere', and that, for instance, geographical barriers exist. That geographical barriers do not necessarily coincide with ecological ones is amply substantiated by the behaviour of plants which become neophytes in a different region. Geographical barriers can presumably also play a role within smaller areas without seas or high mountains, but with a certain variation in the landscape. Although undoubtedly many species are present (in the form of diaspores) in, or have easy access to, sites where they can potentially develop, **primarity** must play a decisive part: if of all the species capable of germinating in a certain place at about the same time, the diaspores of one reach that site before those of the others, this species has a clear advantage over the other species, at least for a certain lenght of time. Only if a later arriving form can get a foothold and, being better adapted to the habitat, can grow faster, or has a strong competitive power, will the first settler be partly or completely ousted. The so-called Law of Beyerinck fits into a more general principle: *"Everything tends to be everywhere"*, in which principle barriers of various kinds are accounted for as well as the selection by the environment (see 3.1.8). This principle also obtains at various other levels of organisation of living matter, and even of the non-living world. That primarity indeed plays a role can be illustrated by means of translocation experiments. Such experiments were carried out on a number of canal sides in Amsterdam with *Linaria cymbalaria* and *Corydalis lutea,* two species which have a weak long-distance dissemination capacity in contrast to an appreciable short-distance dissemination capacity in the local population spectrum (see under 2.3.3). The first-mentioned species is rare in Amsterdam and the second did not occur there at all when the experiment was started.

In 1956 seeds of both species were introduced on canal walls which appeared to be prospective favourable sites judging by the degree of decomposition and by the plant growth already present. *Linaria cymbalaria* became established in two of the three experiments and in one place increased to five individuals. *Corydalis lutea* developed in both places where it was sown, but no new plants appeared, which may be attributable to the absence of ants, the principal disseminators of *Corydalis* seeds. All these experiments, carried out near Weesperstraat, had to be prematurely discontinued on account of large-scale demolitions (the plants would have had to be removed at some time or other anyway to prevent flora adulteration). Attempts to transfer spores of *Asplenium ruta-muraria* did not have the anticipated result, but it is not at all sure if the method was perhaps inadequate. The transplantation of a young plant was successful, but it only started spore production in the year the wall on which it was growing was demolished. Such simple experiments do not provide water-tight arguments to prove the existence of primarity, because the sowing always disturbs the substratum albeit ever so slightly. The mode of 'migration' differs from the normal case in that sown diaspores have a better chance of getting firmly anchored in the substratum.

Pioneer species can only establish themselves after the stone has attained a certain degree of porosity, and most probably lower organisms, such as bacteria, blue-green algae, fungi and possibly also protozoa, have previously already been active. Mosses by no means always precede vascular plants, but ferns often appear in pioneer stages and their prothalli have the growth form of a frondose liverwort. *Asplenium ruta-muraria,* especially, is a wide-spread pioneer. In sites where the substratum has been enriched by nitrogen compounds (for instance washed down by rain), *Sagina procumbens* is often the principal component in early phases. According to Bakker (1961), the capacity of migration and settlement of a species is decided, apart from site accessibility, by the ease of dissemination and dispersal, and the life-span of the diaspores, which may already be present in a dormant condition. The rate of population building is also important; generally speaking it is, like the dissemination capacity, high for pioneer species. In how far the specificity of the environment or the primarity contribute to the development of a stand, can not always easily be established. For instance, on a wall bordering a canal (near Weteringschans, Amsterdam) which had been repaired but not completely renovated in 1954, and on which previously a good stand of *Asplenium trichomanes* had grown, a specimen of this species (which is rare in Amsterdam and western Holland generally) was again discovered in 1967. In between no sporophyte was observed. It is not certain whether that particular site offers favourable conditions for the development of this species of fern (exposure, moisture relations, etc. having remained practically unchanged), or if a locally produced spore remained dormant for a considerable time until the conditions for its germination were favourable.

Mural vegetation is usually pioneer vegetation, because it exhibits the characteristics of initial phases of a succession series: an open stand whose floristic composition and structure is rather simple and which does not show an appreciable integration of its components; a high dominance or subdominance of a single or a few species; a distinct tendency towards the development of vegetation patterns; a predominance of certain categories of life- and growth forms; and the size of the forms present is usually on the small side.

7.4.2 Further development and ultimate stages

Mural vegetation often attains a phase which is presumably capable of remaining constant for decades or even centuries because the environment changes at a very slow rate. Once a species has established itself, aggregation often becomes important. Geophytes are not so often found on walls, but tussock- and cushion formation is exhibited by both hemicryptophytes and therophytes, e.g. by *Linaria cymbalaria, Fumaria officinalis* and *Sagina procumbens*. The extent to which interference plays a part can not easily be established. In cases of codominance of a few species, e.g. of *Parietaria judaica* with *Linaria cymbalaria* and *Asplenium ruta-muraria,* these species often form mosaic patterns, but in the example mentioned *Parietaria* may ultimately gain the upper hand. It is difficult to say in this special case whether this is the result of competition, or of a change in environment unfavourably changing the conditions for *Linaria* and *Asplenium* or (perhaps in connection with this) improving the conditions for *Parietaria*. It is a fact that *Parietaria* vegetation may ultimately become relatively poor in species. A decrease in the number of species in later phases of a succession series (presumably during consolidation) is not at all unusual; it may be associated with the attainment of an equilibrium involving a higher degree of integration of the components and the ousting out of the 'casuals', dominance once more becoming an important feature (compare the development of beech woods in central Europe). In such cases the homogeneity has increased and there is not so much overdispersion but a more normal dispersion.

8 Vegetation of streets: a comparison

In order to acquire a better insight into certain types of mural vegetation it appeared desirable to study vegetation occurring in the cracks and joints of paved roads, as there is a fairly great resemblance in floristic composition between vegetation types with *Poa annua* and *Sagina procumbens* found both on paved streets and on walls. Such vegetation types occur especially on the walls of canals and quay-sides in western Europe along which some substratum of adjacent sites, or rain water containing diaspores from such habitats, runs down. There are several ecological resemblances between walls and paved roadways. In both cases the anthropogenic influence is strong, the available surface for germination of diaspores low, the fluctuation in the temperature often appreciable, and the atmospheric precipitation rapidly removed by run-off and evaporation. To some extent paved streets can be regarded as "horizontal walls" and walls as "vertical streets". However, there are also some important differences, more particularly in the nature of the substratum (mortar, and hence a higher lime content, in walls; usually more nitrogen compounds in the cracks between paving stones), in the inclination (hence: insolation) and, in the case of streets, the factor of being trodden upon or not.

The relationships between vegetation types found on walls and those occurring on paved roadways were discussed in an earlier paper (SEGAL, 1963a). The communities inhabiting cracks between paving stones will not be dealt with in great detail here. In the present context, only their relation with mural vegetation types, possible floristic discontinuities between wall vegetation and vegetation of adjacent paved streets, and the nature of the phytosociological relationships between vegetation units belonging to different 'classes' are relevant.

The floristic composition of the vegetation types present in cracks and joints of paving stones is up to a point dependent on the width of the spaces between the stones or bricks: the wider the spaces, the more the vegetation type approaches that of much trampled gravel paths and road sides, a very wide-spread community in which *Lolium perenne, Plantago major, Capsella bursa-pastoris* and *Matricaria matricarioides* are important species (the *Plantagini-Lolietum* of BEGER 1930). In not excessively trampled sites the vegetation type tends to approach that of more natural habitats, almost always showing transitions to adjacent stands of vegetation on account of resemblances in environmental factors and of vicinism. The smaller the dominant species of the adjacent stand (of usually pioneer vegetation) are, the stronger the resemblance, as, for instance, on roads in open dune terrain.

The width of the voids between paving stones or bricks is also partly determined by the shape and the finishing of these paving materials, by their arrangement in the paving pattern and by the intensity of the traffic. The joints between rather irregularly shaped stones, such as basalt blocks, are usually broad, and they are also wider when much motorised traffic passes over the road, or when they are arranged in a rectangular instead of a zig-zag fashion. In the last instance, on narrow roads, vegetation is best developed in the two rather narrow strips along which the tyres usually roll along since here the stones have been pushed apart most strongly. Treading is another mechanical factor, whose influence on meadow vegetation has been amply discussed by LIETH (1954). Immediate effects of treading are especially the alteration of the soil structure (leading to an increase in compactness), the translocation of parts of the substratum, the direct damage to plants, and the dispersal of diaspores. The unfavourable soil texture is worsened by the disintegration of crumbly soil particles which become pressed, and thus firmly stuck, together: the soil 'closes up'. Generally speaking, the soil becomes firm to solid and has a small air-containing capacity. The moisture content may be high when all the available spaces become filled with water. In paved sites, the joints and cracks frequently collect water and remain inundated for a relatively great lenght of time. Such considerable fluctuations in water content doubtless must exert a strong selective influence. The substratum is usually rich in nitrogen compounds and the amount of organic nitrogen may be considerable where dunging by horses, cattle, rabbits, dogs or other animals plays a role. In coastal dunes rabbit dung favours the incidence of such species as *Sedum acre* and *Senecio jacobea*. The firmly compacted and insufficiently aerated soil is an unfavourable environment for nitrifying bacteria (SISSINGH, 1950), so that organic nitrogen accumulates. The uppermost layer of the substratum in joints of paved surfaces ultimately consists of firmly compacted humus on which species of *Bryum* (mostly *B. argenteum*) and *Ceratodon purpureus* may develop.

Vegetation of strongly trampled sites is subjected to several extreme environmental factors and its structure is simple. Small therophytes prevail, the most successful being prostrate forms and minute rosette-forming plants as well as some rapid growers, such as *Poa annua, Sagina procumbens* and *Capsella bursa-pastoris*. Important hemicryptophytes present are species of *Plantago* (*P. major* and *P. coronopus*). These species are all more or less drought-resistant and they have a long vegetation period with the capacity to flower during a considerable part of the year. According to Blum (1925), such plants exhibit two or more flowering periods e.g. March-May, August-September, and October-December. In the Low Countries and in western Germany this seems to be indeed the rule. However, elsewhere (e.g. in southern England) flowering during the summer season takes place more often, whereas in drier climates the plants are more or less wilted and subdormant in summer Several species (*Plantago, Polygonum aviculare* s.l., *Matricaria matricarioides*) have a long taproot; *Poa annua* and *Lolium perenne* have a dense felt of root-hairs. The adaptation to being trodden upon is usually achieved by the possession of elastic petioles and floral peduncles or pedicels (Pfeiffer, 1937). Progressive succession can only take place if the intensity of treading is moderate or after a street has been closed or abandoned. It usually leads to a *Plantagini-Lolietum* if it does not tend towards some adjacent stand of vegetation.

One could maintain that therophytic vegetation takes up a place of its own among the *Plantagini-Lolietum,* the argument being that it includes the pioneer stands which are, as a rule, particularly well developed towards the edges of vegetation of much-trodden sites where treading is usually the strongest. In later successional stages the therophytes persist but they are far less important, especially their quantitative contribution falling off considerably.

Diemont, Sissingh & Westhoff (1940) distinguish the *Sagino-Bryetum argentei* as a specific vegetation type of joints in paved surfaces, and consider as characteristic several species which feel at home in habitats with a fluctuating water-table and a relatively poor soil, such as *Juncus bufonius* and *Gnaphalium uliginosum.* Later Tüxen (1957) included this association in the alliance of the *Polygonion avicularis* whose ecological species group consists of treading-resistant forms such as *Plantago major, Polygonum aviculare, Lolium perenne, Capsella bursa-pastoris, Poa annua* and *Matricaria matricarioides.* Also included in this alliance is the *Plantagini-Lolietum* and indeed vegetation of paved surfaces shows a marked floristic resemblance to it. The abstract units (associations) can be regarded as being connected by completely continuous transitions, the differences between them being chiefly quantitative ones. Differentiating for the *Sagino-Bryetum argentei* are a few mosses: *Bryum argenteum, B. caespiticium, B. capillare* and *Ceratodon purpureus; Lolium perenne* and *Achillea millefolium* being relatively rare in this habitat. In the most extreme environments, vegetation stands tend to be poor in species and such depauperated stands have been referred to as inops (see 6.1.1). The inops of paved surfaces has as its characteristic species *Poa annua, Sagina procumbens, Plantago major, Polygonum heterophyllum* and *Bryum argenteum,* but in the most extreme environments the number of species may be further reduced. This combination of species may also serve as a starting-point of a series of transitional stages to various other vegetation types. In the terminology of the Franco-Swiss School such transitions towards other higher synsystematic categories can be referred to as subassociations (see under 5.1.3). Transitions of this kind clearly indicate that the system of classification must be visualised as a multi-dimensional structure with the units as points in a space in which clusters of related units constitute the higher synsystematic units, which may overlap in various ways or do not even have sharp boundaries. All units are only abstractions and clear-cut discontinuities are relatively rare. Every association can be extended to an *association assembly* which includes series of transitional cases and may comprise floristically very similar vegetation types referable to different higher units (even to different classes). Fig. 5 is a representation

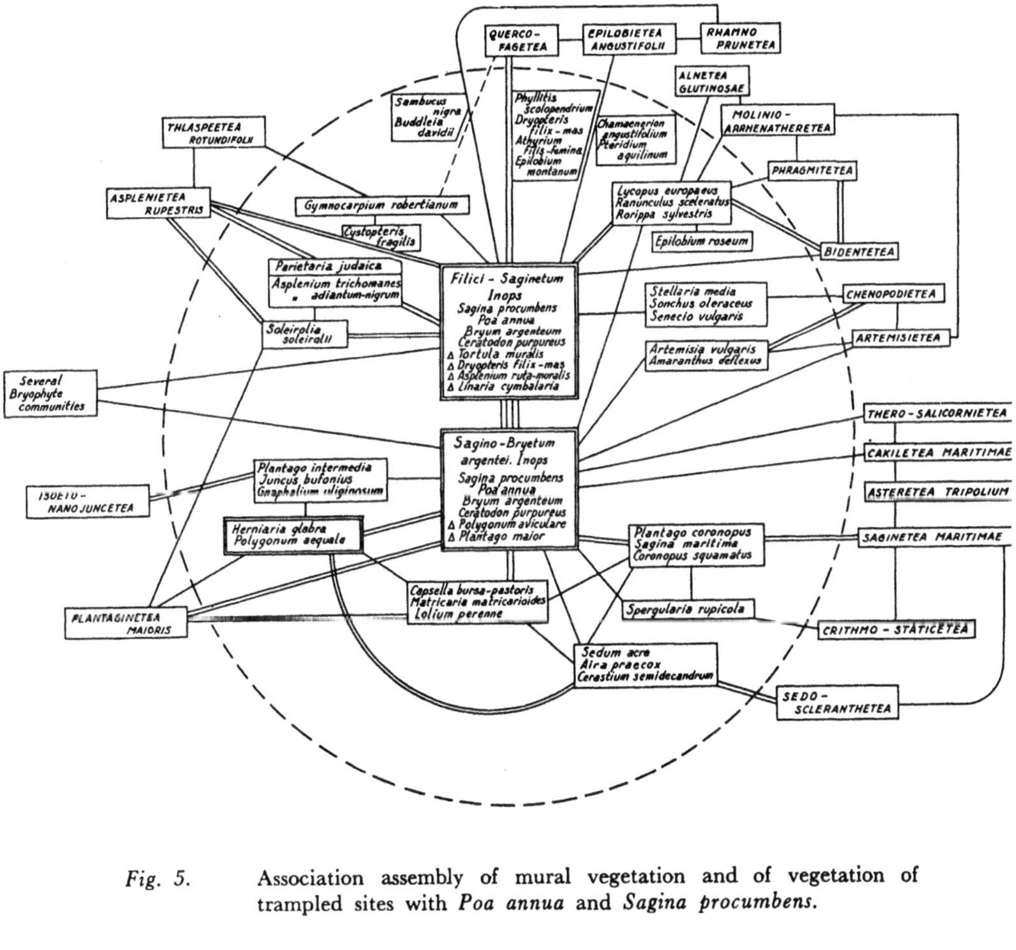

Fig. 5. Association assembly of mural vegetation and of vegetation of trampled sites with *Poa annua* and *Sagina procumbens*.

of an association assembly in which the *Sagino-Bryetum argentei* is combined with the ecologically vicariating *Filici-Saginetum* of walls whose differentiating species are e.g. *Linaria cymbalaria, Asplenium ruta-muraria* and some other ferns, and *Tortula muralis,* and which links up with chasmophytic vegetation types of rock crevices (class *Asplenietea rupestris).* I am convinced that such an approach to phytosociological classification will broaden our insight into the interrelationships among vegetation units and into the burning question of continuity *versus* discontinuity, and may form a bridgehead towards the understanding of ordination of at least coenoclines. Geographically vicariating variants and vegetation types presumably linking up with the *Sagino-Bryetum argentei* through gradual transitions also occur: in continental Europe vegetation of paved surfaces with *Herniaria glabra,* in southern Europe with *Poa infirma, Sclerochloa dura, Cynodon dactylon* and species of *Eragrostis,* in the montane and alpine regions with *Poa supina* (= *P. annua* ssp. *varia*). The structural resemblance (and taxonomic relationship) between *Poa annua, P. supina, Eragrostis, Sclerochloa* and the grass *Catapodium marinum,* which appears in coastal communities, is very striking. The similarity becomes more pronounced as more vicariating species of other taxa occur in related associations, such as different species of *Spergularia,* of *Sagina,* of *Polygonum* sect. *Avicularia,* of *Capsella,* of *Plantago* and of *Taraxacum.* It is even possible to make a comparison of vegetation of trampled sites in Japan as described by MIYAWAKI (1964) in which, apart from a number of Asiatic species of *Eragrostis,* such vicariating forms occur as *Plantago asiatica* and *Sagina japonica,* and *Lepidium virginicum* may perhaps be compared with *L. ruderale* and *L. draba* in similar habitats in Europe. For the rest, a number of species native in Europe are also found in Japan. It would be worth while to verify whether in other parts of the earth (at least in the Holartic region) similar vegetation types occur in corresponding habitats, and to make a comparative analysis of such communities in order to increase our insight into structural problems of at least this group of associations. In Table 50 a survey, undoubtedly incomplete, is given of physiognomically and (at least at the family level)

Table 50 Physiognomic and taxonomic relationships among species of vegetation of trampled sites

Lowlands of W.Europe	Poa annua	Sagina procumbens	Capsella bursa-pastoris Coronopus didymus	Lepidium ruderale	Plantago major	Lolium perenne
Mountains of Europe	Poa supina	Sagina saginoides		Lepidium densiflorum	Plantago alpina	
Warmer regions of Europe	Poa infirmis Sclerochloa dura Eragrostis pilosa poaeoides	Sagina apetala (Spergularia rubra)	Capsella rubella	Lepidium draba	Plantago crassifolia	Lolium rigidum
Costal regions of Europe	Catapodium marinum	Sagina maritima nodosa Spergularia marina media	Cochlearia danica Coronopus squamatus		Plantago coronopus	Parapholis strigosa
Asia		Sagina japonica		Lepidium virginicum	Plantago asiatica	

taxonomically related species. This compilation can easily be augmented with species of *Panicum*, *Cynodon* and *Eleusine,* and with bryophytes. In coastal regions, for instance, *Ceratodon purpureus* ssp. *purpureus* is usually replaced by its ssp. *conicus,* and the more common species of *Bryum* are presumably replaced by *B. marratii* and *B. warneum* (SEGAL, 1968a). A morphological convergence can be observed between *Matricaria matricarioides* and species of *Coronopus.*

8.1 Vegetation analysis

The method of analysis was very similar to that applied to mural vegetation, the percentage of coverage also being related to the substratum proper (the joints and cracks) and not to the total area of the paved surface. The investigation took place in the same regions where mural vegetation was studied. The sampling was not so strictly at random as was the case with the recording of mural vegetation. The simple reason for this was that it appeared very soon that a certain combination of species is extremely common and wide-spread, so that deviating combinations had to be searched for in ecologically more or less different sites. The most general vegetation type agrees with what is called the subassociation of *Capsella bursa-pastoris* of the *Sagino-Bryetum argentei* by TÜXEN (1957), a combination of species including, apart from those constituting the inops, several species of the *Plantagini-Lolietum* (in other words, elements of the alliance *Polygonion avicularis* and units of a higher category). Special attention was paid to deviating vegetation types of coastal regions, and in Czechoslovakia and Poland continental vicariating communities were studied. In the Mediterranean areas and in the alpine zones such sites were only rarely sampled. In all, 177 relevés of vegetation of paved surfaces were made, and in addition a number were made of other trampled sites such as unsurfaced roads and paths. The latter are not discussed in the present paper. For the recording of the data of the relevés standardised forms were used which were rather similar to those used for relevés of mural vegetation (see under 2.2.5.1). Relevés of paved surfaces were used for a comparative study of various methods of data processing, which study is still in progress. Originally a table was drawn up according to the method described by ELLENBERG (1956). The ultimate version took several weeks to prepare and it soon became obvious that various alternative set-ups for a suitable lay-out are possible. The method is very useful for material with a high degree of homotonity, but a selection is always inevitable whenever subunits have to be chosen according to different (floristic) criteria. The same applies to the compilation of tables for subsequent computer analyses. In either case the lay-out can be made as objective as possible by using coefficients of affinity.

It was essential to achieve a satisfactory grading of the groups of species both in the species recorded and in the relevés. The assortment was carried out separately in either case, all species having been taken into account that occurred in more than two of the relevés.

The analysis was started by setting up a matrix in which the frequency of occurrence of each combination of two species was recorded. Along the diagonal the absolute presence of each species was indicated, the chosen starting-point for the sequence being the species with the highest degree of presence (viz. *Poa annua*).

Clusters were found, both along the diagonal and (as far as aggregates of species belonging to several primary clusters are concerned) in other places of the matrix field. The sites were sorted out in a similar fashion. The coefficient of affinity used for both matrices was that of Poore (1956; see under 2.2.5.2).

A matrix based on 2 x 2 convergency tables was prepared too (compare 2.2.5.2). The sequence is an alphabetical one (apart for cormophytes and bryophytes). Finally the information was compared with that of the original table. There proved to be much correspondence, the differences mainly being caused by the bias in the interpretation of the degree of 'weight' of the data which affects the assessment of the table of relevés but not the electronic processing of the data. At a later stage an attempt was made to find a relationship in which the bias of the weighted data could be accounted for. The results of these analyses and the comparison of this form of similarity analysis with other methods of data processing will be the subject of a forthcoming publication.

In the table of relevés an inops could be indicated in a number of cases, but not in the calculated matrix based on Poore's formula. In the latter the cluster of species constituting the inops and at the same time representing the characteristic species of the whole table (i.e. those with the highest degree of presence) can not be distinguished from the group of species of the *Plantagini-Lolietum*. The separation of the two species groups is of course only a relative one and is only purposeful if one strives for the recognition of abstract, delimitable units for classificatory purposes.

8.2 Vegetation of trampled sites as occurring on paved surfaces

The majority of the plant communities found in trampled habitats are in so far related that a number of species of a certain species group appear to be represented in all vegetation types. This species group comprises the following species (the numbers representing the percentage of presence in the 177 relevés):

Poa annua	96%	Polygonum heterophyllum	53%
Sagina procumbens	76%	Bryum argenteum	50%
Plantago major	73%	Ceratodon purpureus	47%
Taraxacum spec. div.	55%	Capsella bursa-pastoris	42%

This species group can be considered to be characteristic of vegetation on paved surfaces in at least large regions of western and central Europe. *Sagina procumbens,* an important species in this community, germinates in damp soil, but can stand considerable fluctuations in moisture relations. In its later stages of development it exhibits optimum growth on a drier soil than during its early stages of development (ELLENBERG & SNOY, 1957). In drier climatic regions the association is only poorly developed or its species are replaced by vicariating forms better adapted to drier conditions such as *Polygonum aequale, Eragrostis poaeoides* and *Herniaria glabra,* which appear to take the place of *Polygonum heterophyllum, Poa annua* and *Sagina procumbens,* respectively. A decrease in presence of *Sagina procumbens* is frequently associated with an increase in numbers of *Spergularia rubra.* The species of *Taraxacum* belong chiefly to the section *Vulgaria,* but in drier habitats other forms, e.g. of the section *Erythrosperma,* are also encountered. Well-developed stands are only present in not excessively dry sites and in places with a high relative humidity. It is not at all surprising that such stands are of more common occurrence in western Europe than elsewhere and that they are relatively rare in eastern Europe. In Holland the vegetation cover of a number of suitable sites was phytosociologically recorded between 1960 and 1967, at least annually, but usually several times per annum. A comparison of the consecutive relevés indicates that the specific composition was very constant in all instances. Some fluctuations in the biomass were noticed, however. *Sagina procumbens* appeared to be well-developed in the wetter years and showed a decrease in drier years, whereas *Polygonum heterophyllum* and *Capsella bursa-pastoris* showed the reverse tendency. The remaining species did not seem to be sensitive during this span of time. Depauperated inops forms of the association occur in sites where treading or motorised traffic is excessive, or other environmental factors are suboptimal e.g. in places where the available surface area for germination of diaspores is small, where the substratum is strongly polluted (e.g. near filling stations) or excessively enriched by dunging, and — in western Europe — where the substratum is too tightly compressed or, relatively speaking, too dry. Such inops communities are very common and the seemingly small number of relevés of such stands in the tables is not at all indicative of the frequency of occurrence of such poorly developed stands. The combination *Poa annua-Sagina procumbens-Bryum argenteum* is especially common and situations in which only the first species is present are not at all rare. Communities poor in species except for the dominance of *Poa annua* have been described as *Poëtum annuae* by KNAPP (1961), who referred vegetation with dominance of *Polygonum aviculare* (sensu stricto?) to the *Polygonetum avicularis* (compare PASSARGE, 1964). *Poa annua* shows optimum growth in more or less sheltered situations e.g. near the foot of a wall, where the effect of treading is least felt. *Polygonum heterophyllum* is more drought-resistant and withstands stronger insolation. A number of relevés of the inops are tabulated in Table 51. The differences in the relevés recorded between 1959 and 1967 in one site are shown in Table 52, which also gives the

Table 51 Community of Poa annua, Sagina procumbens and Bryum argenteum, inops

Nr C	1	2	3	P 13st	U W	MGT	MGP
Date	27-4	11-7	18-7				
Year	62	62	60				
Nr R	232	451	142				
CH	5	30	75				
CM	1	80	15				
Sagina procumbens	1b	2b	4b	*100*	+r-5a	20	20
Poa annua	1b	2b	2a	*100*	1a-4a	15	15
Bryum argenteum	2p	4b	2b	*93*	2p-4b	13	14
Plantago major	-	+p	+p	*75*	+r-1b	0.5	0.6
Polygonum heterophyllum	1a	-	1a	*63*	+a-2b	5	8
Ceratodon purpureus	-	2b	-	*47*	1p-2b	5	11
Taraxacum (div.) spec.	-	-	-	*23*	+r-+p	<0.1	<0.1
Capsella bursa-pastoris	-	-	-	*16*	+r-+p	<0.1	0.1

LEGEND

1. Toulouse (France), street in front of the Basilique St. Sernin.
2. Opladen (North Rhine- Westphalia, Western Germany), terrace near petrol station along the motor way.
3. Wanneperveen (Overijssel, Netherlands), in front of the house nr 218, Veneweg.

Table 52 Dynamics of the community of Poa annua, Sagina procumbens and Bryum argenteum, inops in Wanneperveen (Netherlands)

Nr C	1	2	3	4	5	6	7	8
Year	59	60	61	62	63	64	65	66
Rainfall[1] (mm)								
Dec.-Febr.	183	186	248	268	135	64	217	356
March-May	154	106	136	196	178	125	272	187
June-August	104	279	263	162	304	230	353	410
Sept.-Nov.	111	337	215	166	242	253	200	223
Total annual rainfall	536	929	925	750	777	760	1152	1148
Number of rainy days[2]	175	245	236	225	201	207	257	245
Number of hours of sunshine	1986	1355	1435	1398	1425	1531	1509	1319
Date	8-8	18-7	27-7	1-9	8-8	27-8	10-7	11-8
Nr R	217	142	341	619	350a	536	537	638
CH	70	75	75	70	70	70	85	80
CM	15	15	15	15	15	15	15	15
Poa annua	2a	2a	2a	2a	2a	2a	2a	2a
Sagina procumbens	3b	4b	4b	4a	4a	4a	5a	4-5
Bryum argenteum	2b	2b	2b	2b	2b	2b	2b	2b
Plantago major	+p	+p	+p	+p	+p	+p	+p	+p
Polygonum heterophyllum	1b	1a	1p	1a	1a	+p	+r	+r
Taraxacum spec.	-	-	-	-	-	+r	+r	-
Ceratodon purpureus	-	-	-	-	-	-	-	+p

1) Data from the Royal Dutch Meteorological Institute at De Bilt (Utrecht, Netherlands).
2) Days with 0.1 mm or more precipitation.

mean annual rainfall in the Netherlands during these years. The differences between relevés made in drier sites in Amsterdam were greater, but the stands were richer in species. The figures in the last column of Table 51 of course also refer to other relevés than the three shown *in extenso*. The majority of the relevés were recorded in Holland.

The data shown in Table 52 suggest a relation between the climate and the percentage of coverage of *Sagina procumbens* and of *Polygonum heterophyllum*. In the cases recorded here the differences do not seem to be only the result of differences in the rainfall during the spring, but they may partly be attributable to the weather prevailing during the (early?) summer. Unfortunately, there are as yet no records of the variation in the relevés made in the same spot during the whole growing season.

The first six species in Table 51 may be considered to be characteristic of the inops, the demarcation line for their specificity being drawn at the level of a presence exceeding 40%. It is equally possible that the last two species should be included as well, but the inadequate number of relevés of this depauperated type does not permit a more definite conclusion. The eight species under discussion constitute the basis of the *Sagino-Bryetum argentei* if one wishes to fit in the combination of species as an association.

The distinction of a subassociation with *Ceratodon purpureus* (TÜXEN, 1957) is inane according to these criteria. For the characterisation of the poorest manifestation of the community a smaller species group can be chosen e.g. the first three.

Not only the inops is very common, but also stands richer in species (see e.g. TÜXEN, 1957). Apart from the 'basic' species group of the inops they contain at least one or more species of the following list (the relative presence being mentioned after the names):

Matricaria matricarioides	24%	Trifolium repens	10%
Lolium perenne	20%	Achillea millefolium	9%
Poa pratensis	12%	Cerastium fontanum	6%
Plantago lanceolata	11%		

All these are species frequently encountered in other trampled habitats (such as the *Plantagini-Lolietum)*. The drop in relative degree of presence between *Capsella bursa-pastoris* (42%) and *Matricaria matricariodis* (24%) is indeed striking.

Other species not confined to special types of vegetation of paved surfaces are:

Bryum capillare	(18%)
Bryum caespiticium	(17%)

They occur in various habitats rich in minerals and with rather strong fluctuations in moisture content, including walls. *Spergularia rubra* was recorded in 23 relevés (rel. presence 13%), always in situations where the substratum is relatively pervious to water and the moisture content of the soil shows wide variation, including both sites that are dry for a great length of time and favour the occurrence of *Plantago coronopus* and of *Cynodon dactylon* and sites which remain damp for a long time and support vegetation with e.g. *Gnaphalium uliginosum* and *Juncus bufonius*. The inops agrees with the so-called *Sagino-Bryetum argentei "typicum"* (cf. SEGAL 1943a), but it does not serve a useful purpose if one distinguishes a subassociation for such common forms of appearance of a community, which are in point of fact only impoverished expressions of that community. The community under discussion very closely links up with the *Plantagini-Lolietum,* which prefers, generally speaking, less trampled sites such as road sides. A species often supposed to be characteristic of this community is *Coronopus squamatus* (see e.g. WESTHOFF, DIJK & PASSCHIER, 1946; SISSINGH, 1950), which species presumably occurs most frequently in sites rich in organic nitrogen and/or phosphate, often together with *Lepidium ruderale.* These two species are not common in street vegetation and are predominantly encountered in sea ports. A few relevés relate to sites where the occurrence of *Coronopus didymus* was of a more ruderal nature viz. to paved streets newly laid down on pumped-up sludge in the outskirts of Amsterdam and Utrecht. Table 53 shows examples of relevés of vegetation with species of the *'Matricaria matricarioides* group'.

Table 53 Community of Poa annua and Matricaria matricarioides

Nr C	1	2	3	P 20st	U W	MGT	MGP
Date	19-9	23-8	31-10				
Year	62	67	63				
Nr R	700	432	191				
CH	25	20	15				
CM	45	-	-				
Poa annua	1b	2m	2a	100	1a-3a	19	19
Plantago major	+a	+a	+p	100	+r-2b	1	2
Capsella bursa-pastoris	+p	+p	+p	100	+r-1b	0.5	0.5
Sagina procumbens	2b	-	2m	85	+r-2b	8	9
Polygonum heterophyllum	1p	2b	1p	80	+r-2b	5	6
Taraxacum (div.) spec.	+p	+p	-	55	+r-1p	0.1	0.1
Bryum argenteum	2b	-	-	70	+p-3b	5	8
Ceratodon purpureus	2m	-	-	30	1p-4a	4	12
Matricaria matricarioides	1p	1a	+p	80	+r-2b	3	4
Lolium perenne	-	1p	2m	35	+r-2m	0.4	1
Trifolium repens	-	-	+r	25	+r-1a	0.2	0.9
Coronopus didymus	-	-	1p	15	1p	0.1	0.4
Poa pratensis	-	-	+p	15	+r-+p	<0.1	0.1
Achillea millefolium	-	-	-	15	+r-+p	<0.1	0.1
Cerastium fontanum	1p	-	1a	10	1p-1a	0.1	0.5
Plantago lanceolata	-	+r	-	10	+r	0	0
Agrostis stolonifera	-	-	+p	20	+p-2m	0.1	0.7
Spergularia rubra	-	-	-	15	+r-2b	1	6
Stellaria media	-	-	+r	15	r-+r	0	0
Bryum capillare	2b	-	-	10	2m-2b	1	10
Dactylis glomerata	-	-	+p	10	+r-+p	<0.1	0.1

In addition in 1: Amblystegium serpens +r; in 2: Anagallis arvensis +p, Sonchus oleraceus +r, Polygonum aequale +r; in 3: Polygonum persicaria +p, Elytrigia repens +p.

LEGEND
1. Road from Oosterwierum to Marssum, 0.5 km S. of Jellum (Friesland, Netherlands).
2. Sittard (Limburg, Netherlands), road in front of the abbey of Wintersley. (Much Polygonum aequale in adjacent Plantagini-Lolietum.)
3. Amsterdam-Buitenveldert (Netherlands), pavement of Karel Lotsylaan.

The relevés show quite clearly that the presence of the species of the group including *Matricaria matricarioides* is usually rather limited, but the high degree of presence of *Capsella bursa-pastoris* is striking. Furthermore the relatively low degree of presence of *Bryum capillare, B. caespiticium* (once recorded) and *Ceratodon purpureus* is noteworthy. *Coronopus didymus* is completely absent from all other relevés.

8.2.1 Vegetation of trampled sites in coastal regions

In European coastal areas *Plantago coronopus* is frequently met with. The stands are best developed in heavily treaded, sandy soils which are not so impervious to water as clayey ones; on clay or on loamy soils the stands are depauperated and if *Plantago coronopus* is at all present its vitality is usually much reduced. The coastal variant of street vegetation includes a number of salt-tolerant forms (Table 54).

Table 54 Community of Poa annua and Plantago coronopus

Nr C	1	2	3	P	U W	MGT	MGP
Date	5-8	19-9	3-8	*45st*			
Year	64	62	65				
Nr R	453	704	838				
CH	50	20	70				
CM	-	1	-				
Poa annua	2m	+p	2b	*98*	+r-3a	5	5
Sagina procumbens	2m	1a	2b	*83*	+r-2b	4	5
Plantago major	-	+p	+p	*76*	+r-2a	1	1
Polygonum heterophyllum	+p	-	+p	*71*	+r-2b	2	2
Bryum argenteum	-	-	-	*61*	+p-4a	3	8
Taraxacum (div.) spec.	+p	-	+p	*47*	+p-1a	0.1	0.3
Ceratodon purpureus	-	-	-	*47*	+p-4a	3	7
Capsella bursa-pastoris	-	-	-	*27*	+r-1a	0.1	0.3
Bryum caespiticium	-	-	-	*9*	2m-2a	0.2	2
Bryum capillare	-	-	-	*4*	1p-2a	0.2	5
Matricaria matricarioides	+p	-	-	*29*	+r-2a	0.4	1
Lolium perenne	-	-	2a	*33*	+p-3b	2	8
Plantago lanceolata	+p	-	-	*16*	+r-+p	<0.1	0.1
Trifolium repens	-	-	+p	*13*	+r-1a	0.1	0.6
Bromus mollis	-	-	2a	*9*	+p-2a	0.2	2
Plantago coronopus	3a	1p	2a	*63*	+r-3a	7	8
Coronopus squamatus	1b	-	-	*38*	+r-2a	1	3
Sedum acre	-	-	-	*33*	+r-2b	2	5
Festuca rubra	+p	+p	1p	*33*	+r-2b	0.5	1
Sagina maritima	2a	2m	1p	*31*	+r-2b	0.9	3
Atriplex hastata	-	-	1b	*27*	+r-1b	0.2	0.6
Sonchus oleraceus	+p	-	-	*27*	+r-+p	<0.1	0.1
Spergularia rupicola	+p	-	-	*18*	+p-2a	0.7	4
Spergularia marina	-	2m	-	*18*	+p-2a	0.3	1
Leontodon nudicaulis	-	-	+r	*13*	+r-1p	<0.1	0.1
Senecio vulgaris	-	-	-	*13*	+r-+p	<0.1	<0.1
Agrostis stolonifera	-	+p	1p	*11*	+r-3b	1	11
Streblotrichum convolutum	-	-	-	*11*	+p-2b	0.6	5
Koeleria albescens	-	-	+p	*11*	+p	<0.1	0.1

In addition in 1: Crithmum maritimum +a; Sisymbrium officinale +r; in 2: Spergularia media 2b; Glaux maritima 1b; Plantago maritima 1p; Parapholis strigosa +r; Pottia heimii 2m; in 3: Cerastium arvense, Cochlearia danica 2m; Cirsium vulgare +b; Poa pratensis, Cerastium fontanum 1p; Trifolium dubium, Bellis perennis, Achillea millefolium, Atriplex littorale +p; Daucus carota +r.

LEGEND

1. Porthleven (Cornwall, England), slope 18° S. towards harbour.
2. Holwerd (Friesland, Netherlands), parking area near harbour.
3. Closing dike of former Zuiderzee between Holland and Friesland, near the monument.

The community was recorded in all coastal parts of both the Atlantic and the Mediterranean areas studied. A number of the salt-tolerant species are characteristic for vegetation of the bottom parts of dune slopes alongside tidal marshes, in the splashing zone of the highest water-levels (highest spring tides), i.e. in the zones of contact of the halo- and the xeroseres e.g. *Plantago coronopus, Sagina maritima, Spergularia marina* and *Cochlearia danica*. These, and other species are the specialists of a singular environment in which appreciable fluctuations in the moisture- and the salt content of the soil take place. Vegetation stands of this kind are referred to a separate class *Saginetea maritimae*. Other species supposed to be characteristic are: *Parapholis strigosa, Catapo-*

dium marinum, Pottia heimii and *Amblystegium serpens* var. *salinum,* all of which were also found in vegetation of trampled sites with *Plantago coronopus.* Most probably *Cerastium diffusum, Ceratodon purpureus* ssp. *conicus, Bryum marratii, B. warneum* and *Rhynchostegiella compacta* var. *salina* can be added to this group of species. The last variety was recorded from some localities in the Netherlands and from one in Devon (S. England) in heavily treaded sites, and furthermore quite frequently at the foot of dunes in the Island of Terschelling (the Netherlands). In such habitats this species is usually much more common, but is presumably often mistaken for *Amblystegium serpens* (Segal, 1968a). Other species occurring both in such habitats and in trampled sites are *Glaux maritima* and *Sagina nodosa* var. *moniliformis.* In a few instances it could be established that the stands are sometimes flooded by sea water (e.g. near Katseveer, Zeeland, Netherlands) and if this is the case such species as *Salicornia europaea* and *Suaeda maritima* may be present. The salinity is usually brought about by splashes of salt water, by the spreading of fishing nets for drying, or by transport of materials from which salt water drips down. *Honckenya peploides* and *Elytrigia pungens* are especially found in sites where there is an accumulation of blown-in sand, but only as depauperated specimens. The presence of *Atriplex hastata,* a species frequently found near the flood mark of beaches, and of other species of ruderal sites, such as *Sonchus oleraceus, Senecio vulgaris* and *Lepidium ruderale,* is indicative of the high nitrogen content. This vegetation type is rather variable and the variants will be discussed in some more detail.

In the phytosociological literature, vegetation with *Plantago coronopus* has been described by e.g. Horvatic (1934; 1963: *Lolio-Plantaginetum commutatae*) and Tüxen (1950: *'Cynodon dactylon-Plantago coronopus*-association') and by Tüxen & Oberdorfer (1958: *'Plantago Coronopus-Trifolium fragiferum*-association'). Their relevés refer to stands in the Mediterranean and south Atlantic regions of Europe. Boerboom (1960) speaks of a littoral subassociation or variant of the *Sagino-Bryetum argentei* in the dunes near Wassenaar (Netherlands), but his relevés were recorded more inland and show transitions to dunal vegetation (with e.g. *Hippophaë rhamnoides*). Such transitions are of common occurrence, and in southern England and western France there are also transitions to vegetation of open sites with *Sedum anglicum.*

Of special interest is the occurrence of *Coronopus squamatus* in vegetation on trampled sites in coastal ports. This species is in the Netherlands restricted in its distribution to coastal regions and if it occurs elsewhere (e.g. along the great rivers) it is presumably adventitious in most cases (compare Sissingh, 1950). In the northern parts of the Netherlands this species appears in meadows on heavy clay, but as a rule in heavily treaded sites, e.g. near gates between meadows where the plant cover is locally reminiscent of an impoverished *Plantagini-Lolie-*

tum and includes species equally characteristic of the *Sagino-Bryetum argentei* (e.g. much *Poa annua*). Transitions towards vegetation with *Sedum acre* are common and they are in many cases not attributable to vicinism, but to a distinct ecological relationship. Most probably the soil type and the high organic nitrogen content, among other things, explain the correspondence, the high nitrogen content in fishing ports *argentei* (e.g. much *Poa annua*). Transitions towards vegetation with *Sedum acre* is separately discussed.

8.2.2 Vegetation with Sedum acre and Streblotrichum convolutum

Between paving stones and bricks on sandy soil rich in nitrogen in western Europe *Sedum acre* and *Streblotrichum convolutum* are frequently encountered in a plant community, which I found chiefly in the Netherlands, in the coastal dune area, but sometimes more inland. I recorded it only once in Italy, but I expect that it is present also in other parts of Europe. *Sedum acre* is also well represented in variants of the community with *Plantago coronopus* (see 8.2.1), but *Streblotrichum convolutum* seems to be much more characteristic of the species group of the vegetation under discussion. There is, furthermore, a certain correspondence with the community of *Herniaria glabra* (see 8.2.3). The development of vegetation with *Sedum acre* is probably furthered by organic fertilisation with the dung of rabbits, sheep, dogs or horses. The stands are predominantly encountered on sandy soils with an admixture of sea shells or rich in lime. As a rule the habitat is somewhat unstable, which is specially noticed in the motility of the substratum caused by transport of blown sand. The upper layer of the soil is, therefore, less solid than in other trampled habitats and the optimum development of the community can only take place in the broader joints between paving stones in more untrodden places, often in strips of pavement at the foot of walls where the shifting sand accumulates and turbulent air currents may play an important role. In Table 55 a few representative examples are given.

The proportional contribution of *Bryum caespiticium* and *B. capillare* is fairly large, that of the group of species including *Matricaria matricarioides* small; the salt-tolerant species of the community with *Plantago coronopus* are almost entirely lacking and also the differentiating species of the *Herniaria glabra* association. The resemblance with the community of *Plantago coronopus* is chiefly determined by *Sedum acre* and *Festuca rubra.*

The species of *Taraxacum* occurring in *Sedum acre* vegetation include, apart from some representatives of the *Vulgaria* group, *T. tortilobum* and *T. erythrospermum* (s.l.), which presumably represent good differentiating taxa.

Table 55 Community of Poa annua and Sedum acre

Nr C	1	2	3	P 22st	U W	MGT	MGP
Date	31-10	17-10	27-6				
Year	62	63	62				
Nr R	743	502	393				
CH	60	40	30				
CM	30	20	45				
Poa annua	2a	1p	2b	100	+p-2b	5	5
Sagina procumbens	2b	3a	2b	86	+p-3a	7	8
Bryum argenteum	+p	2m	+p	77	+p-4a	5	6
Plantago major	+p	+p	1p	68	+r-2a	0.9	1
Taraxacum (div.) spec.	1p	-	+p	68	+r-2b	0.2	0.3
Ceratodon purpureus	2b	+p	3b	55	+p-4a	11	17
Polygonum heterophyllum	-	+p	+r	50	+r-2a	0.8	2
Capsella bursa-pastoris	+p	-	-	46	+r-2a	0.7	2
Bryum caespiticium	-	2m	-	41	1p-3b	5	11
Bryum capillare	-	-	-	36	+p-2b	2	4
Lolium perenne	-	-	+r	23	+r-2a	0.4	2
Plantago lanceolata	-	-	+r	23	+r-+p	<0.1	0.1
Matricaria matricarioides	-	-	-	18	+r-1p	<0.1	0.1
Cerastium fontanum	-	-	-	14	+a-1b	0.3	2
Achillea millefolium	-	-	+r	14	+r-2m	0.1	0.8
Sedum acre	3a	2a	2a	86	+r-3b	15	17
Streblotrichum convolutum	2a	2a	+p	59	+p-2b	3	5
Festuca rubra	-	-	1b	36	+p-2m	0.5	1
Erigeron canadensis	-	+p	-	27	+r-1p	<0.1	0.1
Senecio vulgaris	-	-	-	23	+r-2a	0.4	2
Cerastium semidecandrum	1p	-	+p	18	+p-1p	<0.1	0.3
Veronica officinalis	-	+p	-	18	+p-1p	<0.1	0.2
Geranium molle	+p	-	-	18	+r-+p	<0.1	0.1
Aira praecox	-	-	1a	14	+p-2m	0.2	1
Rumex acetosella	1b	-	+p	14	+p-1b	0.2	1
Hypochaeris radicata	-	-	+r	14	+r-+a	<0.1	0.1
Koeleria albescens	-	-	-	14	+r-+p	<0.1	<0.1
Sonchus oleraceus	-	-	-	14	+r-+p	<0.1	<0.1

In addition in 2: Barbula unguiculata 2m; Poa trivialis, Veronica officinalis +p; in 3: Cephaloziella hampeana 2m; Agrostis stolonifera 1a; Viola tricolor ssp. curtisii, Hieracium umbellatum +p; Cerastium diffusum, Jasione montana, Corynephorus canescens, Elytrigia pungens, Galium verum, Holcus lanatus, Empetrum nigrum, Bromus mollis, Cladonia spec. +r.

LEGEND

1. Noordwijk aan Zee (Zuid-Holland, Netherlands), pavement of Emmaweg near Pieternelweg.
2. Mechelen (Limburg, Netherlands), garden path of parish of H. Johannes de Doper.
3. Oosterend (island of Terschelling, Netherlands), Badweg about 50 m S. of the cycle path to Boschplaat.

The only more or less similar vegetation type previously described in the phytosociological literature is found in a relevé made by Boerboom (1960).

8.2.3 Vegetation with Herniaria glabra and Polygonum aequale

Herniaria glabra was most frequently encountered in paved roads in central and southern Europe. In drier sites this species, often growing together with *Spergularia rubra,* seems to replace *Sagina procumbens. Spergularia rubra* is rather rare in vegetation of paved sites in western Europe, but it is already much more common in W. Germany than in the Netherlands (compare Tüxen, 1957). *Herniaria glabra* is a species which, especially in central Europe, occurs frequently in vegetation types referable to the *Sedo-Scleranthetea,* but shows transitions towards more or less ruderal vegetation (sometimes in trampled sites) which is usually furthered by anthropogenic or sometimes also zoogenic disturbance. In such sites, recorded in Poland and in Czechoslovakia, but also on riverine dunes in the Netherlands, in addition to *Herniaria glabra,* the following species are especially characteristic: *Potentilla argentea, Artemisia campestris, Scleranthus annuus, S. perennis, Brachythecium albi-*

cans and other species of the *Sedo-Scleranthetea*. Such a type was also described by WALDHEIM (1947) from southern Sweden.

Herniaria glabra often grows among paving stones in somewhat exposed sites, most frequently sloping to the south. Such sloping localities are, for instance, the squares before the dome church in Bamberg (Germany) and before the St. Veit's church in Prague. When traffic increases in intensity *Herniaria* gradually disappears, as I could observe quite clearly in Prague during repeated visits between 1962 and 1967. In the street-dwelling variant the occurrence of *Polygonum aequale* is striking (it is not, however, restricted to habitats where *Herniaria glabra* is found). *Polygonum aequale* often seems to replace *P. heterophyllum* in drier sites and appears to become more common towards the more continental (and Mediterranean?) parts of Europe. On sandy soils in western Europe the species was frequently encountered, occasionally in mass stands. The resemblance between vegetation with *Herniaria glabra* and *Polygonum aequale* follows from Table 56.

Table 56 Community of Poa annua and Herniaria glabra (1-3) and community of Poa annua and Polygonum aequale (4-6)

Nr C	1	2	3	P	U W	MGT	MGP	4	5	6	P	U W	MGT	MGP
Date	19-4	30-4	7-6	*16st*				9-7	15-10	2-9	*6st*			
Year	62	65	63					67	67	63				
Nr R	123	107	055					237	684	439				
CH	30	40	30					15	15	40				
CM	-	-	-					-	1	80				
Poa annua	1a	2a	1p	*100*	+p-2b	15	15	2m	2m	1p	*100*	1p-2a	3	3
Plantago major	-	+p	-	*75*	+p-2b	2	2	1p	1p	-	*83*	+p-2a	2	2
Capsella bursa-pastoris	+r	+p	+p	*63*	+r-2m	0.6	0.9	-	+p	+r	*67*	+r-+p	<0.1	0.1
Sagina procumbens	2m	2a	-	*56*	+r-2a	1	3	-	1p	1p	*70*	1p-2b	3	5
Polygonum heterophyllum[1)]	-	-	-	*38*	+r-2a	2	5	-	-	+r	*60*	+r-1p	0.1	0.2
Taraxacum (div.) spec.	-	-	+r	*38*	+r-+p	0.2	0.5	+p	+p	+p	*83*	+r-1p	0.1	0.1
Ceratodon purpureus	-	-	-	*31*	2m-2b	2	5	-	+p	1b	*60*	+p-3b	7	17
Bryum argenteum	-	-	-	*25*	+p-2b	<0.1	0.3	-	2m	3b	*67*	2m-4a	21	31
Matricaria matricarioides	-	-	-	*19*	1p-2m	4	2	-	-	-	*1x*	+p		
Herniaria glabra	2b	3a	2b	*100*	+p-3a	8	8	-	-	-	-	-	-	-
Spergularia rubra	2a	+p	1p	*60*	+p-2a	0.8	2	-	-	-	*1x*	1b		
Potentilla argentea	-	-	2b	*25*	1b-2b	3	15	-	-	-	-	-	-	-
Agrostis tenuis	+r	-	1p	*25*	+r-2m	<0.1	0.2	-	-	-	-	-	-	-
Veronica arvensis	-	-	1p	*19*	+p-1p	<0.1	0.3	-	-	-	-	-	-	-
Rumex acetosella	-	-	+p	*19*	+r-+p	<0.1	0.2	-	-	-	-	-	-	-
Polygonum aequale	-	-	1p	*38*	+p-2a	0.6	2	2b	2b	2b	*100*	1a-2b	14	14
Erigeron canadensis	1p	-	-	*13*	1p-2m	0.2	1	1p	2m	1b	*83*	1p-2a	2	3
Sonchus oleraceus	+r	-	1p	*13*	+r-1p	<0.1	0.2	-	-	2a	*33*	+p-2a	1	4
Sedum acre	-	-	[illegible]	*18*	1p-3a	0.6	4	-	-	-	-	-	-	-
Achillea millefolium	-	-	+p	*13*	+p-2a	0.5	4	-	-	+p	*1x*	+p		
Trifolium repens	-	-	-	*13*	+p-2a	0.5	4	-	-	-	-	-	-	-
Streblotrichum convolutum	-	-	-	*13*	2m	0.3	3	-	-	-	-	-	-	-
Scleranthus annuus	-	+r	-	*13*	+r-1b	0.2	2	-	-	-	-	-	-	-
Cerastium semidecandrum	-	-	-	*13*	+p-2m	0.2	1	-	-	-	-	-	-	-
Bromus mollis	-	1p	-	*13*	+r-1p	<0.1	0.3	-	-	-	*1x*	+p		
Lolium perenne	-	-	-	*13*	+r-+p	<0.1	0.1	-	-	-	*1x*	+p		

[1)] In two cases this may have been Polygonum aequale.

In addition in 1: Stellaria media var. apetala +p; Senecio vulgaris +r; in 3: Trifolium dubium, Carex hirta, Antennaria dioica +p; Artemisia campestris +r; in 6: Poa trivialis 2a; Bryum capillare +p.

LEGEND
1. Le Grau du Roi (Gard, France), road alongside the harbour, W.side.
2. Siena (Tuscany, Italy), Via di Monna Agnese near the dome church.
3. Road 2 km S. of Rajgród (Poland).
4. Hildesheim (Niedersachsen, Western Germany), pavement near the dome church.
5. Amsterdam (Netherlands), pavement in front of nrs 219 and 221, Sarphatistraat.
6. The Hague (Netherlands), pavement of Prinsessegracht near bridge to Korte Voorhout.

The variability of the stands with *Herniaria* is relatively low. The representation of the species of the group including *Matricaria matricarioides* is small, but nevertheless there are apparently distinct transitions to *Plantagini-Lolietum* vegetation. Occasionally vegetation was found showing transitions towards vegetation of paved surfaces with *Sedum acre* and *Streblotrichum convolutum.*

Related *Herniaria* vegetation, sometimes showing a clear affinity with the *Plantagini-Lolietum,* has been recorded by BRAUN-BLANQUET (1949: Davos, Switzerland), FRÖDE (1958), and PASSARGE (1963; 1964: Table 3, no. 6; 1965). Fröde referred to it as *'Herniaria glabra-Spergularia rubra*-Gesellschaft' (!) which he considers to be equivalent to the *Sagino-Bryetum argentei.* Passarge distinguished a *'Spergularia echinospora-Herniaria glabra*-Gesellschaft' of river banks, a vegetation type related to vegetation with *Corrigiola littoralis.*

The number of relevés of *Polygonum aequale* vegetation is too restricted to permit any conclusions. The rather consistent occurrence of *Erigeron canadensis* in the relevés is striking. The representation of the latter species in related vegetation types partly referred to the *Polygonetum avicularis* by PASSARGE (1963) strongly suggests that *Polygonum aequale* was represented in at least some of his relevés, instead of *P. aviculare*; the former species is fairly common in eastern Germany for that matter and also appears in the community of *Eragrostis poaeoides* which becomes gradually more common from W. to S.E. Europe and is somewhat allied to the community with *Herniaria.* According to FOERSTER (1968), *Polygonum aequale* grows in *Plantagini-Lolietum* meadows in Nortrhine-Westphalia, *P. heterophyllum* being restricted to fields under crops. Although the species found among field crops in western Europe is presumably often *P. heterophyllum,* it is, however, also quite common in the *Plantagini-Lolietum* and the *Sagino-Bryetum.* On the more lime-containing soils of the S.E. Netherlands I found *P. aequale* both in the *Plantagini-Lolietum* and in its transitions towards ruderal vegetation with *Urtica dioica,* where it was strongly developed. In the more continental parts of Europe *P. aequale* probably occurs in such types of vegetation on soil rich in alkaline substances, but also in joints of paved roads. In western Holland *P. aequale* was occasionally recorded as growing on paved streets, but it was absent in immediately adjacent stands of the *Plantagini-Lolietum.*

8.2.4 Vegetation with Stellaria media and Artemisia vulgaris

On a substratum rich in nitrogen, especially alongside the foot of a wall, a vegetation type is often encountered which contains nitrophilous species such as *Stellaria media, Chenopodium album, Artemisia vulgaris, Amaranthus deflexus, Hordeum murinum* and *Galinsoga parviflora* and clearly represent transitions towards ruderal vegetation but always includes a few species resistent to trampling. Characteristically *Urtica dioica,* which is sensitive to treading, is usually lacking and *U. urens* only appears on paved surfaces carrying very little traffic or on deserted streets, which show a stronger transition towards ruderal communities. The incidence of these nitrophilous vegetation types is the result of the activities of domestic animals and of the accumulation of discarded waste matter. A number of species represented in such sites are cosmopolitans or followers of human culture, which have thus acquired (or eventually will acquire) a world-wide distribution.

The most frequently encountered type is vegetation with *Stellaria media*, found in urban areas throughout western and central Europe. Vegetation with elements of the much more tread-sensitive ruderal *Hochstauden* vegetation is much rarer or it is already much degenerated to become a more typical ruderal type as soon as the treading factor becomes negligible. A special type, repeatedly found in W. France, especially in Brittany and the western Loire basin, is characterised by the presence of *Amaranthus deflexus* (see Table 57).

Table 57 Community of Poa annua and Stellaria media (1-3) and community of Poa annua and Artemisia vulgaris (4-6)

Nr C	1	2	3	P	U W	MGT	MGP	4	5	6	P	U W	MGT	MGP
Date	9-7	28-9	2-9	*9st*					11-8	13-9	*7st*			
Year	67	66	63					63	62	62				
Nr R	239	666	440					461	597	595				
CH	70	20	40					12	15	20				
CM	2	5	5					8	1					
Poa annua	3a	2a	1p	*100*	1p-3a	11	11	+p	1a	2m	*100*	+p-3a	8	8
Taraxacum (div.) spec.	2m	+p	+p	*89*	+p-2a	3	3	1a	+r	+r	*100*	+r-1a	0.3	0.3
Sagina procumbens	+p	2m	1p	*78*	+p-3a	4	5	1p	-	2m	*57*	1p-4a	7	15
Plantago major	-	+r	-	*67*	+r-2a	3	4	-	-	+p	*71*	+p-1b	0.6	1
Ceratodon purpureus	2m	-	-	*56*	2m	1	3	2a	1p	-	*71*	1p-2b	4	9
Bryum argenteum	-	-	+p	*44*	+p-2m	0.3	1	2m	-	-	*29*	2m	1	3
Polygonum heterophyllum	-	-	-	*33*	+p	<0.1	0.1	-	1b	2a	*43*	1b-2a	2	5
Capsella bursa-pastoris	2a	-	-	*67*	+r-2a	1	2	-	-	-	*1x*	+p		
Bryum capillare	-	-	-	*44*	+p-2m	0.6	1	2a	-	-	*29*	+p-2m	1	4
Bryum caespiticium	-	-	2m	*33*	+p-2m	0.3	1	1p	-	-	*43*	1p-2m	0.5	1
Poa pratensis	+p	+p	-	*44*	+r-1a	0.2	0.5	-	-	-	-	-	-	-
Lolium perenne	+p	-	-	*22*	+p-1p	0.1	0.3	-	-	-	*29*	1p	0.1	0.4
Trifolium repens	-	-	-	*1x*	+p			-	+p	+p	*29*	+p	<0.1	0.1
Matricaria matricarioides	-	+p	-	*1x*	+p	<0.1	0.1	+r	-	+p	*43*	+r-+p	<0.1	0.1
Sonchus oleraceus	1a	+r	1a	*67*	+r-2a	1	2	1a	+r	+r	*86*	+r-2a	1	1
Stellaria media	3a	2a	2a	*100*	2a-3b	18	18	-	-	-	-	-	-	-
Epilobium montanum	-	-	-	*22*	+r-+p	<0.1	0.2	-	-	-	-	-	-	-
Senecio vulgaris	-	1a	-	*44*	+r-1a	0.2	0.4	-	-	-	*29*	+p	<0.1	0.1
Chenopodium album	-	-	+p	*33*	+r-+p	<0.1	0.1	+r	-	+r	*57*	+r-+p	<0.1	<0.1
Artemisia vulgaris	-	-	-	*22*	+r	0	0	+a	-	+r	*71*	+r-+a	0.3	0.4
Erigeron canadensis	-	-	-	*1x*	+p			+b	-	-	*43*	+p-1b	0.6	1
Galinsoga parviflora	-	-	-	*1x*	+p			-	-	-	*29*	+b-2a	2	6
Amaranthus deflexus	-	-	-	-	-	-	-	-	2a	-	*57*	1a-2a	4	6
Hordeum murinum	-	-	-	-	-	-	-	-	+r	-	*43*	+r-1p	0.1	0.2

In addition in 2: Glechoma hederacea +r; in 3: Urtica urens 2b; Poa trivialis 1b; Amaranthus cf. retroflexus 1a°; Plantago lanceolata +r; in 4: Agrostis gigantea 2a; Lycopys europaeus, Tussilago farfara +r; in 5: Cynodon dactylon 1p.

LEGEND

1. Königslutter (Niedersachsen, Western Germany), pavement along the Elm church.
2. Woudenberg (Utrecht, Netherlands), pavement near the wind mill near the road to Zeist.
3. The Hague (Netherlands), pavement of Prinsessegracht near Korte Voorhout, alongside a wall.
4. Rotterdam (Netherlands), pavement near bridge of Spaanse kade opposite the "White House".
5. Chinon (Indre-et-Loire, France), street alongside steps of the ramparts.
6. Angers (Maine-et-Loire, France), pavement of Rue Michelet.

Jovet (1957) described a corresponding type of vegetation with *Amaranthus deflexus* from Angers, and Tüxen & Oberdorfer (1958) referred a related community from N.W. Spain to a *Spergularia rubra-Amaranthus deflexus* association supposed to be distinguishable by the presence of *Spergularia rubra, Sagina apetala* and *Malva* cf. *parviflora*. In England (e.g. in the Port of London) I encountered a similar community which included, among other species, *Senecio squalidus, Lactuca serriola* and *Chamaenerion angustifolium*.

Such communities not infrequently resemble vegetation of trodden sites alongside walls which includes wall-dwelling species (see 8.2.6).

8.2.5 Vegetation of paved surfaces in moist sites

Vegetation of moist paved surfaces can be of at least two types: a type rich in bryophytes with *Marchantia polymorpha* on sites rich in nitrogen, and a type of a relatively infertile substratum subjected to great changes in moisture content. This second type contains species of the *Nanocyperion* alliance. A special variant of this is a community with *Corrigiola littoralis,* which is of rare occurrence among paving stones, however. The last-mentioned species is on the increase in western Europe on drier substrata, especially in vegetation developing in gravel- and cinder-paths near railway stations and harbours. Such occurrences have an adventitious character. Vegetation with species of the *Nanocyperion* has been described by DIEMONT, SISSINGH & WESTHOFF (1940) and is mentioned by TÜXEN (1947). The differentiating species include *Juncus bufonius* and *Gnaphalium uliginosum,* species whose seeds can only germinate in damp soil and exhibit a decrease in abundance in dry years, especially when the germination period in spring is dry (compare ELLENBERG & SNOY, 1957). This holds for *Sagina procumbens* too (see 8.2) but to a lesser extent. However, after their germination and juvenile phases these species can withstand a moderate and not too prolonged desiccation. Apart from the species mentioned, the following are also characteristic of this kind of habitat: *Veronica serpyllifolia, Plantago intermedia, Pohlia camptotrachela* and *Atrichum tenellum,* and, furthermore, *Spergularia rubra* is quite common.

In vegetation with *Corrigiola littoralis, Polygonum lapathifolium* is of frequent occurrence. The relation with *Herniaria* vegetation has been mentioned (see 8.2.3), the affinity being especially striking in the communities described by PASSARGE (1965) from the banks of the river Elbe near Magdeburg.

For descriptions of such vegetation types I refer to the relevant literature, because I have too few relevés at my disposal (which were exclusively recorded from paved sites, for that matter). It is likely that the communities under discussion will become rarer as pollution increases. In a number of localities mentioned by Diemont et al., there was no trace of vegetation related to the *Nanocyperion* and elsewhere it was only fragmentarily developed, but on a number of sand and gravel paths well-developed stands of this community were found.

In sites rich in nitrogen which remain damp for considerable lengths of time, *Marchantia polymorpha* or *Lunularia cruciata* frequently occur. The latter species is especially common in southern Europe, but is also found in western and central Europe in localities where the effect of frosts is negligible e.g. in greenhouses and in gardens. *Lunularia* was found twice together with *Marchantia,* but it is possible that *Lunularia* forms a community of its own which is physiognomically, ecologically and floristically related to the *Marchantia* community. Such communities particularly occur in places where much rain water drains off (e.g. near leaking rain pipes) or where much water is sprayed (in gardens and parks). Some examples are shown in Table 58.

Table 58 Communities of Poa annua and of Marchantia polymorpha and/or Lunularia cruciata

Nr C	1	2	3	P	U W	MGT	MGP
Date	19-9	26-9	17-10	*10st*			
Year	62	63	63				
Nr R	703	478	498				
CH	15	20	15				
CM	20	80	80				
Sagina procumbens	2a	2a	2b	*100*	+p-3a	10	10
Poa annua	1b	2a	1p	*100*	+r-2a	4	4
Ceratodon purpureus	2m	2b	2p	*90*	+p-3a	8	9
Bryum argenteum	+p	+p	-	*80*	+p-2a	2	2
Plantago major	+r	-	-	*50*	+r-2a	0.8	2
Taraxacum (div.) spec.	+p	-	-	*40*	+r-+a	0.2	0.4
Bryum capillare	-	-	-	*70*	2a-3a	9	13
Bryum caespiticium	-	-	-	*30*	+p-2m	0.4	1
Marchantia polymorpha	2b	2a	-	*90*	2a-4a	20	22
Barbula unguiculata	1p	3b	+p	*70*	+p-3b	6	9
Epilobium roseum	1b	-	-	*40*	+r-2m	1	2
Lunularia cruciata	-	-	3b	*30*	2m-3b	9	30
Stellaria media	-	-	-	*20*	+p-2b	2	18
Senecio vulgaris	-	-	-	*20*	+r-2a	3	16
Chamaenerion angustifolium	+r	-	-	*20*	+r-+a	0.2	0.8
Bryum creberrimum	1p	-	-	*20*	1p	0.1	0.4

In addition in 1: Epilobium parviflorum +r; in 2: Rhynchostegium confertum 1a; Poa pratensis 1p; Amblystegium serpens 1p; in 3: Rhynchostegium murale 3a; Epilobium montanum +a; Oxyrrhynchium praelongum +p; Hypnum cupressiforme var. tectorum +p.

LEGEND

1. Leeuwarden (Friesland, Netherlands), alley leading off from the Grote Kerkstraat, near spout of rainpipe.
2. Maastricht (Limburg, Netherlands), path in St. Servaas churchyard at Vrijthof.
3. Hoensbroek (Limburg, Netherlands), path in churchyard.

The characteristic group of species includes *Barbula unguiculata* and *Epilobium roseum,* but also other species of *Epilobium* and *Chamaenerion angustifolium* may occur, the latter in particular in England and in Scotland e.g. on not excessively trampled approaches to basement-stories. It is possible that *Bryum creberrimum* is also a characteristic species, This moss, previously considered to be extremely rare in the Netherlands, was twice recorded with capsules. The total absence of *Polygonum heterophyllum* is striking. The stands, therefore, have but little pioneer character, *Capsella bursa-pastoris* having only been recorded once in a transitional stage towards a *Poa annua-Stellaria media* community (which may be more or less related). There is also an affinity with trampled vegetation in which *Soleirolia soleirolii* occurs and which was recorded as relevés in two localities in southern England, the sites being kept moist by rain water flowing down (see Table 59).

Possibly such trampled communities are more common in the eu-atlantic parts of western Europe. It will be clear that these relevés show a strong resemblance to those of mural vegetation with *Soleirolia* (see 6.10).

Table 59 Community of Poa annua and Soleirolia soleirolii

Nr C	1	2		1	2
Date	3-8	4-8	Trifolium repens	-	+p
Year	64	64			
Nr R	404	440			
CH	40	60	Soleirolia soleirolii	2b	3b
CM	50	20	Brachythecium rutabulum	2a	2b
			Barbula unguiculata	1p	+p
Poa annua	2m	2a	Epilobium montanum	+r	+p
Sagina procumbens	2a	2m	Tortula muralis (var. aestiva)	2a	-
Bryum argenteum	2p	+p	Rhynchostegium muralis	2a	-
Taraxacum (div.) spec.	+p	+p	Linaria cymbalaria	+a	-
Plantago major	-	+a	Lunularia cruciata	1p	-
Ceratodon purpureus	-	+p	Hypnum cupressiforme	+p	-
			Veronica officinalis	-	1p
Bryum caespiticium	2m	1p	Oxyrrhynchium praelongum	-	1p
Bryum capillare	+p	-	Prunella vulgaris	-	+r

LEGEND
1. Totnes (Devon, England), Atherton Lane.
2. Falmouth (Cornwall, England), garden path of Melvill Road opposite Sea View Road.

8.2.6 Vegetation of paved surfaces containing wall-dwelling plants

Apart from vegetation which includes *Soleirolia,* sometimes, and especially close to the base of walls, communities are encountered which contain species that predominantly grow on walls, such as *Linaria cymbalaria* and *Parietaria judaica.* Such stands with *Parietaria judaica* are much rarer outside the Mediterranean region, vicinism nearly always playing a role (presumably on account of the dispersal mechanism of the two species mentioned). All seven relevés from the Mediterranean area are from Italy and they consistently include *Parietaria judaica,* whereas this species is absent from the three relevés from western Europe (Table 60).

Table 60 Communities of Poa annua and Parietaria judaica and/or Linaria cymbalaria

Nr C	1	2	3	P 10st	U W	MGT	MGP
Date	27-9	5-5	5-5				
Year	65	65	65				
Nr R	1263	145	144				
CH	60	25	25				
CM	-	s	2				
Poa annua	3a	2m	2a	100	1a-3a	13	13
Sagina procumbens	-	2m	2m	70	+p-2b	2	3
Taraxacum (div.) spec.	2a	-	+p	60	+p-2a	0.9	1
Ceratodon purpureus	-	+p	2m	60	+p-3a	4	8
Plantago major	+a	-	-	50	+r-1b	0.4	0.8
Bryum argenteum	-	-	-	20	2p	0.1	0.4
Capsella bursa-pastoris	+p	-	-	20	+r-+p	<0.1	0.1
Polygonum heterophyllum	2a	-	-	1x			
Bryum caespiticium	-	-	1p	30	+p-2m	0.3	1
Parietaria judaica	-	2a	2a	70	2m-4a	11	16
Stellaria media	+p	-	2a	70	+p-2a	2	3
Linaria cymbalaria	2a	1b	-	60	+p-3a	9	15
Sonchus oleraceus	+p	+a	-	60	+p-+a	0.2	0.3
Oxalis corniculata	-	-	1a	40	1a-2a	1	3
Veronica cymbalaria	-	1b	1p	20	1p-1b	0.4	2
Senecio vulgaris	1b	-	-	20	+r-1b	0.4	2
Euphorbia peplus	-	-	+p	20	+r-+p	<0.1	0.1
Centranthus ruber	+p	+r	-	20	+r-+p	<0.1	0.1

In addition in 3: Polycarpon tetraphyllum, Erigeron naudini +p.

LEGEND
1. Autun (Saône-et-Loire, France), nr 6, Rue St. Andoche, alongside a wall.
2. Florence (Tuscany, Italy), Belvédère.
3. Florence, near nr 50, Via di S. Leonardo, alongside a wall.

The resemblance to Mediterranean *Parietaria* vegetation is manifest, but there is also a strong affinity with the community of *Poa annua* and *Stellaria media.*

8.2.7 Other communities of paved surfaces

In a number of localities in southern and central Europe up to about 2000 m alt., relevés of other tread-resistant communities were made. The total number of relevés recorded is too low to permit any conclusions other than those that can be gleaned from the phytosociological literature. The records include relevés with *Eragrostis* from Czechoslovakia, with *Sclerochloa dura* and *Sagina apetala* from the Mediterranean area, and with *Poa supina* from the montane regions. All these undoubtedly represent distinct vegetation units. Apart from various species of *Eragrostis*, *Cynodon dactylon*, *Lepidium ruderale* and *Euphorbia humifusa* are characteristic of the S. and E.-European vegetation types. There are diverse *Eragrostis* communities, e.g. vegetation with *E. pilosa* (compare BRAUN-BLANQUET, 1949), and with *E. poaeoides* (with, *inter alia*, *Euphorbia maculata*, see OBERDORFER, 1957); the latter perhaps 'merges' in the Mediterranean area with *Sclerochloa dura* communities, which are related to communities with *Eragrostis* anyhow (compare e.g. the relevés in PIGNATTI, 1953). In these vegetation types *Sagina apetala*, *Capsella rubella*, *Crepis bursifolia*, *Polycarpon tetraphyllum*, *Euphorbia chamaesyce*, *Eu. humifusa* and *Cynodon dactylon* are differentiating species in respect of the *Sagino-Bryetum argentei* and allied vegetation types.

Of the *Sclerochloa dura* communities, one type occurring in coastal areas shows a strong resemblance to the community of *Poa annua* and *Plantago coronopus*. It contains, apart from *Plantago coronopus*, *Coronopus squamatus* and perhaps also *Euphorbia peplus* (compare BRAUN-BLANQUET et al., 1952; BLONDEL, 1941).

Vegetation types reported in the phytosociological literature, but which I did not come across, are the *Plantagini-Sporoboletum* and the community of *Tillaea muscosa* and *Spergularia capillacea* of Atlantic Spain (BRAUN-BLANQUET, 1967). The latter community is structurally very similar to the vegetation types previously discussed.

Vegetation with *Poa supina* and *Plantago alpina* I found in the French Alps. Such communities have also been described from Davos by BRAUN-BLANQUET (1949) and from S.W. Germany by OBERDORFER (1957), who speaks about a *Poa variae-Saginetum* prov., and by LIPPERT (1966).

8.3 Vegetation with Poa annua - a summary

If one attempts to classify the communities of paved sites using the system of Braun-Blanquet, one runs into the same inadequacy of the hierarchic and monolithic system of classification. When striving for a synthesis, the drawing of border lines and the recognition of vegetation units prove to be arbitrary. One can, for instance, take a very broad concept of the *Sagino-Bryetum argentei* as a starting-point and include all types enumerated in this chapter, except those mentioned in the last paragraph, in this association.

One could also go so far as to distinguish certain communities on dominance such as a *Polygonetum avicularis*, a *Poetum annuae* and a *Saginetum procumbentis*, but these assemblies have in this case a meaning totally different from that of the 'associations'. (However, if one defines the concepts of a faithful species and a differentiating species in a very special and restricted sense, these communities can be fitted into Braun-Blanquet's system: compare the definition of WESTHOFF, 1965. Although Westhoff would himself undoubtedly apply the concept of a faithful species in a different sense tot that used here, his definition implies such a fit). If it is permissible to base associations on groups of differentiating species, one can, for instance, regard the community of *Poa annua* and *Plantago coronopus* as an association or as a subassociation. The alternative interpretations are as follows:

Sagino-Bryetum argentei	Sagino-Bryetum argentei
inops	inops
	matricarietosum (matricarioidis)
stellarietosum (mediae)	stellarietosum (mediae)
artemisietosum (vulgaris)	artemisietosum (vulgaris)
var. with Amaranthus deflexus	amaranthetosum deflexi
gnaphalietosum (uliginosi)	gnaphalietosum (uliginosi)
var. with Corrigiola littoralis	corrigioletosum (littoralis)
marchantietosum	marchantietosum
? var. with Lunularia cruciata	? lunularietosum
parietarietosum (judaicae)	parietarietosum (judaicae)
sedetosum (acris)	sedetosum (acris)
plantaginetosum coronopidis	Coronopi squamatae-Poëtum annuae
herniarietosum (glabrae)	Herniario-Poëtum annuae
polygonetosum aequale	? Polygonetum aequale
soleirolietosum	referable to Soleirolietum
dryopteridetosum	referable to Filici-Saginetum

Together with the related types of mural vegetation these vegetation types can be united into an 'association assembly' (compare SEGAL, 1963a), the types treated in this chapter being referable to the *Polygonion avicularis* and the mural types to the *Tortulo-Linarion cymbalariae*. A diagrammatic representation is shown in Fig. 5. The phytosociological affinities between species and species groups have been estimated, but will be discussed in a forthcoming paper in which it is compared with affinity coefficients obtained by means of other methods. In the figure one can visualise transitional cases between at least all groups of species connected by lines. In reality the number of phytosociological relations between these groups of species is of course appreciably larger.

The association assembly can be extended to include e.g. other vegetation types belonging to the *Polygonion avicularis*, such as those discussed in the previous paragraph. It may include also vegetation of damp forest paths with *Glyceria declinata*, or vegetation types referable to other classes which are at least structurally related and possess combinations of species of the same genera, such as the various types of the *Saginetea maritimae*, and, finally, also transitions to other communities, (e.g. to the *Centaurio-Saginetum saginetosum danicae*). Transitions of the *Sagino-Bryetum argentei* to other vegetation types also exist, e.g. to *Puccinellia distans* vegetation, to *Bidention* vegetation, and even to vegetation types of hygroseres. In principle the number of 'subassociations' can be extended to include transitions to a considerable number of higher units of the classification system. The allocation of the *Sagino-Bryetum argentei* in the *Plantaginetea majoris* is of course floristically in order. However, a floristic system can conceivably also be based on taxa higher than the species, the structure being taken into account as well. Consequently the ecology can clearly be defined even if the focal point may be subject to shifting. In such a system at least the *Polygonion aviculare* and the *Saginetea maritimae* communities could be united. Apart from a number of subgenera or sections of other genera such as *Poa, Plantago, Sagina, Spergularia, Coronopus* and *Catapodium* would constitute the characteristic taxa. Such a vision of a system of classification might result in a more thorough integration of the greatest possible number of criteria which is not the case with the Franco-Swiss School of phytosociology.

The matrix set up for the affinities of the various species is shown in Table 61, from which arguments can be found pleading in favour of a separate status of stands with *Herniaria glabra* apart from some deviating spectra.

The ploidy spectrum of stands of vegetation of streets and other paved surfaces is as follows:

	unweighted values	weighted values
2n	36	39
2n + 4n	6	1
4n	28	29
2n (+ 4n) + ⩾ 6n	10	10
⩾ 6n	10	8
4n + ⩾ 6n	10	14

These values do not vary substantially among the various communities, only the representation of the category 4n + ⩾ 6n being appreciably higher in the stands with *Sedum acre* than in the remainder of the communities (the scores being U 17% and W 48% as against U 3-11% and W 3-12%, respectively).

The life-form spectrum according to Raunkiaer yields the following results:

	unweighted values	weighted values
therophytes	44	43
geophytes	1	0.3
hemicryptophytes	49	45
chamaephytes	6	12
phanerophytes	0.3	0.0

The contribution of the therophytes is most substantial in the stands with *Herniaria glabra* (their score: U 13%, W 46%).

The life-form classification according to Iversen yields the following spectrum:

	unweighted values	weighted values
seasonal xerophytes	3	9
euxerophytes	2	1
hemixerophytes	11	9
mesophytes	71	58
hygrophytes	14	22

Table 61 Matrix of species recorded from trampled sites
See text under 8.3. and 6.13.

1	2	3	4	5	6	7	8	9	10	11	12	13	14	15	16	17	18	19	20	21	22	23	24	25	26	27	28	29	30	31	32	33	34	35	36	37	38	39	40	41	42	43	44	45	46	47	48	
				X			+									x							+																									1 Achillea millefolium
				+		+				X						x							x																									2 Agrostis stolonifera
				X					x				X																																			3 Aira praecox
													X							X							x			+																		4 Atriplex hastata
							x									x							x			X			X																			5 Bromus mollis
									o			+				x	X			-	X	+		+	+						x					x												6 Capsella bursa-pastoris
																																					x							X				7 Cerastium fontanum
														+							o					x			x	x													+			x		8 Cerastium semidecandrum
																X	x			x					x		x																					9 Coronopus squamatus
																																		x														10 Cynodon dactylon
																X		+					X					o				+			+													11 Dactylis glomerata
																									-																	X	x		X			12 Epilobium roseum
																							+									x									+							13 Erigeron canadensis
																x				X							X			X			x													+		14 Festuca rubra
																				o				X		X		O	X						X					o		-						15 Herniaria glabra
																			X		-											X																16 Linaria cymbalaria
																							X					-								x						o						17 Lolium perenne
																					X				X												x											18 Matricaria matricarioides
																					O							-																				19 Medicago lupulina
																									O							+			x													20 Parietaria judaica
																								o			X			x			X									O						21 Plantago coronopus
																						x			+																-						O	22 Plantago major
																							-			o		X	o	o										x								23 Poa annua
																												o	x						X					o								24 Poa pratensis
																										X		O	+	-													-					25 Polygonum aequale
																																			-		+		o			o			O		-	26 Polygonum heterophyllum
																												O	X					+						o								27 Potentilla argentea
																																	X							O	-							28 Sagina maritima
																													o			o		-						X		+	X			x		29 Sagina procumbens
																																		x						-								30 Scleranthus annuus
																																		o	-								x			X		31 Sedum acre
																																x			x					o								32 Senecio vulgaris
																																								o								33 Sonchus oleraceus
																																				O							-					34 Spergularia marina
																																					X					o	o					35 Spergularia rubra
																																				x												36 Stellaria media
																																																37 Taraxacum (div.) spec.
																																										-						38 Trifolium repens
																																											x		X			39 Barbula unguiculata
																																								+	+	+	+				X	40 Brachythecium rutabulum
																																									X	X	X		+	x		41 Bryum argenteum
																																										X	X				X	42 Bryum caespiticium
																																											X	+	X			43 Bryum capillare
																																													X	x	x	44 Ceratodon purpureus
																																																45 Funaria hygrometrica
																																																46 Marchantia polymorpha
																																																47 Streblotrichum convolutum
																																																48 Tortula muralis

Seasonal xerophytes are only of great significance in stands with *Sedum acre* (their score being U 8% and W 44%) and are for the rest exclusively found in stands with *Plantago coronopus* (score: U 3%, W 5%) and with *Herniaria glabra* (score: U 1%, W 2%). In inops stands and in stands with *Parietaria judaica* euxerophytes are also absent. In stands with *Plantago coronopus* hemixerophytes are more numerous than in all other communities (the score being: 19% U, 25% W against 1-7% U and 0.0-5% W, respectively). Hygrophytes attain high scores in the inops stands (27% U, 49% W), and low scores in stands with *Herniaria* (6% U, 5% W).

The growth form spectrum shows the following picture:

	unweighted values	weighted values
trees and shrubs	0.2	0.0
liliids	8	3
echiids	32	12
poids	28	34
Asplenium-type	1	2
fumariids	2	2
fragariids	14	17
orthotropic bryophytes	15	27
plagiotropic „	2	4

Liliid forms are most important in stands with *Herniaria glabra* (their score being: U 12%, W 15%). The *Asplenium*-type is only represented in stands with *Parietaria judaica* (the score being: U 9%, W 23%), in which stands the fumariids are also well represented (their score is 10% U and 21% W as against 0-3% and 0-2% W in the rest of the stands).

Orthotropic mosses are particularly numerous in the inops stands (representation: 10% U, and 21% W, as against 0-3% U and 0-2% W in better developed stands), plagiotropic bryophytes in stands with *Marchantia polymorpha* (representation: 6% U and 17% W).

The dissemination spectrum is as follows:

	unweighted values	weighted values
vegetative reproduction	0.1	0.0
autochory	3	7
anemochory, dissemination capacity large	0.4	0.0
„ „ „ medium	24	7
„ „ „ low	13	8
hydrochory	2	1
endozoochory	1	0.3
epizoochory	0.4	0.0
myrmecochory	3	8
anthropochory	53	69

Autochory is especially common among the species of *Parietaria* communities (the score being 20% U and 36% W), and so is myrmecochory (the score: 20% U, 36% W). Human interference is always very considerable. Diaspores of many species can adhere to footwear and thus be dispersed, as was demonstrated by e.g. CLIFFORD (1956). Dispersal by birds feeding on the achenes of grasses (*Poa annua!*) has possibly been underestimated in this spectrum (compare GREEN, 1961).

The formation spectrum yields high scores for species of communities of dry, open sites (U 14%, W 12%), especially of stands with *Herniaria glabra* (U 35%, W 44%) and with *Sedum acre* (U 24% and W 26%), but such species are not represented in inops stands. Halophytes are restricted to the community with *Plantago coronopus* (their score being: 3% U and 3% W). Ruderal species also attain high scores (U 17%, W 7%), but especially the category "vegetation of contact zones and various" is very well represented (the score being: U 36%, W 42%). The majority of the bryophytes can grow both terrestrically and epilithically (W 10%, and U 17%, respectively), or they are indifferent in respect of the substratum (W 7%, U 13%; in the spectrum bryophytes and vascular .plants have been taken together).

The distribution spectrum shows that by far the greatest number of species have a large area of distribution:

	unweighted values	**weighted values**
Cosmopolitan	35	35
Supraholarctic	4	3
Holarctic	24	30
Eurasiatic	17	10
European	10	11
Mediterranean	1	1
Mediterranean-Atlantic	3	3
Mediterranean-Atlantic-C. European	5	7

All other categories have scores under 1%. The Mediterranean and Mediterranean-Atlantic categories include mostly species of the communities with *Parietaria judaica* (U 9% and W 3% for Mediterranean species, U 13% and W 26% for Mediterranean-Atlantic taxa). The distribution inside Europe yields scores of 86% (U and W) for species with a pan-european distribution, and, furthermore, fairly high scores for the Mediterranean-Atlantic-C. European distribution pattern (U 4%, W 9%) and the Mediterranean-Atlantic distribution (U 4%, W 3%).

9 Some phytogeographical and taxonomic aspects of mural vegetation

The attention of many a worker has been drawn to plants found growing on walls. Incidental floristic inventories have been made in the past in a number of towns, mainly in France and in Italy. One of the first reports was from SEBASTIANI (1815) who studied the Flavian amphitheater in Rome. Older studies in Italy include a publication of DEAKIN (1855) and several publications by MAZZINI (1875-1877) on the Colosseum in Rome, by DAMANTI (1903) of walls in Palermo, by DA ROSA (1905) of walls in Naples, by BEGUINOT (1911-1915) of walls in Padua and by GABELLI (1915) of walls in Siena.

From France reports are available concerning the area of the Vosgues (CHATIN, 1861; KIRSCHLEGER, 1862), Gisors (LEPAGE, 1861), Paris (VALLOT, 1884; 1887), Poitiers (RICHARD, 1888), and Cercy-la-Tour (GAGNEPAIN, 1897). In Britain, RISBETH (1948) compiled a list of wall-dwelling plants from Cambridge, WOODELL & ROSSITER (1959) of mural vegetation at Durham, and KENT (1961) of species in Middlesex.

BARNEWITZ (1898) compiled a list of wall-dwelling plants in Brandenburg (Germany) and JOURDAN for Oran (1867) and Algiers (1872). Floristic surveys in the Netherlands were made of the quasy-sides of Amsterdam (MEYER, 1943; VAN KONINGSDAAL et al., 1956), of Dordrecht (GUITTART, 1957), of Utrecht (C.J.N., 1959) and of Amersfoort (BEYLSMIT & MATEN, 1965). Previous incidental records from the Netherlands likewise concern only the floristic aspects (DE WEVER, in at least 12 lists of records between 1911 and 1943). Most of the authors mentioned did not restrict their inventories to plants growing on more or less vertical wall surfaces and it must be borne in mind that a large number of species which may become established on horizontal ledges and wall tops are of rare occurrence, or lacking altogether, on vertical mural sites. Nevertheless, the total number of vascular plants recorded from vertical walls in the area studied amounts to at least 1,200 and the number of bryophytic species is at least 300. A complete list is not given here because the purpose of the present investigation is not an inventory of this kind but an ecological and phytosociological study. In a forthcoming publication all species recorded from walls in the Netherlands will be enumerated. The list of species constituting Appendix III contains only those species which appeared in the published relevés.

For each taxon listed supplementary data are provided which concerns the taxonomic position, the degree of ploidy, the life form and growth form, the floral colour (if any), the mode of dissemination, the distribution and the area occupied in Europe, and the landscape element in which it usually occurs. The various classificatory criteria are mentioned in 2.3 and in 4.6. The various codes (and code symbols) are given in Appendix II. The respective data are combined into spectra for each community (or other syntaxonomic unit) and into spectra for the different geographical areas (the delimitation of these areas is shown in Fig. 1, Appendix I). The spectra will be discussed in this chapter.

During the field studies, in a number of sites notes were made for certain taxa, concerning the ecology, and sometimes also the method of dispersal and the taxonomy. Some of the more interesting taxonomic and ecological aspects, mostly relating to "critical" species and aggregate species or species groups, and to the role of naturalised plants and neophytes in mural vegetation, will be discussed in more detail.

9.1 Taxonomic spectrum

In Table 62 a comparison is made between the taxonomic spectra of the eu- and subatlantic regions combined and the eu- and submediterranean regions combined. Only those taxonomic ordines are included whose score in at least one of the columns is at least 1.0% (Conifers excepted).

Table 62 Taxonomic spectra of mural vegetation of (1) the eu- and subatlantic regions (111, 112, 120, 131, 132), of (2) the eu- and submediterranean regions (310, 320, 331, 332), and of (3) all data (inclusive 200 and 400), compiled from unweighted numbers (U) and weighted numbers (W)

	1		2		3	
taxon	U	W	U	W	U	W
Marchantiales	0.4	2.8	0.3	0.2	0.4	2.5
Dicranales	3.4	4.1	0.0	0.0	2.5	3.2
Grimmiales	1.1	1.2	0.7	0.2	1.0	1.0
Pottiales	12.2	13.7	8.0	8.0	11.2	12.3
Bryales	8.7	6.1	1.6	0.6	7.0	4.1
Leucodontales	0.1	0.3	1.3	8.2	0.4	1.6
Hypnales	6.7	12.6	2.3	6.9	5.5	10.8
Aspidiales	13.5	14.3	5.4	7.1	11.7	12.9
Pteridales	0.0	0.1	0.6	1.6	0.2	0.4
Polypodiales	1.7	1.6	1.0	1.1	1.5	1.8
Bryophyta	34.3	40.9	15.0	21.4	29.6	36.6
Pteridophyta	15.5	16.7	7.2	9.9	13.6	14.7

	1		2		3	
taxon	U	W	U	W	U	W
Poales	9.9	2.4	9.9	0.8	9.9	2.1
Urticales	2.1	8.8	9.2	37.6	3.8	14.9
Caryophyllales	4.8	2.6	3.3	0.4	4.4	2.1
Geraniales	0.7	0.2	3.5	0.5	1.4	0.3
Apiales	1.4	3.9	1.5	3.8	1.4	3.8
Rubiales	1.6	1.7	2.7	3.3	1.9	2.1
Rosales	3.3	1.0	10.5	6.1	5.0	2.1
Myrtales	3.1	0.3	0.1	0.0	2.4	0.2
Brassicales	3.8	6.6	6.2	5.3	4.3	6.5
Solanales	7.0	11.7	12.3	6.3	8.3	10.5
Asterales	9.0	2.9	13.5	3.8	10.3	3.1
Coniferopsida	0.2	0.0	0.1	0.0	0.2	0.0
Monocotyledones	10.0	2.4	10.7	0.9	10.1	2.1
Dicotyledones	39.5	40.0	67.1	67.7	46.1	46.0

Some salient differences are noticeable in the groups of the *Dicranales, Leucodontales* and other bryophytic ordines; and of the *Urticales, Geraniales, Apiales, Rosales, Myrtales* and *Solanales.* The differences between unweighted and weighted values are particularly large in the *Leucodontales (Scorpiurium circinatum), Hypnales (Homalothecium sericeum* and *Hypnum cupressiforme), Poales (Poa* spec. div.), *Urticales (Parietaria judaica, Soleirolia soleirolii), Caryophyllales (Cerastium), Geraniales (Geranium), Rosales (Sedum, Rubus), Myrtales (Epilobium), Solanales (Linaria)* and *Asterales.* The spectra not included in the present report show relatively high scores for the *Hypnales* in area 131 (U 8.9; W 17.8) and 132 (U 8.9; W 12.8), the *Polypodiales (Poly-*

podium) in areas 111 (U 3.7; W 1.7) and 112 (U 2.8; W 0.9), and the *Asterales* in area 331 (U 18.4; W 6.8). Furthermore, the *Dicranales (Ceratodon)* are strongly represented in area 120 (W 7.8), and the *Rosales* in area 310 (U 14.5), and the *Brassicales* are poorly represented in areas 111, 112, 120 and 200 (U 2.0-3.4).

The discontinuities observed in these and in other distributional spectra are of course not quite real but nevertheless indicate a certain trend. The areas included in the investigation are more or less isolated from one another by geographical barriers (mountain chains, etc.). A large number of species not restricted in their distribution to a Mediterranean, a eu-atlantic or a central European region exhibit certain distributional trends, e.g.

Asplenium ruta-muraria becomes more frequent from the Mediterranean region through the submediterranean areas to the more Atlantic ones;

Phyllitis scolopendrium is increasingly more common from a continental climate through a subcontinental one to the eu-atlantic zone;

Ceterach officinarum exhibits the same trend as *Phyllitis,* but also increases in number from a Mediterranean to a submediterranean climatic area.

Gymnocarpium robertianum increases in numbers from the eu-atlantic parts to the continental regions and from the lowlands to the montane zones, and the same applies to *Cystopteris fragilis*;

Poa compressa increases from the eu-atlantic and submediterranean through subatlantic to continental areas;

Parietaria judaica increases from continental areas towards Atlantic and towards Mediterranean zones;

Cheiranthus cheiri becomes gradually more common from continental regions towards Atlantic ones and from continental towards submediterranean ones;

Corydalis lutea is more frequently recorded from subatlantic areas than from eu-atlantic or continental ones;

Leptobryum pyriforme increases going from eu-atlantic areas towards more continental ones, and *Bryum murorum* exhibits the reverse trend.

It follows that communities in which these (and similarly behaving) species are conspicuous elements likewise shift and peter out.

Each of the climatic regions also have a number of more or less unique taxa e.g. area 111: *Buddleja davidii* and *Senecio squalidus*; area 112: *Linaria purpurea*; 111 and 112 together: *Spergularia rupicola* and particularly *Asplenium adiantum-nigrum;* 120: many riparian forms such as *Lycopus europaeus* and species of *Rorippa*; 131: *Campanula rotundifolia, Cardaminopsis arenosa, Erysimum cheiranthoides, Saxifraga rosacea* ssp. *sponhemica, Rumex scutatus* and *Rhynchostegium murale;* 112 and 132 together: *Petrorhaga prolifera*; 131 and 132 together: *Ctenidium molluscum, Encalypta streptocarpa* and *Tortella tortuosa*;

200: *Asplenium viride, Polystichum lonchitis* and *Distichium capillaceum*; 310: *Chaenorrhinum origanifolium, Valerianella eriocarpa, Arabis hirsuta* and *Pylasia polyantha*; 320: *Capparis spinosa, Helichrysum italicum, Hyoseris radiata, Senecio lividus, Urtica atrovirens, Hypericum perforatum* and (in a coastal strip a few dozens of kilometres wide) *Lobularia maritima*; 331: *Centaurea aspera, C. solstitialis* and *Cheilanthes fragans*; 320 and 331 combined: *Antirrhinum latifolium, Parietaria lusitanica, Anogramma leptophylla* and *Selaginella denticulata*; 332: *Calamintha nepeta*; and 400: *Leontodon hispidus, Plantago media, Diplotaxis erucoides, Orthotrichum lyellii* and *Leucodon sciuroides*. Some species appear regularly on walls in certain areas, but are elsewhere frequently found in different habitats, and often more or less exclusively so. This is, for instance, the case with *Veronica hederifolia, Erophila verna* and *Cardamine hirsuta* (wall-dwelling in the Mediterranean area), *Crepis vesicaria* ssp. *taraxacifolia* (in mural vegetation in S. Europe and in Brittany), and *Geranium molle* (found on walls in Tuscany). Conceivably different intraspecific taxa are involved (ecological races?).

From a floristic point of view, of the Mediterranean areas in relatively close proximity of one another, 331 is more or less intermediate between 320 and 332. A corresponding combination of species is often encountered in 320 at higher altitudes than in the cooler and wetter climate area 332.

9.2 Ploidy spectrum

Some ploidy spectra are shown in Table 63. Apart from the various categories of the degree of ploidy, in imitation of PIGNATTI (1960), the ploidy index *d/p* is given in which *d* is the percentage of diploids (symbol "2") and *p* the percentage of polyploids (symbols "4", "6" and "7" combined).

Table 63 Ploidyspectra from different geographical regions, sampled from unweighted numbers (U) and weighted numbers (W)
For a detailed explanation of the symbols see under 2.3.2.

	geographical region	400		131+132		120		111+112		310		331+332		320		total	
symbol	ploidy	U	W	U	W	U	W	U	W	U	W	U	W	U	W	U	W
2	2n	26.2	18.6	31.8	41.0	37.0	55.7	34.9	57.4	36.4	51.4	43.7	63.5	45.9	82.2	36.7	54.7
3	2n+4n	15.2	16.2	18.1	12.7	11.5	12.5	16.1	10.6	11.6	9.8	11.8	7.3	9.5	1.8	14.3	10.4
4	4n	24.1	6.7	23.2	28.5	25.6	22.5	28.1	25.0	25.9	25.7	19.2	20.0	18.6	10.4	23.9	23.3
5	2n(+4n)+≥6n	10.3	48.2	5.6	11.2	4.8	4.2	2.7	3.4	4.7	0.5	2.7	1.0	5.2	0.0	4.5	5.8
6	≥6n	19.3	9.5	14.3	4.8	12.8	3.7	10.6	1.7	5.3	5.0	10.5	7.1	8.5	2.3	11.6	3.9
7	4n+≥6n	4.8	0.9	6.8	1.8	8.3	1.4	7.4	1.9	16.1	7.6	12.0	1.9	12.1	2.4	8.9	2.0
index: d/p		0.54	1.09	0.72	1.17	0.80	2.02	0.76	2.01	0.77	1.34	1.05	2.17	1.17	6.44	0.83	1.88

The category of the hybrids is negligible and was omitted. The data indicate that the percentages of weighted and unweighted values of the same category may be widely divergent. In all cases where either relatively high or relatively low values appear, in a horizontal column of unweighted values, this effect is often amplified in the weighted ones (compare, for instance, the category "2" for the areas 400 and 320, category "7" for area 400, and the categories "3", "5" and "6" for area 320). There are, furthermore, some indications of the following trends:

1. Going from E. to W. in Europe, the relative contribution of diploids in stands of mural vegetation increases and that of polyploids falls off;
2. Going from W. and C. Europe to S. Europe, the percentage of diploids increases and of polyploids decreases in mural stands.

In area 120 the ploidy index is a little higher than in the eu-atlantic regions. The spectrum of this area deviates the least from the spectrum of the whole of the material, and the spectrum of area 310 is not very much different either. The greatest deviation is found in the spectrum of area 400 and to a less extent in area 320.

The agreement with the overall spectrum can be evaluated by means of Poore's index. This yields the following values:

Geographical region	Unweighted numbers	Weighted numbers
120	0.97	0.97
310	0.91	0.91
320	0.87	0.69
400	0.85	0.46

It goes without saying that these figures are only very roughly indicative of a geographical variation. The values obtained may perhaps be interpreted as meaning that at the same latitude the degree of ploidy in mural vegetation increases as the climate is more extreme. In this connection a drier climate, which is associated with greater fluctuations in the temperature, seems to have a stronger influence than a higher annual atmospheric precipitation or a higher relative humidity. Walls are by their very nature extreme environments, where these factors are always very important, so that it is not at all surprising that in the truly continental regions good stands of mural vegetation are rare or absent. It would be interesting to compare the ploidy spectrum of the whole flora of e.g. Ireland, England, the Low Countries, W. Germany, E. Germany, Poland and parts of Russia of corresponding latitudes in order to verify if the trends are the same as those shown by the wall-dwelling taxa alone. It is feasible that the effect of extreme habitats is greater than that of more stable environments. The increase in the number of diploids from N. to S. agrees with Hagerup (1931), if the degree of ploidy of mural vegetation is indeed representative of the whole flora of a certain area.

In addition, a sequence is clearly discernible in the weighted values of category "3", these values decreasing regularly from a continental climate to an atlantic one, and from a more northern to a more southern occurrence. The unweighted values of this category of the areas 112 and 132 (18.9 and 18.5, respectively) are perhaps somewhat higher than those of the areas 111 and 131 (14.6 and 18.0, respectively). Of the categories "4" and "7" of area 400, the relative contribution of the total number of the species present exceeds by far the relative contribution of their value; the reverse is the case with category "5". Similar differences exist in e.g. area 320 (for the categories "3", "5", "6" and "7", respectively).

The incidence of tetraploids seems to be the greatest in C. Europe. Category "6" again shows a gradual decrease from E. to W. The various indices presumably give an approximation of the total situation. It is striking in this connection that the indices of weighted values are always much higher than those of unweighted ones. The quantitative share of the diploids usually predominates more strongly than their qualitative contribution and it is possible that this also holds true for natural vegetation in the same regions.

It is not at all surprising that the number of species whose degree of ploidy is unknown is greater in E. and S. Europe than in W. Europe. The general rule is that, as far as mural vegetation is concerned, the weighted contribution of this category is always considerably lower than its unweighted share.

9.3 Spectrum of life forms according to Raunkiaer

In Table 64 the biological spectra of some of the areas are shown. It is clear that on the whole mural vegetation consists predominantly of hemicryptophytes. The therophytes are relatively numerous in the

Table 64 Life form spectra according to Raunkiaer of mural vegetation in (1) Atlantic regions (111, 112, 120, 131, 132), (2) E. European regions (400), (3) C. European montane and alpine regions (200), and (4) Mediterranean regions (310, 320, 331, 332), compiled from unweighted numbers (U) and weighted numbers (W)

		1		2		3		4		total	
		U	W	U	W	U	W	U	W	U	W
symbol	life form										
1	therophytes	14	3	11	8	10	2	35	4	20	3
2	geophytes	1	0.7	4	4	4	10	1	0.3	1	0.8
3	hemicryptophytes	69	60	76	75	71	66	41	67	61	62
4	chamaephytes	12	29	5	10	12	19	17	20	13	26
5	nanophanerophytes	3	8	-	-	1	0.7	6	9	3	7
6	macrophanerophytes	1	0.2	5	8	2	3	0.2	0.0	1	0.2

Mediterranean area, but their weighted contribution is small, many species apparently having a relatively unimportant degree of coverage. Geophytes are only important in the montane and alpine parts of C. Europe, and they are still of some significance in E. Europe. The role of chamaephytic forms is the least in E. Europe, the greatest in S. Europe. In the Atlantic areas some differentiation can be observed:

Geographical region	Unweighted numbers	Weighted numbers
111 + 112	15	41
120	7	15
131	10	24
132	21	35
310	22	33

Where the chamaephytes are more numerous, the number of hemicryptophytes is lower, and *vice versa.* The proportional contribution of chamaephytes in area 310 also differs somewhat from that in the eumediterranean areas and agrees with the values in area 132. The relatively high percentage of nanophanerophytes in the Mediterranean region is attributable to the frequent occurrence of *Capparis spinosa* in area 320. ANZALONE (1951) gives the following biological spectrum of 385 wall-dwelling species found in Italy (unweighted values):

T 42 — G 10 — H 28 — Ch 4 — P 17

His spectrum refers to both vertical and horizontal wall surfaces, but there is a reasonably close agreement with column 4 of Table 64, the greatest deviations occurring in the quantities of geophytes and phanerophytes, which are much higher in Anzalone's spectrum.

The spectra of mural vegetation at Cambridge (compiled by RISBETH, 1948) and at Durham (WOODELL & ROSSITER, 1958) are also in reasonably good agreement with column 1 of Table 64, but also in these cases the spectra were compiled from both vertical and horizontal wall surfaces and the same categories as in Anzalone's spectrum are relatively high (unweighted values):

Risbeth: T 27 — G 5 — H 49 — Ch 4 — P 15
Woodell & Rossiter: T 16 — G 4 — H 55 — Ch 8 — P 17

Therophytes are often abundant on horizontal wall surfaces.

9.4 Spectrum of life forms according to Iversen

Some life form spectra according to Iversen are shown in Table 65. Generally speaking, the spectra of unweighted and weighted values do

Table 65 Life form spectra according to Iversen of mural vegetation in (1) eu-atlantic regions (111, 112), (2) area 120, (3) subatlantic regions (131, 132), (4) E. European regions (200), (5) C. European montane and alpine regions (400), (6) Mediterranean regions (310, 320, 331, 332), compiled from unweighted numbers (U) and weighted numbers (W)

		1		2		3		4		5		6		total	
symbol	life form	U	W	U	W	U	W	U	W	U	W	U	W	U	W
	Vascular plants														
1	saison xerophytes	2	*2*	0.3	*0.1*	3	*2*	2	*0.1*	0.6	*0.8*	8	*7*	4	*3*
2	euxerophytes	0.2	*0.1*	0.1	*0.1*	0.2	*0.0*	0.5	*0.7*	-	-	0.6	*0.2*	0.3	*0.1*
3	hemixerophytes	17	*19*	10	*23*	24	*34*	23	*31*	12	*10*	24	*21*	19	*24*
4	mesophytes	64	*65*	52	*51*	60	*58*	49	*47*	53	*70*	54	*68*	57	*61*
5	hygrophytes	15	*15*	33	*18*	13	*6*	20	*21*	33	*18*	13	*4*	18	*10*
6	telmatophytes	0.5	*0.1*	5	*7*	0.4	*0.3*	-	-	1	*1*	0.1	*0.0*	1	*2*
	Bryophytes														
7	xerophytes	42	*53*	48	*58*	43	*49*	26	*23*	26	*61*	52	*41*	44	*51*
8	mesophytes	29	*27*	12	*13*	27	*25*	44	*33*	29	*11*	30	*50*	24	*25*
9	hygrophytes	3	*6*	4	*7*	4	*13*	8	*16*	4	*16*	6	*6*	4	*9*
0	poikilhygrophytes	25	*15*	37	*23*	26	*14*	22	*28*	42	*22*	13	*4*	27	*16*
relative portion of Bryophytes (%)		28	*30*	36	*48*	38	*44*	36	*34*	34	*40*	15	*21*	30	*37*

not differ so much as they do in the case of a life form spectrum according to Raunkiaer. The number of seasonal xerophytes is highest in the Mediterranean areas, which is of course associated with the prevailing periodicity of the seasons, and the number of therophytes is also large for the same reason. Hemixerophytes are most numerous in mural vegetation in E. and S. Europe. In all areas the contribution of the mesophytes is the highest, and apparently mural vegetation is not adapted to conditions of extreme drought, hygrophytes even playing an important role in areas 120 and 400. The first conclusion is feasible (compare 6.9.1.4), the second at first sight incomprehensible, but understandable if one bears in mind that mural vegetation often prefers the immediate vicinity of water.

9.5 Growth form spectrum and sociability spectrum

The growth forms of herbaceous plants being connected with their sociability (see 4.6.4), these particulars were united in the same table (Table 66). Among the herbs the differences are not very large.

Table 66 Growth form and sociability spectra of mural vegetation in (1) Atlantic regions (111, 112, 120, 131, 132), (2) E. European regions (400), (3) C. European montane and alpine regions (200) and (4) Mediterranean regions (310, 320, 331, 332), compiled from unweighted numbers (U) and weighted numbers (W)

symbol	growth form and sociability class	1 U	1 W	2 U	2 W	3 U	3 W	4 U	4 W	total U	total W
11-14	trees and shrubs	2	0.2	3	4	2	2	4	0.7	2	0.4
21	liliids	11	2	12	5	17	13	23	5	14	3
22	echiids	11	2	13	3	11	2	13	2	11	2
23	poids	21	12	24	8	9	4	14	7	19	11
24	Asplenium-type	12	21	9	10	17	24	19	53	14	28
25	fumariids	6 -	15	3	30	6	20	8	8	7	14
26	fragariids	3	4	1	0.5	1	0.1	4	4	3	4
27-28	corynephorids and androsaciids	0.2	0.0	0.4	0.0	2	0.0	0.1	0.0	0.2	0.0
29	soleiroliids	0.2	2	-	-	-	-	-	-	0.1	1
32-35	orthotropic bryophytes	27	25	29	28	29	26	11	9	23	22
41-46	plagiotropic bryophytes	7	16	5	12	7	8	4	13	7	15
	Vascular plants										
1	sociability 1	35	8	44	20	44	24	45	9	38	9
2	2	65	88	55	80	54	76	55	91	61	89
3	3	0.3	0.1	0.6	0.0	3	0.1	0.5	0.3	0.4	0.1
4	4	0.3	3	-	-	-	-	-	-	0.2	2
5	5	0.3	0.7	-	-	-	-	0.0	0.1	0.2	0.5
	Bryophytes										
1	sociability 1	0.1	0.3	-	-	-	-	-	-	0.1	0.2
2	2	21	21	29	14	33	22	18	13	21	20
3	3	74	67	66	54	59	63	76	82	73	69
4	4	6	11	5	32	8	15	6	5	6	11
5	5	-	-	-	-	-	-	-	-	-	-

The weighted contribution of rosette-plants and other species of "solitary" habit (degree of sociability 1) is always much smaller than the unweighted share, but with poids the reverse is the case. In Mediterranean communities particularly, the chasmophytes are conspicuous components. Soleiroliids are represented by *Soleirolia soleirolii* in areas 111 and 112 (U 0.7, W 7).

Mosses do not play such an important role in mural vegetation in the Mediterranean area as they do elsewhere. Of the orthotropic forms the bryids are always the most important elements (total of U 21, of W 20), but in area 400 the representation of the grimmiids is also appreciable (W 10). The vascular plants mostly belong to types with a degree of sociability (=DS) 1 or 2. The weighted percentages of coverage of taxa with DS 1 are considerably lower than the unweighted values. A DS of 1 is of rare occurrence among bryophytes, a DS of 3 being most common.

It goes without saying that the weighted values yield the best expressions of the structure (as they do in the life form spectra). The unweighted values provide interesting additional information concerning the relative amount of species of each category.

9.6 Spectrum of floral colours

Some spectra of floral colours are shown in Table 67, and also the corrected spectra in which the categories "Pteridophytes" and

Table 67 Spectra of floral colours for (1) Atlantic regions (111, 112, 120, 131, 132), (2) E. European regions (400), (3) C. European montane and alpine regions (200) and (4) Mediterranean regions (310, 320, 331, 332), compiled from unweighted numbers (U), weighted numbers (W), corrected unweighted numbers (CU) and corrected weighted numbers (CW)

symbol	colour of flowers	1		2		3		4		total		1		2		3		4		total	
		U	W	U	W	U	W	U	W	U	W	CU	CW	CU	CW	CU	CW	CU	CW	CU	CW
00	Pteridophytes	24	27	29	23	33	47	8	13	19	23	-	-	-	-	-	-	-	-	-	-
01	membranaceous or scarious-white	8	16	6	1	5	3	10	48	8	25	-	-	-	-	-	-	-	-	-	-
02	white	8	8	10	7	8	3	16	11	11	9	12	14	16	9	13	5	20	30	15	18
03	purple	7	18	3	9	7	16	5	7	7	15	11	32	5	12	11	33	7	19	10	30
04	blue	3	2	3	2	8	6	6	1	4	2	4	4	5	2	12	12	8	2	5	4
05	green	20	8	22	8	13	4	16	2	19	6	30	14	33	10	21	8	20	5	27	12
06	yellow	18	15	17	60	18	18	19	6	18	12	27	27	27	65	29	37	25	16	26	24
07	pink and mauve to lilac	8	1	9	2	8	1	12	6	9	2	12	2	14	2	12	1	15	16	13	4
08	red	3	3	-	-	2	2	4	5	3	4	4	6	-	-	2	4	5	13	4	8

"flowers membranaceous or scarious-white" are disregarded. These spectra indicate that in mural vegetation in montane and alpine regions of C. Europe especially the ferns are numerous, whereas they are not so well represented in the Mediterranean. Plants with membranaceous or scarious-white petals are more important, quantitatively speaking, in the Mediterranean than elsewhere *(Parietaria!)*. White-flowered taxa are also most frequently represented in the Mediterranean mural communities, whereas in C. Europe they are the least represented. Purple-flowered species appear to be more numerous in mural vegetation in W. and in C. Europe than they are in S. and E. Europe. Blue-flowered ones are most frequently found in C. Europe. Graminoids seem to be most commonly found in W. and E. Europe. Yellow-flowered species are conspicuous in all areas but best represented in E. Europe. Pink, mauve and lilac flowers are more important in the Mediterranean region than elsewhere, at least quantitatively. Red-flowered taxa are, generally speaking, rare, and only in S. and W. Europe of some importance *(Centranthus ruber!)*.

No opinion can be given regarding the possible significance of the differences between the various areas.

9.7 Dissemination spectrum

Of the possible dispersal mechanisms, the categories "vegetative reproduction" and "barochory" have not been taken into account in Table 68, their percentages being negligible, and the categories "polychory", "sterile" (= non-reproductive in mural stands) and "unknown" have not been included in the calculations. For a discussion of the various categories the reader is referred to 2.3.3 and 3.1.9.

Autochory is relatively unimportant in area 400, and is of greater significance in the Atlantic areas than in the Mediterranean ones. Anemochory is of relatively frequent occurrence in all areas, but less so

Table 68 Dissemination spectra of the various geographical regions based on unweighted numbers (U) and weighted numbers (W)
DC = dissemination capacity; "great" means: diaspores as fine as dust, "moderate" means: diaspores provided with a pappus, "small" means: diaspores winged

symbol	geographical region / dissemination type	111 U	111 W	112 U	112 W	120 U	120 W	131 U	131 W	132 U	132 W	400 U	400 W	200 U	200 W	310 U	310 W	320 U	320 W	331 U	331 W	332 U	332 W	total U	total W
20	autochory	12	38	11	24	8	19	10	18	14	21	4	9	10	18	10	9	13	8	8	8	10	7	10	17
	anemochory																								
51	DC great	35	35	28	19	28	35	30	43	28	30	35	33	36	60	20	27	8	6	8	10	10	21	24	37
52	DC moderate	23	5	[illegible]	20	21	3	22	5	11	0.4	24	4	17	4	13	6	14	4	20	5	25	0.8	19	5
53	DC small	13	8	8	14	12	5	17	14	27	21	13	9	11	4	31	26	24	7	29	20	20	13	19	12
60	hydrochory	2	1	1	0.3	16	13	2	0.8	0.2	0.0	4	4	0.6	0.0	0.2	0.0	-	-	0.5	0.0	-	-	4	3
	zoöchory																								
71	endozoöchory	2	0.4	4	0.5	1	0.1	4	0.1	3	0.1	4	6	8	3	2	2	8	11	4	0.8	6	0.3	4	2
72	epizoöchory	-	-	0.2	0.0	0.2	0.0	0.2	0.0	-	-	-	-	1	0.0	0.2	0.0	0.2	0.0	0.7	0.0	1	0.0	0.3	0.0
73	myrmecochory	7	33	9	28	5	24	12	19	15	28	8	43	10	19	16	31	30	65	27	67	33	57	14	34
80	anthropochory	6	2	5	1	9	2	4	2	3	0.1	9	3	7	3	4	0.2	3	0.4	3	0.1	5	0.3	5	1

Table 69 Formation spectra of the various geographical regions based on unweighted numbers (U) and weighted numbers (W)

symbol	geographical region / formation type	111 U	111 W	112 U	112 W	120 U	120 W	131 U	131 W	132 U	132 W	400 U	400 W	200 U	200 W	310 U	310 W	320 U	320 W	331 U	331 W	332 U	332 W	total U	total W
	Vascular plants																								
10	rocks, cliffs, walls and slopes of rock debris	26	63	28	61	12	27	22	38	35	62	16	43	28	54	39	66	33	75	35	64	29	67	25	48
20	pioneer vegetation of dry sites	5	1	7	2	3	0.6	9	4	9	2	8	4	6	1	12	0.8	11	2	10	0.8	9	0.6	7	2
30	shallows and mud flats	0.3	0.3	0.6	0.3	-	-	-	-	-	-	-	-	-	-	-	-	-	-	-	-	-	-	0.1	0.1
40	arable land	0.2	0.3	1	0.3	0.4	0.0	0.6	0.0	2	0.1	-	-	-	-	6	0.1	4	0.4	7	0.1	11	0.2	2	0.1
50	ruderal terrains	7	1	7	0.6	7	1	7	1	6	1	9	3	4	2	9	0.3	18	1	18	3	13	0.8	9	1
60	banks, reed lands, marshes and peat bogs	0.4	0.0	0.3	0.3	4	4	0.4	0.2	0.1	0.0	0.9	0.7	-	-	0.1	0.0	-	-	0.3	0.0	0.1	0.0	1	1
71	hay-fields, meadows and other grassy sites	4	0.7	3	3	4	0.6	2	0.1	2	0.1	2	0.1	4	0.1	2	0.4	0.6	0.0	2	0.1	3	0.1	3	0.3
72	vegetation of tall suffruticose plants	0.5	0.3	-	-	1	0.3	0.2	0.0	0.1	0.0	0.9	0.0	-	-	0.1	0.0	-	-	-	-	-	-	0.4	0.1
80	dwarf shrub, scrub and forest vegetation	4	3	5	4	6	3	5	1	1	0.1	12	5	4	4	3	0.5	6	2	4	0.3	3	0.4	5	2
90	contact zones and various	23	14	17	5	24	18	14	5	9	8	16	6	16	16	9	3	13	6	12	9	10	5	17	9
	Bryophytes																								
06	wide-spread not only in pioneer vegetation	-	-	0.1	0.0	0.3	0.8	0.8	0.4	0.3	0.1	0.4	0.0	0.7	0.4	0.7	0.2	0.2	0.1	0.3	0.0	0.6	0.1	0.4	0.3
26	open, generally dry habitats	0.0	0.0	0.2	0.4	0.1	0.1	-	-	0.3	0.2	-	-	0.3	0.0	-	-	-	-	-	-	-	-	0.1	0.1
96	humid habitats	-	-	0.2	0.0	0.0	0.0	0.0	0.0	-	-	-	-	0.3	0.3	0.1	0.0	-	-	0.3	0.0	-	-	0.1	0.0
16	epilithes	10	8	12	7	10	16	13	18	14	9	7	14	14	12	8	5	5	6	5	9	8	6	10	12
07	epiphytes	0.0	0.0	-	-	0.0	0.0	-	-	0.1	0.0	0.4	0.5	-	-	-	-	-	-	-	-	0.1	0.0	0.0	0.0
17	terrestrial and epilithic	10	8	10	17	12	11	13	15	8	4	19	17	17	17	5	10	6	6	4	8	8	16	10	11
18	epilithic and epiphytic	3	5	4	14	2	3	3	8	6	12	0.9	4	0.7	1	2	2	0.9	2	0.6	3	1	2	2	6
19	epilithic and lignicole	0.1	0.1	-	-	0.2	0.6	-	-	-	-	-	-	-	-	-	-	-	-	-	-	-	-	0.1	0.2
98	indifferent	7	7	4	3	13	17	11	11	5	2	7	4	5	4	3	4	1	0.4	2	4	2	2	7	8

in the Mediterranean areas where myrmecochory is, from a quantitative point of view, one of the principal methods of dispersal (stands with *Parietaria!*). Myrmecochory is also important in all other areas. Hydrochory is mainly found in the Atlantic regions, especially in area 120, and in area 400. Endozoochory is relatively important in the montane and E. European regions, but not of great significance in the densely populated area 120. Anthropochorous taxa seem to be most favoured in areas 200 and 400.

9.8 Formation spectrum

The formation spectrum (shown in Table 69) indicates that the contribution of the chasmophytic forms is always appreciable, especially in a quantitative respect, but is highest in the Mediterranean region and lowest in area 120. Other important categories are: species of pioneer of mud flats and shallows only appear in the eu-atlantic areas. In the bryophyte layer, obligatory or facultative epiliths predominate. Species of mud flats and shallows only appear in the eu-atlantic areas. In the Mediterranean region agricultural weeds and ruderal taxa are relatively numerous. In area 120 riparian forms and tall suffrutices are of relatively common occurrence. Generally speaking, forest-dwellers are more frequently encountered than elements of scrub vegetation, moors or heaths. In the Mediterranean zone, however, some species of maquis and garigue vegetation are represented. The category "contact vegetation and various" mainly includes forms not belonging to one of the groups previously mentioned. In respect of the sum total, the scores are very low for cultivated plants occasionally run wild (U 0.3; W 0.0) and for adventitious taxa (U 0.0, W 0.0). The scores for forms of contact zones are only appreciable in area 120 in the contact zone of fresh water and a drier substratum (U 6; W 2), and to a lesser extent in area 111 (U 1; W 0.1) and area 400 (U 0.9; W 2). The inclusion of a species in one formation class only means that the species in question is most frequently encountered in that formation class. However, a given species may behave rather differently in different areas, its requirements often being more specialised towards the limits of its area of distribution than they are elsewhere.

9.9 Distributional spectra

The spectra of the total areas of distribution and of the distribution within Europe were separately compiled for the vascular plants and the bryophytes (Table 70). The distribution of vascular plants inside Europe is much more often of a local nature than that of mosses and among the latter certain categories are even completely lacking (such

Table 70 Global distribution spectra and local distribution spectra of the various regions, based on unweighted numbers (U) and weighted numbers (W), for both Vascular plants (Va) and Bryophytes (Br)

	geographical region	111				112				120				131				132				
symbol	global distribution	UVa	WVa	UBr	WBr	UVa	WVa	UBr	WBr	UVa	WVa	UBr	WBr	UVa	WVa	UBr	WBr	UVa	WVa	UBr	WBr	UVa
10	Cosmopolitan	11	9	31	24	10	5	28	14	17	10	38	42	20	24	27	24	14	15	30	18	24
20	Supraholarctic	13	5	4	4	8	3	5	12	11	6	9	11	9	7	8	10	4	3	6	1	11
30	Holarctic	20	9	60	70	19	7	56	69	29	27	51	44	22	15	60	57	15	7	56	75	30
40	Eurasiatic	26	26	0.3	0.8	26	24	-	-	27	26	0.4	0.0	28	21	-	-	30	24	-	-	20
50	European	7	1	-	-	9	1	-	-	8	2	-	-	9	2	0.2	0.8	4	0.2	-	-	8
51	Alpine	-	-	-	-	-	-	-	-	-	-	-	-	0.4	0.0	-	-	-	-	-	-	0.7
52	Atlantic	0.2	0.0	-	-	0.4	0.0	-	-	0.2	0.0	-	-	0.4	0.1	-	-	-	-	-	-	-
53	Mediterranean	0.3	0.1	-	-	2	0.0	-	-	0.1	0.0	-	-	1	3	-	-	3	0.0	-	-	-
54	C. European	0.1	0.0	-	-	-	-	-	-	0.1	0.0	0.6	0.6	1	0.2	-	-	0.3	0.0	-	-	1
55	Medit.-Atl.	11	53	1	0.1	18	43	6	3	3	17	0.3	0.0	4	11	0.2	0.3	17	38	2	3	-
56	Atl.-C. Eur.	0.1	0.0	-	-	0.1	0.2	-	-	-	-	0.3	0.0	0.1	0.0	-	-	-	-	-	-	-
57	Medit.-C. Eur.	0.3	0.8	1	1	0.6	1	0.9	0.0	0.1	0.0	0.3	0.6	2	3	1	0.6	3	1	0.9	0.5	1
58	Boreal-C. Eur.	0.1	0.1	-	-	0.1	0.2	-	-	0.2	0.0	-	-	0.3	0.3	-	-	0.9	0.1	-	-	-
59	Med.-Atl.-C. Eur.	6	14	2	0.7	4	1	5	2	3	11	1	3	4	14	4	7	7	13	5	2	2
70	Neophytic	5	16	-	-	4	14	-	-	0.8	0.6	-	-	0.1	0.0	-	-	0.3	0.0	-	-	1
80	Alien	-	-	-	-	0.1	0.0	-	-	-	-	-	-	0.1	0.0	-	-	-	-	-	-	0.7
90	Naturalised	0.2	0.0	-	-	0.4	0.3	-	-	0.8	0.1	-	-	0.4	0.0	-	-	0.6	0.1	-	-	-
	local distribution																					
50	European	60	24	87	88	53	17	74	81	81	54	96	95	74	51	92	89	47	26	85	85	79
51	Alpine	-	-	-	-	0.1	0.0	-	-	-	-	-	-	0.4	0.0	-	-	-	-	-	-	0.7
52	Atlantic	0.2	0.0	-	-	0.5	0.0	-	-	0.2	0.0	-	-	0.4	0.1	-	-	-	-	-	-	-
53	Mediterranean	0.7	0.2	-	-	2	0.0	-	-	0.1	0.0	-	-	1	3	-	-	5	0.7	0.6	0.7	-
54	C. European	0.1	0.0	-	-	-	-	-	-	0.1	0.0	0.6	0.6	1	0.2	-	-	0.3	0.0	-	-	1
55	Medit.-Atl.	17	23	2	2	24	46	8	15	3	3	0.4	2	4	11	0.8	2	21	41	2	3	-
56	Atl.-C. Eur.	0.4	0.4	-	-	0.3	0.2	-	-	0.4	0.1	0.3	0.0	0.1	0.0	0.1	0.0	+	-	-	-	1
57	Medit.-C. Eur.	1	0.8	2	2	1	1	2	0.3	0.8	0.4	0.3	0.6	3	4	1	0.8	5	2	0.9	0.5	4
58	Boreal-C. Eur.	0.5	0.7	-	-	0.1	0.2	-	-	2	0.4	0.3	0.0	0.9	0.7	0.4	0.1	0.9	0.1	-	-	5
59	Med.-Atl.-C. Eur.	15	36	9	9	14	21	11	5	10	27	2	3	14	30	6	9	20	31	12	11	7

as "alpine" and "Atlantic"). Neophytes, adventitious forms and other naturalised species are mainly found among the vascular plants.

Cosmopolitan species are, comparatively speaking, numerous and especially of importance in area 400; their numbers increase in Europe from W. to E. In the Mediterranean area the scores for the vascular plants are lower than anywhere else in Europe. The high percentages of the mosses recorded from area 120 are perhaps not quite comparable with .the scores recorded elsewhere, because a number of stands without any vascular plants recorded from area 120 were used for the calculations. Species with a supraholarctic and holarctic distribution are relatively important in the areas 200 and 400 and it is not surprising that their representation is lowest in the Mediterranean region. Alpine taxa have only been encountered in C. Europe, but one should take the fact into account that it was only in this area that mural vegetation of alpine or montane zones was studied (and only to a lesser extent). Area 200 does not only comprise the alpine regions of C. Europe but also the lowlands and the river valleys in that area (e.g. the Inn valley in Austria). The contribution of species with a restricted distribution is unimportant: Atlantic (also poorly represented in W. Europe), C. European (the relative contribution is also low in C. Europe), and also Atlantic-C. European and Boreal-C. European species. Species with a Mediterranean distribution are principally of importance in the Mediterranean and their representation increases in a southward direction. However, mosses with a typical Mediterranean area of distribution are relatively unimportant in Mediterranean mural communities. The category "Mediterranean-Atlantic" is only poorly represented in areas 200 and 400; it is more important in area 112 than in area 111, and the importance increases in a southernly direction. The contribution of this

Guss. (= *A. serpyllifolia* ssp. *tenuior* (M. et K.) Arc.), so that for the purpose of data processing all subsequent records of the two were combined. *A. leptoclados* is fairly common in mural stands, particularly on N.-facing walls in the Mediterranean region.

9.10.2 Asplenium ruta-muraria L.

Since LOVIS & REICHSTEIN (1964) distinguished a diploid form of the wall-rue as its ssp. *dolomiticum,* the more common form has to bear the name ssp. *ruta-muraria.*

The wall-rue is one of the very few European ferns hardly ever found on any other substratum than rock crevices or walls. BARKMAN (1958) reports its occurrence on *Sambucus nigra* as an epiphytic plantlet of 0.5 cm high, apparently developed in the excessively wet summer and autumn of 1954-1955, but the specimen died after a short dry spell. *Asplenium trichomanes* and *A. adiantum-nigrum* are frequently encountered on old tree stumps, earth walls, etc. In the northern part of the Netherlands (in the provinces of Friesland, Drente and Groningen) the wall-rue grows on many old churches but nearly always on their S.- or W.- facing sides, as was noticed as early as 1912 by VELDHUIZEN. Conceivably this phenomenon is associated with the more rapid decomposition on the walls facing S. and W. On N.-facing walls, and especially in overshadowed places, *A. ruta-muraria* exhibits a somewhat different growth habit, its leaflets being usually longer and narrower and not so rigid as they are in specimens of S.-facing surfaces. Some authors (e.g. DE WEVER, 1942) report that the species is sciophilous, but in some places (e.g. in Amsterdam) specimens occurring in the immediate vicinity of street lights appear to be thriving. At any rate, a "long-day" periodicity seems to have a favourable effect on its development.

A distinct preference for a certain exposure was not noted elsewhere in Europe.

Especially in W.- and in C. Europe, the wall-rue is *the* pioneer species of stands of mural vegetation. It is quite common in cracks and slits in walls and of wall-joints and can already become established when the pH is still as high as 8.7.

9.10.3 Ficus carica L.

The fig-tree is generally reputed to be a truly Mediterranean species. According to MEIER & BRAUN-BLANQUET (1934) it is a faithful species of the *Asplenietea rupestris* i.e. of a certain kind of chasmophytic vegetation. Outside the Mediterranean area the fig-tree was recorded in mural vegetation in Brittany and in S. England several times and also once at Veere (Walcheren, Netherlands). In the area of Greater London it was repeatedly found, and apparently thriving, on rubble heaps of bombed sites (LOUSLEY, 1949). Presumably these extra-mediterranean records are all cases of a subspontaneous establishment of the fig, the seeds of cultivated specimens having been dispersed by birds. According to CLAPHAM, TUTIN & WARBURG (1962), the fig is "occasionally self-sown and ± naturalised in England". The possibility of the seeds being carried across considerable distances must not be excluded, however, and at least some localities in the eu-atlantic region lie close to the natural area of distribution of the species.

9.10.4 Fumaria officinalis L.

Included in this taxon is *F. densiflora* DC. which was found in mural stands in the Mediterranean region.

9.10.5 Galium mollugo L.

In mural vegetation, and especially in association with *Parietaria judaica,* the ssp. *erectum* Syme (= *G. capsiriense* Jeanbert ex Timbal-Lagrave, see SHAW, 1960), is not infrequently encountered. Between this subspecies and the "typical" one transitions occur, so that a specific distinction is not warranted. *G. mollugo* ssp. *erectum* is presumably a calciphile.

9.10.6 Gymnocarpium robertianum (Hoffm.) Newm. and G. dryopteris (L.) Newm.

Gymnocarpium robertianum and *G. dryopteris* are treated as separate species, although some authors (e.g. LAWALREE, 1950) consider them to be only subspecies or varieties of one single species. Some of the main distinctive morphological characters indeed overlap. Glandular hairs are not always absent in *G. dryopteris,* but they are always less developed. Hybrids found on two occasions, viz. in Amsterdam and at Lautaret (France), were completely sterile. According to LOVE (1967) the valid name for the genus is *Carpogymnia*, but MORTON (1965, 1967) adduces cogent arguments to conserve the name *Gymnocarpium*. This name is, therefore, maintained here.

9.10.7 Juncus bufonius L. and J. hybridus Brot.

The differences between these taxa have been discussed in a previous note (SEGAL, 1960). Taxonomic synonyms of *J. hybridus* are, inter alia, *J. ambiguus* Guss. and *J. bufonius* L. ssp. *ranarius* Song. et Perr.

9.10.8 Linaria cymbalaria (L.) Mill.

Linaria cymbalaria (syn. *Cymbalaria muralis* Gaertn., Mey. & Scherb.) belongs to a group of species generally considered to be neophytic in large areas of Europe. According to Cufodontis (1947), it was originally a native of Italy (of the Apennines and Istria), but according to Hegi's "Flora" it is a native of S. Europe, N. Africa and W. Asia. In the famous *Cruijdtboeck* (herbal) of Dodoens (Dodonaeus) of 1644 the species is reported to have become much more common in the Low Countries and the expectation is expressed that this plant would eventually have to be regarded as an indigen. Kirschleger (1862) recorded the ivy-leaved toadflax from many towns along the Rhine in C. Europe and pointed to the interesting fact that it is not mentioned in Bauhin's Pinax of 1622.

Linaria cymbalaria occurs here and there in Italy and Yugoslavia in rock crevices. It presumably commenced its spread to other parts of Europe in historical times, because it found a suitable site much more often on walls than in rock crevices. Perhaps, the erection of stone walls and buildings by the ancient Greeks and Romans provided the first potential habitats. The gradual expansion of its natural area of distribution need not necessarily have been caused by human interference. *Linaria cymbalaria* must have followed a distinct pattern of migration along the valleys of the great rivers and their tributaries, and along the W. coasts of Europe, considering the density of its localities in these areas. Also in C. Europe the species is especially seen along the greater rivers (such as the Vistula in Poland).

Dispersal is frequently carried out by ants, as I have repeatedly observed. This myrmecochory does not explain the long-distance translocation of the seeds. The species can increase locally by autochory, the fruiting pedicels becoming negatively heliotropic and pushing the capsules into the substratum. The species is reported to be generally found in damp situations. Cufodontis is of the opinion that it is clearly intolerant of drought and of strong temperature fluctuations. The species is indeed rare in the more continental parts of Europe and restricted in its occurrence to more or less moist sites, whilst in the Mediterranean zone it becomes more common at higher altitudes, avoiding the very dry areas. In the eu-atlantic areas, and also in area 120, *Linaria cymbalaria* is often found on S.-facing walls and seems to be thermophilous. In the Netherlands it always grows more lushly and usually more abundantly in hot than in cool summers, its coverage sometimes varying appreciably from year to year at the same site. In the climate prevailing in area 120 the species is often perennial, but it is not frost-resistant and during severe winters many plants die off. Flowering may commence early in the season: in Holland from mid-March onwards; in Paris sometimes as early as February; in the Mediterranean region presumably in winter. Flowering may continue till late autumn. The impression was gained that there is a form which flowers later in the year and forms denser clumps, but the number of observations is as yet insufficient.

A white-flowered form was occasionally recorded (Amersfoort, Buren and Windesheim, Netherlands, and Roquevaire, France).

9.10.9 Parietaria judaica L. and P. officinalis L.

Parietaria judaica and *P. officinalis* are quite distinct. The differences in morphology, chromosome number, distribution and ecology have been discussed in detail by Mennema & Segal (1967).

9.10.10 Poa annua L.

This street grass is now a typical cosmopolite through human agency and it was the first cormophyte to be recorded as a weed from Antarctica (Skottsberg, 1954). The species has a remarkable adaptive capacity in respect of climatic influences. In spite of its name it is by no means always annual. According to Magrou (1950), the annual habit is genetically determined and the perennial form (ssp. *varia* Gaudin) has a mycorrhiza. In W. Europe, annual plants predominate in the lowlands, to be replaced by the perennial (hemicryptophytic!) form in the alpine region, but perennial forms are also found in the plains. Biannual plants are also frequently encountered, and they presumably dominate in the montane part of S. Europe and elsewhere in sites with a high relative humidity. The tendency towards a biannual life-cycle is strongest in damp places (at least in the Low Countries).

Magrou substantiated his conclusions by translocation experiments with plants both from the mountains and from the plains, but this still does not provide cogent proof of the incidence of genetic differences between all perennial individuals from the W.- and S. European lowlands and the annual forms. The impression was gained by the present author that all sorts of transitions exist between age classes and that, apart from morphological features directly associated with the perennial habit (such as intravaginal shoots), there are no clear-cut morphological differences between them (at least in the Netherlands, where both the annual and the perennial forms are quite common). In the Netherlands plants of the annual habit form appear especially in relatively dry places and the perennial ones in relatively moist sites, e.g. in trampled stands on peaty soil. In some places, within short distance of a few metres, specimens transitional between the annual and the perennial habit are encountered, e.g. in streets and roadsides in the marshy region of N.W. Overijssel, where for roadmaking a body of sand is always deposited first. No observations have been made concerning the presence of a mycorrhiza.

9.10.11 Poa bulbosa L.

In the Mediterranean area, viviparous forms of ***Poa bulbosa*** are frequently encountered, particularly in the stands with ***Asplenium trichomanes*** and *Ceterach officinarum* which usually appear on N.-facing walls.

9.10.12 Poa compressa L. and P. nemoralis L.

Poa compressa is a species of open sites and occurs in dry grasslands and on rocky slopes in lime-rich areas. It also occurs in mural habitats, and appears at sites strongly influenced by Man, such as dikes, quarries, clay pits and stone- and slag heaps of (coal) mines. The species is somewhat nitrophilous and this partly explains why it is a follower of human culture.

In C. Europe it exhibits a considerable degree of variation, most peculiar being a form which in several respects seems to be intermediate between *P. compressa* and *P. nemoralis* and was mainly recorded at mural sites in N.E. France, S. Germany and Luxemburg (but also from Belgium and the southern part of the Netherlands). The intermediate specimens often attain a greater size than either *P. compressa* or *P. nemoralis* and they appear to remain morphologically constant after having been transplanted in an experimental garden. The seeds are well-developed.

Instead of being a hybrid, the intermediate form may well prove to be a separate taxon, presumably a good species putatively considered to be of hybrid origin and possibly an allopolyploid of *P. compressa* and *P. nemoralis*. The chromosome number has, as yet, not been counted. It may not, however, be so easy to solve the problem by simply counting the chromosomes, a large number of different chromosome counts of *Poa compressa* having already been reported. According to CLAUSEN (1961) the latter species is apomictic, but we may assume that, as is often the case with apomicts, sexual reproduction also takes places to some extent so that hybridisation and introgression may occur.

Poa nemoralis is hardly ever found growing on walls, but occurs mainly in open spaces in deciduous forests. However, it sometimes appears in massive stands in not very dry and not too exposed sites created by Man, such as the N. slopes of waste rock heaps of mines. In S. England it is occasionally seen in mural vegetation and in the London area in flower beds and on waste land (ALLEN, 1964). Glaucescent forms of *P. nemoralis* somewhat resembling *P. compressa* occur in C.

Europe in mixed *Quercus-Carpinus* forests. The intermediate form of *P. compressa* and *P. nemoralis* frequently encountered in mural vegetation in some areas has for the purpose of this study been provisionally referred to as *"Poa pseudocompressa"*.

9.10.13 Poa pratensis L. s.l.

In contradiction to a number of recent floras a distinction is made between *P. pratensis* L. (sensu stricto), *P. angustifolia* L. and *P. subcoerulea* Smith, which are not only morphologically but also ecologically distinguishable. *P. pratensis* s.s. is a species of well-developed meadowland. *P. angustifolia* often grows in open and relatively dry habitats, e.g. in open grassland on lime-stone, in agricultural fields, on more or less nitrogen-enriched places on rocky slopes or on escarpments of rock debris, in quarries, etc. It is of common occurrence in mural vegetation, particularly in C. Europe and in area 120, often associated with *Chelidonium majus*. *P. subcoerulea* is usually found on a poorer type of soil than the other two species e.g. on sandy soils or on leached-out soils of open slopes, and its appearance seems to be favoured by a source of organic nitrogen (such as grazing by sheep). It was only occasionally recorded on top of walls where a great quantity of sand and humus had accumulated.

9.10.14 Polygonum aviculare L. s.l.

In 1912 LINDMAN published an impressive paper concerning the variability of *Polygonum aviculare,* which species he subdivided into three taxa viz. *P. heterophyllum* (= *P. aviculare* s.s.), *P. calcatum* and *P. aequale* (the latter with two subspecies). In England, STYLES (1962) apart from *P. heterophyllum* and *P. aequale* (which he, presumably in error, calls *P. arenastrum* Boreau, cf. LAMBINON, DUVIGNEAUD & LAWALREE, 1965), also distinguished the taxon *P. rurivagum* Jordan on the basis of a biometric approach. SCHOLZ (1960) apart from all taxa mentioned, recognised *P. monspeliense* Thiébaud and *P. neglectum* Besser, both presumably habitat modifications of *P. heterophyllum.* According to Styles, in a number of characters *P. rurivagum* is morphologically more or less intermediate between *P. aequale* and *P. heterophyllum,* but is supposed to differ from both by the altogether different length of the ochrea. The fruit is said by Styles to be (2.5-)2.9-3.3(-3.8) mm long, but according to Scholz it is only about 2 mm long. In the Netherlands, the segregation of *P. rurivagum* can neither satisfactorily be achieved by means of the criteria given by Styles, nor by means of those mentioned by Scholz, and a preliminary survey of the Dutch material not clearly referable to either *P. aequale* or *P. calcatum* yielded an approximately binomial distribution of the same characters reported by Styles to exhibit variation curves with two maxima. The recognition of *P. rurivagum* was, accordingly, not considered recommendable.

The distinction between *P. aequale* and *P. calcatum* is likewise difficult. In S. Germany and in Czechoslovakia forms were encountered, sometimes in large numbers, that seemed to be intermediate between the two. Dr. Scholz, who kindly examined some sheets of this material upon my request and named it "*P. calcatum*", mentioned in a letter to me that in Berlin also such intermediate specimens are not at all rare. He is apparently somewhat at a loss when it comes to interpreting such specimens and wonders if they can possibly be hybrids. The seeds are always well-developed, however.

Such observations necessitate the consideration of a different treatment of the aggregate species *P. aviculare* s.l. in Europe, the best solution presumably being the recognition of "*P. aequale*" and "*P. calcatum*" as subspecies of a single taxon. In any event, experimental taxonomic studies are required to procure more insight into the taxonomy of the whole complex.

P. heterophyllum is commonly found in trampled sites and in agricultural fields. Forms corresponding with "typical" *P. aequale* (and presumably also specimens agreeing with the concept of "*P. calcatum*") are especially common in ruderal communities (the *Artemisietea*) and usually occur at drier sites than *P. heterophyllum*. Cohabitation of the two taxa is not at all a rare phenomenon.

9.10.15 Rubus spec. div.

Determination of the *Rubi* in mural stands was mostly impossible because specimens of the brambles usually remain sterile when growing on walls. One of the most frequent species in mural vegetation in S.- and W. Europe is *R. ulmifolius* Schott.

9.10.16 Sonchus oleraceus L.

Sonchus oleraceus belongs to the most common elements of mural vegetation and is particularly at home in communities with *Parietaria judaica*. This sow-thistle is reputed to be characteristic of communities of agricultural weeds, but a number of its features render the mural habitat a suitable environment for the germination of its seeds: the species grows readily on a substrate rich in basic minerals and in nitrogen, is heliophilous and withstands appreciable changes in temperature (compare Lewin, 1948). It also thrives on the rubble heaps of bombed buildings. Nevertheless, the mural habitat does not seem to be suitable altogether, because the specimens found growing on walls usually remain small and only on rare occasions reach the flowering stage.

9.10.17 Taxus baccata L.

The yew is especially at home in not very dense beech forests or in ashwood growing on steep slopes in areas with an oceanic climate, e.g. in W. Europe and in the lower mountains of Switzerland. It is a frequent invader of chalk scrub in England, but it seldom forms pure woods except in the western South Downs in West Sussex and the adjoining part of Hampshire (TANSLEY, 1939). *Taxus baccata* prefers a lime-containing soil, but WEBB & GLANVILLE (1962) recorded *Taxus* from forests on acid rocks in some localities in W. Ireland. (Perhaps the influx of minerals by the winds blowing from the sea have something to do with this somewhat unusual occurrence.) In Switzerland the yew is found growing in crevices of steep rocky cliffs (VOGLER, 1904). The distributional records compiled by MEUSEL (1939) show that in Germany this tree is most frequently encountered along the rivers Weser and Elbe and their tributaries.

Taxus baccata is by no means rare on walls in W.- and in C. Europe with a predominantly N.- or E. aspect, being particularly common in Luxemburg and much rarer in Holland. It is ornithochorous and at mural sites the seeds have presumably always been introduced by birds. The wall-dwelling specimens usually attain only modest dimensions and do not often bear fruit, but the sterility need not necessarily be the result of an inadequate environment and may be explained by the fact that the partly decomposed walls supporting stands with *Taxus* are often demolished or renovated so that the yews simply do not get a chance to live long enough to attain sexual maturity. The occurrence in some places (such as the Netherlands) may be subspontaneous from seeds of cultivated yew-trees or -hedges, but this is not at all certain.

9.10.18 Barbula vinealis Brid.

In the beginning of the present study no distinction was made between "typical" *Barbula vinealis* and *B. vinealis* var. *cylindrica* (Tayl.) Boul., and all data are recorded under *B. vinealis*. *B. vinealis* var. *vinealis* is, generally speaking, found in a drier environment, usually in stands with *Parietaria judaica*, and its var. *cylindrica* in more or less shady or moderately humid places.

9.10.19 Hypnum cupressiforme Hedw.

Recently, BARKMAN (1966) published a key for the determination of the intraspecific taxa of *Hypnum cupressiforme*. By far the most commonly encountered one is its f. *tectorum* (Brid.) C. Jens. (= var. *tectorum* Brid.), but also its var. *resupinatum* (Tayl.) Schimp. and var.

lacunosum Brid. have repeatedly been recorded. The various taxa sometimes seem to be linked by transitional forms. The var. *resupinatum* is usually found growing in fairly damp situations, especially at N.-facing sites, but in the eu-atlantic areas sometimes also on S.-facing walls.

9.10.20 Tortula muralis Hedw.

This is a variable species, its variability especially showing up in the colour and in the size and length of the vitreous hairs. A whole series of different chromosome counts, lying between n=28 and n=56, has been reported (VAARAMA, 1956; RICHARDS, 1959), but an unequivocal correlation between morphological features and chromosome number has never been demonstrated. In dry habitats the specimens are usually of a greyish-green colour and in moister situations they are a brighter green. In moister growing places the vitreous hairs are usually short or completely absent. In many places all possible transitions between the extreme colours and the extreme lengths of the vitreous hairs can be observed, often quite clearly so on S.-facing walls with an overshadowed, damp base. The extreme forms with short vitreous hairs or with none at all (described as *T. obtusifolia* Schleich. and *T. muralis* var. *aestiva* Brid. = *T. aestiva* Pal. Beauv.) and the extreme colour types (described as e.g. var. *incana*) are, in my opinion, no more than habitat forms modified by the environment.

9.11 The part played by alien taxa in mural vegetation

Several workers have paid attention to non-indigenous species and to their classification in a number of categories e.g. RIKLI (1903), THELLUNG (1912), SCHEUERMANN (1948) and SUKOPP (1966). Partly after their suggestions, the following categories of autochthone and alien elements of a certain area are distinguished:

1. autochthone species of natural habitats;
2. autochthone species also cultivated as ornamentals (e.g. species of *Saxifraga* and *Sedum*);
3. indigenous species of which a number of individuals have migrated from their original natural habitats to artificial habitats and are more or less adapted to these "secondary" habitats (e.g. *Parietaria judaica* in S. Europe): apophytes of Thellung;
4. species naturalised in prehistoric times and not deliberately imported by Man, usually growing in semi-natural or artificial sites, e.g. many ruderals and argicultural weeds: archeophytes of Thellung;
5. species which extended their original area of distribution in historic times;
6. species introduced in historic times and now behaving as if they are

indigenous to the area i.e. they reproduce themselves successfully and constitute an element of certain plant communities: neophytes (they could be divided into escapes from cultivation e.g. *Robinia pseudacacia:* ergasiophygophytes of Thellung, and species that originally appeared adventitiously, e.g. *Erigeron canadensis;* furthermore, a distinction can be made, according to Sukopp, between taxa occurring in stands of natural or semi-natural vegetation, such as communities found on recently felled woodland, and taxa only found in artificial habitats such as arable land and walls);

7. garden escapes maintaining themselves in artificial habitats, e.g. *Tulipa sylvestris* in W. Europe: epokophytes of Thellung;
8. garden escapes locally maintaining themselves in small areas without extending their distributional area: ergasiolipophytes of Thellung;
9. cultivated plants not normally capable of maintaining themselves under natural conditions: ergasiophytes of Thellung;
10. adventitious taxa, usually only capable of maintaining themselves for a short time and as a rule appearing near loading and offloading zones and along main traffic routes: ephemerophytes of Thellung.

The qualification "indigenous" is applicable to categories 1 to 5 (or possibly 1 to 6) inclusive. The differences between the diverse categories are mostly not very sharp and in the sequence of enumeration followed here the delimitations between the preceding and the following category are usually the vaguest. However, (10) can also be placed next to (6), and (1) is almost as close to (3)-(6) as it is to (2); a monolithic "hierarchy" does not exist.

Apart from representatives of category (1), mainly species of the categories (3), (4), (6) and presumably also (5) constitute elements of mural vegetation. In a number of cases, however, it is difficult to decide to which category a characteristic wall-dwelling taxon must be referred *in a certain area* (examples: *Linaria cymbalaria, Parietaria judaica, Cheiranthus cheiri* and *Corydalis lutea.* In the following discussion these four species will be considered together for the sake of convenience).

For the possibility of a plant becoming a neophyte, the following factors are a prerequisite (see EGLER, 1961): (1) availability of a suitable site; (2) agencies of introduction and dissemination; (3) type of reproduction and number of generations involved; (4) age of individual plants and thus its individual persistence; (5) actual time involved.

Suitable habitats for neophytes are usually open sites, such as pioneer stands, unnatural habitats and disturbed environments, i.e. unstable habitats in which the autochthone floral elements mostly have but little competitive capacities and which may contain certain micro-habitats unsuitable for or inaccessible to indigenous species. In such environments therophytes often predominate, so that in the course of time there are more opportunities for settlement than in more stable environments. The diaspores of neophytes are often capable of bridging great distances by being anemochorous, anthropochorous or hydrochorous and the individuals often grow fast and do not live long. According to Sukopp (1966) the proportion of the therophytes among the neophytes is small, but the arborescent forms are still scarcer. Colonisation may proceed with considerable rapidity, so that some plants originally introduced as ornamentals or aquarium plants and some not intentionally introduced forms (such as species of *Erigeron*) have spread all over Europe in less than a century. Other species took centuries to become naturalised and this presumably also holds true for a number of wall-dwelling taxa in W. Europe.

Summary

CHAPTER 1 — Introduction

Little is known of the ecology of mural vegetation. The present study, begun in the Netherlands, was later extended to include localities elsewhere in Europe e.g. in Great Britain, Belgium, Luxemburg, France, Italy and W. Germany. In nearly 1,200 sites relevés were made of stands of vegetation on more or less vertical wall surfaces. In addition, vegetation of cracks and joints in paved surfaces (streets, pavements, squares, etc.) was studied in 177 sites, because the communities occurring in this particular habitat often resemble certain types of mural vegetation.

CHAPTER 2 — Statement of the problem and methods

Site selection was not at random but it was not altogether subjective either. Mural stands of long-inhabited regions were predominantly studied, the purpose of the present investigation not being restricted to the mere description of mural plant communities, but also embracing the variability of mural vegetation types and the recognition of possible continuities in the composition of mural stands in geographical sequences. For the visual estimations of abundance and percentage of coverage of the individual species, the scale of BARKMAN, DOING & SEGAL (1964) was used. The relevés thus obtained were sequentially arranged according to the affinity index of POORE (1956). Subsequently matrices were set up on the basis of 2 x 2 convergency tables in which the affinity (both in a positive and a negative sense) of the most frequently represented species had been calculated. For all species appearing in the relevés, in the form of separate lists (Appendices II and III), the following data are enumerated: idiosystematic (natural) order (according to PULLE, 1952), degree of ploidy, life form according to Raunkiaer and according to IVERSEN (1936), growth form, sociability, floral colour, dissemination type, formation group, general distribution, and occurrence in Europe. By means of these data, spectra (of the relative values of each category in per cents.) were calculated, both for the various communities and for each of the 11 geographical areas distinguished (and for some combinations of the latter), and for both unweighted and weighted values. For the necessary calculations electronic computers were employed. The ecological environment was studied by means of e.g. chemical analyses of the substratum and measurements of temperatures and light intensities.

CHAPTER 3 — Ecology of mural vegetation

Special attention was paid to topographical and to historical factors (ageing, decomposition, etc.). In central Europe and the subatlantic areas well-developed stands of mural vegetation are more or less clearly restricted to the basins of the greater rivers and their principal tributaries. Certain species exclusively occur in mural habitats, at least in certain regions; examples are *Linaria cymbalaria, Corydalis lutea, Cheiranthus cheiri, Parietaria judaica* and *Tortula muralis.* Whenever the inclination of a mural site decreases, the number of species increases as a rule and the stands of vegetation become less "characteristic" by the admixture with numerous "incidentals". The exposure of the sites is very important with regard to the representation of many species and of several of the mural communities. A number of taxa showing a marked preference for north-facing wall surfaces in the Mediterranean region do not exhibit such a clear preference in the eu-atlantic parts of their distributional area.

The deleterious effect of atmospheric pollution is greatest on lichens, but also considerable in the case of bryophytes. Sequences of an increasing toxitolerance were drawn up.

The pH values of the substrata supporting well-developed stands of mural vegetation usually lie between pH 7 and pH 8, those of the substrata carrying pioneer phases of mural communities mostly between pH 8 and pH 9. Decomposition of the masonry and/or accumulation of extraneous dust and soil particles constitute important requisites for the development of a plant cover. Water is often the limiting factor. Many species of mural communities possess adaptations to appreciable fluctuations in the moisture content of the substratum and the surrounding atmosphere.

CHAPTER 4 — Structure

Vertical patterns (stratification) and horizontal ones are discussed, special attention being paid to the limits of the stands. The degree of sociability was determined by means of objective values. The quantitative minimum area and its qualitative counterpart are defined. They both appear to be nearly constant for stands (with optimum development) with a certain structure. The minimum area is, ultimately, dependent on the density of representation of the aspect-forming growth form. A system of growth forms is proposed.

CHAPTER 5 — Classification of mural communities

The starting-point of any evaluation of a stand should preferably be the ecosystem, which can, if necessary, be subdivided into subsystems (biotic community, stenocoenoses). However, the interdependence of living elements of an ecological system, and the relation between each constituting element and its environment are insufficiently known. In pioneer stands of vegetation a system of classification is not so clearly recognisable as it is in more differentiated and more intricate communities. In mural vegetation the layer of herbs and the layer of bryophytes are to a large extent independent of one another.

The concept of a "faithful species" can not very well be applied objectively. Communities can be distinguished on the basis of *differentiating combinations of species* and/or *ecological groups*. A satisfactory system of classification should not be exclusively based on floristic criteria. A certain sequence in the application of the different criteria to be used might further the uniformity of treatment e.g. in the following order: (1) floristic, (2) structural, (3) ecological, and (4) geographical criteria.

CHAPTER 6 — A survey of mural communities

The various mural communities are amply discussed, their variants transitional towards other vegetation types and their changing specific composition along geographical gradients being taken into account. Stands poor in species, and often showing dominance of only a single species, are referred to as 'inops' stands (or as 'inops' stages of a community). Communities transitional towards communities of other 'higher syntaxonomic groups' may, together with the allied vegetation types of those higher groups, be united into an 'association assembly'. A number of coenoclines could be recognised.

In Mediterranean mural communities with *Parietaria judaica*, myrmecochory appeared to be of considerable importance. In the eu-atlantic regions studied the contribution of ferns in mural stands is fairly large. In the medio- and subatlantic areas stands with *Sagina procumbens* and *Poa annua* are of frequent occurrence.

CHAPTER 7 — Succession in mural vegetation

The ultimate stage of the successional seres depends to a large extent upon the degree of alteration of the environment by the plant cover. For the development of mural vegetation the following factors are important: the progressive decomposition, the accumulation of fine-grained particles in the substratum (including the sedimentation of extraneous matter), a lowering of the pH and a decreasing inclination.

The occurrence at mural sites of Fungi, Algae and Lichens is briefly surveyed. Algal forms of the *Protococcus* (sensu lato) type are often represented on walls which remain rather moist for long periods of time.

During the succession of bryophytic forms, initially acrocarpic mosses play the most important role, to become gradually replaced by pleurocarpic ones. In western- and central Europe *Asplenium ruta-muraria* often acts as a pioneer, to be followed by *Linaria cymbalaria.* From stands containing these two species diverse communities may originate, e.g. communities with *Corydalis lutea,* with *Cheiranthus cheiri,* or with *Parietaria judaica.* In the eu-atlantic areas studied *Asplenium adiantum-nigrum* may act as a pioneer species, and in southern Europe *Parietaria judaica.*

CHAPTER 8 — Vegetation of streets and other paved surfaces: a comparison

Vegetation occurring in the cracks and joints of paving stones shows a certain degree of resemblance to mural communities with *Poa annua* and *Sagina procumbens* and all these vegetation types can be united in an association assembly. The various communities of paved surfaces occurring in different parts of Europe, or even outside Europe, exhibit a remarkable resemblance in their structural features, a number of the constituting genera contributing vicariating species in the different regions. The very common inops stands chiefly contain species with a large area of distribution including several cosmopolitan taxa. In other communities, also, the role of such wide-spread forms is relatively important. Anthropochory is one of the major types of diaspore dispersal among all these communities.

CHAPTER 9 — Some phytogeographic and taxonomic aspects

The spectra of the communities of the different geographical areas is amply discussed. Polyploidy becomes relatively more common in mural vegetation from southern Europe to the north, and from western Europe towards the east.

The contribution of hemicryptophytes is always substantial and in southern Europe the representation of therophytic forms is also considerable, at least qualitatively. Mural vegetation consists mainly of mesophytes. Among the vascular plants, the degree of sociability '2' is of frequent occurrence, among the bryophytes the degree of sociability '3'. Anemochory is important in many communities, but in vegetation with *Parietaria judaica*, particularly in southern Europe, myrmecochory plays an important part in diaspore dispersal. Many characteristically wall-dwelling taxa have a Mediterranean-Atlantic or a Mediterranean distribution.

Of a number of species, ancillary data regarding their taxonomy and/or ecology are given. The significance of neophytes is discussed in some detail.

Résumé

CHAPITRE 1 — Introduction

L'écologie des végétations murales est encore peu connue. Nos recherches ont débuté aux Pays-Bas et ont été étendues plus tard à d'autres régions d'Europe, e.a. en Angleterre, en Belgique, au Luxembourg, en France, en Italie et en Allemagne fédérale. La végétation des murs verticaux a été analysée dans environ 1.200 localités. En outre nous avons étudié les végétations d'entre pavés, de 177 endroits. Celles-ci ont en effet très souvent des traits similaires à certains types de végétations murales.

CHAPITRE 2 — Orientation du problème et méthodes

Le choix des milieux n'a pas été fait au hasard, mais pas non plus de manière subjective. Les murs situés dans des régions habitées de longue date ont été principalement l'objet de nos recherches. Notre but a été non seulement la description des types de végétation, mais surtout l'étude de la variabilité des végétations murales et la distinction de continuité éventuelle dans la composition de séries géographiques. Nous nous sommes servis de l'échelle de BARKMAN, DOING & SEGAL (1964) pour l'estimation de l'abondance et des degrés de couverture. Les relevés obtenus ainsi ont été classés d'après la formule d'affinité de POORE (1956). Ensuite des matrices ont été élaborées sur la base des tables de 2 x 2 convergence, dans lesquelles avait été calculée l'affinité (aussi bien dans le sens positif que négatif) des espèces les plus fréquentes. Pour toutes les espèces nous avons mentionné séparément (voir Appendices II et III) les données suivantes: l'ordre idiosystématique (selon PULLE, 1952), le degré de ploïdie, les types biologiques d'après Raunkiaer et d'après IVERSEN (1936), les types morphologiques, la sociabilité, la couleur de la fleur, les types de dissémination, les groupes de formation, l'aire et la répartition en Europe. A l'aide de ces données nous avons calculé des spectra (valeurs relatives pour chaque catégorie en pourcentage) aussi bien pour les types de végétation que pour les 11 régions géographiques préalablement distinguées, et pour un certain nombre de ces dernières combinées, tant pour les valeurs pesées que non-pesées.

Pour les calculs nous nous sommes servis d'ordinateurs. Les recherches écologiques ont été effectuées, entre autres, à l'aide d'analyses chimiques du substrat et des mesures de température et de lumière.

CHAPITRE 3 — Ecologie des végétations murales

Nous nous sommes intéressés particulièrement aux facteurs topographiques et historiques. Dans l'Europe centrale et dans les régions subatlantiques, les végétations murales bien développées se manifestent surtout dans les bassins des grands fleuves. Certaines espèces se trouvent être spécifiques des végétations murales, p.e.: *Linaria cymbalaria, Corydalis lutea, Cheiranthus cheiri, Parietaria judaica* et *Tortula muralis.* Lorsque l'inclination diminue, le nombre d'espèces augmente généralement et la végétation devient moins "spécifique". Pour beaucoup d'espèces et de types de végétation l'exposition est d'une grande importance. Un certain nombre d'espèces ayant une préférence pour les murs exposés au nord dans la région méditerranéenne ne manifestent pas une préférence marquée dans la région eu-atlantique. L'influence de la pollution de l'air est très grande sur les lichens, mais également considérable sur les muscinées. Nous avons élaboré des séquences afin de mesurer la toxitolérance. Le pH du substrat des végétations murales bien développées se trouve généralement entre 7 et 8; le pH des stades pionniers se situe souvent entre 8 et 9. Pour le développement des espèces la décomposition des particules de poussière ou de sol, accompagnée ou non de sédimentation, constitue une condition importante. L'eau est souvent un facteur limitant. Beaucoup d'espèces sont adaptées à de grandes fluctuations de l'humidité du substrat et de l'atmosphère.

CHAPITRE 4 — Structuration

Nous avons discuté les structures verticales (la stratification) et horizontales et nous nous sommes intéressés spécialement aux limites de la végétation. Nous avons déterminé la sociabilité à l'aide de valeurs objectives.

Nous avons défini l'aire minimum qualificative et représentative. Elles s'avèrent presque constantes pour des végétations au développement optimum ayant une certaine structure. L'aire minimum s'avère finalement être dépendante de la densité de la forme habituelle déterminant l'aspect. Un système de formes habituelles a été élaboré.

CHAPITRE 5 — Classification des végétations murales

Le point de départ de l'étude d'une végétation devrait être l'écosystème. Celui-ci peut être subdivisé éventuellement en subsystèmes (biocénose et sténocénoses). Cependant on connaît peu l'interdépendance des organismes et de leur milieu.

Pour les végétations pionnières il s'agit probablement moins d'un système que dans le cas des communautés plus différenciées et structurées. Dans les végétations murales les plantes vasculaires et les bryophytes sont vraisemblablement, dans une large mesure, indépendants.

Il est difficile d'appliquer, objectivement, le concept des caractéristiques. On peut distinguer les végétations d'après les combinaisons d'espèces différentielles et ou d'après les groupes écologiques. Un système satisfaisant ne devrait pas être basé exclusivement sur les critères floristiques. Un ordre fixe dans l'application des critères pourrait contribuer à l'unité de traitement; p.e.: 1. critères floristiques; 2. critères structuraux; 3. critères écologiques; 4. critères géographiques.

Des transitions de types de végétations à d'autres "unités supérieures" peuvent être réunies avec des types de végétations affiliés, issus de ces "unités supérieures" dans un "cercle d'associations".

CHAPITRE 6 — Aperçu des végétations murales

Nous avons amplement traité des végétations murales, de leurs transitions vers d'autres types de végétation et des modifications de leur composition suivant les régions géographiques. Nous avons mentionné des végétations pauvres avec souvent une dominance d'une seule espèce, comme des végétations "inops". Nous avons pu déterminer un certain nombre de coenoclines. Dans les végétations méditerrannéennes à *Parietaria judaica,* la myrmécochorie s'est avérée importante.

Dans les régions eu-atlantiques la participation des fougères est assez grande. Dans les régions médio- et sub-atlantiques les végétations à *Sagina procumbens* et *Poa annua* sont fréquentes.

CHAPITRE 7 — Succession des végétations murales

Le stade final de la succession dépend de l'aptitude de la végétation à modifier le milieu. Pour le développement des végétations murales les facteurs suivants sont importants: décomposition progressive, augmentation des particules fines du substrat (avec apport ou non par sédimentation), baisse du pH et diminution de l'inclinaison.

Nous avons donné un bref aperçu de la végétation murale des *Fungi, Algae* et *Lichenes.* On trouve souvent des *Protococcus* sensu lato sur des murs humides durant de longues périodes. Dans la succession bryophytique, les mousses acrocarpes jouent au début un rôle important; dans un stade ultérieur, il s'agira plutôt des mousses pleurocarpes. En Europe occidentale et centrale *l'Asplenium ruta-muraria* agit souvent comme pionnier, suivi par *Linaria cymbalaria.* A partir des communautés de ces espèces d'autres végétations peuvent se développer, p.e. avec *Corydalis lutea, Cheiranthus cheiri* ou *Parietaria judaica.* Dans les régions eu-atlantiques *Asplenium adiantum-nigrum* agit parfois comme pionnier; de même *Parietaria judaica* en Europe méridionale.

CHAPITRE 8 — Comparaison avec les végétations de pavés

Les végétations d'entre pavés montrent une certaine ressemblance avec les végétations murales à *Poa annua* et *Sagina procumbens* et toutes deux peuvent être réunies dans un même cercle d'association. On peut observer une ressemblance de structure remarquable entre les végétations de plusieurs parties de l'Europe ou même extra-européennes.

Dans quelques genera on peut distinguer des espèces vicariantes. Les végétations de type "inops", très communes, se composent généralement d'espèces ayant une grande aire, dont beaucoup de cosmopolites. Dans les autres types de végétation aussi le rôle de ces espèces est important. Pour tous les types de végétation l'anthropochorie a une grande influence.

CHAPITRE 9 — Quelques aspects phytogéographiques et taxonomiques

Nous avons discuté les spectres de plusieurs régions géographiques. On peut constater une polyploidie progressive pour les végétations murales allant de l'Europe méridionale vers le Nord et allant de l'Europe occidentale vers l'Est.

La contribution des hémicryptophytes est toujours importante et en Europe méridionale la contribution des thérophytes est également considérable, au moins sur le plan qualitatif.

Les végétations murales se composent surtout de mésophytes. Parmi les plantes vasculaires on trouve fréquemment la sociabilité 2, et parmi les bryophytes la sociabilité 3. L'anémochorie est importante pour beaucoup de types de végétation, mais dans les végétations à *Parietaria judaica* la myrmécochorie joue un rôle important, surtout en Europe méridionale. Nombre d'espèces murales caractéristiques ont une distribution méditerranéenne-atlantique ou méditerranéenne. Pour quelques espèces nous avons mentionné des données écologiques et, ou taxonomiques complementaires. Nous avons étudié de près la signification des néophytes.

Zusammenfassung

1. KAPITEL — Einleitung

Von der Oekologie von Vegetationen auf Mauern ist nur wenig bekannt. Die Untersuchung fing in den Niederlanden an, dehnte sich aber später u.a. aus über England, Belgien, Luxemburg, Frankreich, Italien und Westdeutschland. An annähernd 1200 Stellen wurde die Vegetation vertikaler Mauern analysiert. Auch wurden an 177 Stellen Vegetationen zwischen Strasensteinen untersucht, weil diese Vegetationen oft Uebereinstimmungen mit gewissen Typen von Mauervegetationen aufweisen.

2. KAPITEL — Problemstellung und Methoden

Die Wahl der Standorte wurde weder dem Zufall überlassen, noch geschah sie subjektiv. Hauptsächlich wurden Mauern in alten Kulturgebieten untersucht. Ziel war nicht nur Vegetationstypen zu beschreiben, sondern vor allem die Variabilität von Mauervegetationen zu untersuchen und eventuelle Kontinuitäten in der Zusammensetzung in geographischen Reihen zu unterscheiden. Verwendet wurde für das Abschätzen der Abundanz und dem Bedeckungsprozentsatz der Masstab von BARKMAN, DOING & SEGAL (1964). Die Vegetationsaufnahmen wurden nach der Affinitätsformel von POORE (1956) sortiert. Ferner wurden Matritzen ausgearbeitet als Resultat von 2 x 2 Tabellen, in welchen die Affinität (sowohl positiv als negativ) zwischen den Arten, die am meisten vorkommen, berechnet wurde. Von allen Arten wird in einer separaten Liste Folgendes angegeben (Anhang I und II): idiosystematische Ordnung (nach PULLE, 1952), Ploidiegrad, Lebensformen (nach Raunkiaer und IVERSEN, 1936), Wuchsformen, Soziabilität, Blumenfarbe, Disseminationstyp, Formationsgruppe, Areal und Verbreitung innerhalb Europas.

Mit diesen Angaben wurden Spektra (relative Werte für jede Kategorie in Prozenten) berechnet sowohl für die Vegetationstypen als für die 11 unterschiedenen geographischen Gebiete und für eine Anzahl Kombinationen sowohl für ungewogene wie für gewogene Werte. Für die Berechnungen wurden Komputer eingeschaltet. Die ökologische Untersuchung war gerichtet auf chemische Analysen vom Substrat und auf Messungen von Temperatur und Licht.

3. KAPITEL — Oekologie von Mauervegetationen

Besondere Aufmerksamkeit ist topographischen und historischen Faktoren gewidmet worden. In Mitteleuropa und in subatlantischen Gebieten sind gut entwickelte Mauervegetationen vor allem gebunden an Stromgebiete grosser Flüsse. Es zeigte sich bei manchen Arten, das sie spezifisch für Mauervegetationen sind, z.B. *Linaria cymbalaria, Corydalis lutea, Cheiranthus cheiri, Parietaria judaica* und *Tortula muralis.* Wenn die Inklination abnimmt, nimmt in der Regel die Zahl der Arten zu, und die Vegetation wird weniger 'spezifisch'. Die Exposition ist von grosser Bedeutung für viele Arten und Vegetationstypen. Eine Anzahl von Arten, die eine Vorliebe haben für Nordmauern im mediterranen Gebiet, zeigen keine Vorliebe dafür im atlantischen Gebiet.

Der Einfluss von Luftverunreinigung ist auf Flechten sehr gross, aber auch erheblich auf Moose. Für den jeweiligen Grad von Toxitoleranz wurden Reihen aufgestellt.

Der pH des Substrates von gut entwickelten Mauervegetationen liegt meistens zwischen 7 und 8, von Initialstadien häufig zwischen 8 und 9. Die Verwitterung und Sedimentation von Staub- und Bodenteilchen ist eine wichtige Vorbedingung für die Entwicklung der Arten. Der beschränkende Faktor ist oft das Wasser. Viele Arten sind grossen Schwankungen im Feuchtigkeitsgehalt von Substrat und Atmosphäre angepasst.

4. KAPITEL — Struktur

Die vertikalen Gefüge (Schichtung) und die horizontalen Gefüge werden erörtert. Spezielle Aufmerksamkeit wird den Grenzen der Vegetation gewidmet. Die Soziabilität ist mit Hilfe objektiver Werte bestimmt worden. Der qualitative und repräsentative Minimalraum wurde definiert. Dabei stellte sich heraus, dass sowohl ersterer wie letzterer für (optimal entwickelte) Vegetationen mit einer bestimmten Struktur ungefähr konstant ist. Der Minimalraum ist schliesslich von der Dichte der aspektbestimmenden Wuchsform abhängig. Ein System von Wuchsformen wurde ausgearbeitet.

5. KAPITEL — Klassifikation von Mauervegetationen

Ausgangspunkt für die Betrachtung einer Vegetation müsste das Oekosystem sein. Dies wäre eventuell in Subsysteme aufzugliedern (Biozoenose und Stenozoenosen). Von der Abhängigkeit zwischen den Organismen untereinander und ihren Standorten ist jedoch wenig bekannt. In Pioniervegetationen ist vermutlich in geringerem Masse die Rede von einem System als in stärker differenzierten und strukturierten Gesellschaften. Auf Mauern sind vermutlich Gefässpflanzen und Bryophyten oft in hohem Masse unabhängig. Der Begriff Charakterart ist schwierig objektiv zu hantieren. Vegetationen werden auf Grund von differenzierenden Artenkombinationen unterschieden und/oder von ökologischen Gruppen. Ein gutes System müsste nicht nur allein floristisch fundiert werden. Eine gewisse Rangordnung in den Kriterien könnte möglicherweise die Einheit in der Betrachtung fördern, z.B. 1. floristische, 2. strukturelle, 3. ökologische und 4. geographische Kriterien.

6. KAPITEL — Uebersicht von Mauervegetationen

Die Vegetationstypen auf Mauern werden ausführlich besprochen, mit ihren Uebergängen zu anderen Vegetationstypen und ihrer Veränderung in der Zusammenstellung in anderen geographischen Gebieten. Uebergänge von Vegetationstypen zu anderen 'höheren Einheiten' können zugleich mit verwandten Vegetationstypen aus den höheren Einheiten in einen 'Assoziationskreis' zusammengebracht werden. Artenarme Vegetationen, oft mit Dominanz einer Art, sind als 'Inopsvegetation' beschrieben worden. Eine Anzahl von Coenoklinen konnte abgeleitet werden. In mediterranen Vegetationen mit *Parietaria judaica* zeigte es sich, dass Myrmekochorie wichtig ist. In den mittel- und subatlantischen Gebieten kommen viele Vegetationen mit *Sagina procumbens* und *Poa annua* vor.

7. KAPITEL — Sukzession von Mauervegetationen

Das Endstadium der Sukzession ist abhängig von dem Grad, in welchem die Vegetation das Milieu verändern kann. Für die Entwicklung sind die zunehmende Verwitterung, das Ansteigen von feinen Substratteilchen (auch durch Sedimentation), sinkender pH und abnehmende Inklination wichtig. Eine kurzgefasste Uebersicht enthält Fungi, Algae und Lichenes auf Mauern. *Protococcus* sensu lato ist oft auf Mauern vertreten, die auf lange Dauer ziemlich feucht bleiben.

Bei der Sukzession von Moosen spielen anfangs akrokarpe Moose eine Rolle, später mehr pleurokarpe Moose. In West- und Mitteleuropa tritt *Asplenium ruta-muraria* oft als Pionier auf, nachgefolgt von *Linaria cymbalaria.* Aus der Gemeinschaft dieser Arten können sich andere Vegetationen entwickeln, in denen z.B. *Corydalis lutea, Cheiranthus cheiri* oder *Parietaria judaica* vorkommen. In den euatlantischen Gebieten kann *Asplenium adiantum-nigrum* als Pionier auftreten und in Südeuropa *Parietaria judaica.*

8. KAPITEL — Strassenvegetationen: ein Vergleich

Vegetationen zwischen Pflaster zeigen gewisse Uebereinstimmungen mit Mauervegetationen mit *Poa annua* und *Sagina procumbens,* und beide sind in einem Assoziationskreis zu vereinigen. Eine auffallende Uebereinstimmung in Struktur gibt es zwischen Vegetationen von verschiedenen Teilen Europas und ausserhalb davon, wobei vikariierende Arten in einer Anzahl von Genera anzuzeigen sind. Die sehr allgemeine Inopsvegetation besteht hauptsächlich aus Arten mit einem grossen Areal. Darunter befinden sich viele Kosmopoliten. Auch in anderen Vegetationstypen ist der Anteil dieser Arten gross. Anthropochorie spielt bei allen Vegetationen eine wichtige Rolle.

9. KAPITEL — Einige phytogeographische und taxonomische Aspekte

Die Spektra der verschiedenen geographischen Gebiete werden besprochen. Es hat sich herausgestellt, dass Polyploidie auf Mauern zunimmt von Süd- nach Nordeuropa und von West- nach Osteuropa.

Wichtig ist immer der Anteil der Hemikryptophyten und in Südeuropa ist, jedenfalls qualitativ, auch der Anteil der Therophyten erheblich. Mauervegetationen sind vor allem aus Mesophyten aufgebaut. Bei den Gefässpflanzen kommt Soziabilität 2 sehr viel vor, bei den Bryophyten Soziabilität 3. Anemochorie ist in vielen Vegetationstypen wichtig, aber in Vegetationen mit *Parietaria judaica* spielt, vor allem in Südeuropa, Myrmekochorie eine wichtige Rolle. Viele typische Mauerarten haben eine mediterran-atlantische oder eine mediterrane Verbreitung.

Von einer Anzahl Arten sind ergänzende Angaben von Oekologie und/oder Taxonomie gegeben worden. Die Bedeutung von Neophyten ist näher besprochen worden.

ADDENDA ET CORRIGENDA

p. 9, 2nd par.,	7th line:	*read* was described *instead of* cover was recorded
p. 12,	28th line:	*add* or rather frost sensitivity *behind* temperatures
p. 15, 2nd par.,	1st line:	*read* sample plot surface *instead of* experimental surface area
	3rd " :	*read* uncompletely developed vegetation *instead of* vegetation fragments
	6th " :	*add* rather *behind* be
	14th " :	*read* stands *instead of* experimental areas
	23rd " :	*read* stand *instead of* experimental area
p. 23, 6th par.,	4th line:	*read* cover degree *instead of* presence
	10th " :	*read* cover degree *instead of* degree of presence
p. 25, 1c":		*strike* and S.W.
p. 30, 2.Autochory:		*read* (bolochory) or disengaging tissue (blastochory) *instead of* or
p. 31,	14th line:	*read* bolochory *instead of* autochory
p. 34,	21st line:	*read* calciphilous and calcifugous *instead of* calciphile and calcifuge
p. 47, 3.1.3,	2nd line:	*read* stand *instead of* vegetation fragment
	3rd " :	*read* being *instead of* recording
p. 48, 3.1.3:		*strike* last line (and also...case)
p. 53,	6th line from below:	*add* solar *behind* amount of
p. 59,	24th line:	*read* indigenous *instead of* autochthone
p. 60,	14th line:	*read* decreases *instead of* increases
p. 78,	last line:	*read* S.E. *instead of* S.W.

p. 95, 3rd par.,	7th line:	*strike* ecological
p. 96, 4.4,	1st line:	*strike* causal
p. 98,	last line:	*read* stands *instead of* vegetation fragments
p.101, 4.5,	3rd line from below:	*read* could have *instead of* has
	2nd " " " :	*read* separate *instead of* minimum
p.106, 4th par.,	4th line:	*strike* ecological
	12th " :	*read* hygrophytic, aerohygrophytic *instead of* aerohydrophytic
p.111, B:		*read* Leskeids ("*Leskea* type"of Barkman) *instead of* Radulids ("*Radula*-form" of Ochsner)
p.112, B:		*read* Lecanorids *instead of* Calopacids
p.116, 4th par.,	3rd and 5th line:	*read* stands *instead of* experimental areas
p.120, 3rd par.,	6th line:	*read* maximum *instead of* minimum
p.122,	16th line:	*read* *topocline* *instead of* *topoclin*
p.125, 3rd par.,	2nd line:	*strike* fragments
p.128,	2nd line from below:	*read* synthesis *instead of* analysis
p.130, 4th par.,	3rd line:	*read* floristic *instead of* biological
	last line:	*place* 4. Geographical criteria *below* 3. Ecological criteria
p.225,		*add to* 6.14: Summarizing may be said that *Parietarion judaicae* vegetation generally may be found in drier habitats (climates) than *Cymbalario-Asplenion* vegetation, and often occur on southward exposition. *Cymbalario-Asplenion* rather prefer habitats with small temporal differences in humidity and temperature, and often occur on north- and eastward exposition. *Parietarion judaicae* vegetation show relatively more chamaephytes, hemixerophytes, sociability 3 and myrmecochory. The distribution is mainly Mediterranean. *Cymbalario-Asplenion* vegetation show rela-

tively more ferns, geophytes and helophytes, telmatophytes, sociability 2, anemochory and anthropochory, forest and riparian plants. The distribution is more restricted to atlantic, continental and N.Europe. *Parietaria judaica* behaves as a pioneer in the Mediterranean region. Occasionally this helds in W. Europe, but never in continental Europe.

p.318: *add* MEYER DREES, E. 1951 - Verklarende lijst van termen uit de plantensociologie en synoecologie. Rapport Bosbouwproefstation Bogor 48. 140 p.

" : *read* NORDHAGEN *instead of* NCRDHAGEN

p.319: *read* ORDNUFF, R. 1967 *instead of* ORDNUFF, R. 1965

References

References with an asterisk indicate: not seen by the author.

ALLEN, D.E. 1964 - Poa nemoralis (in Plant notes). Proc. Bot. Soc. Brit. Isles 5: 233.

ALLORGE, P. 1941 - Essai de synthèse phytogéographique du Pays basque. Bull. Soc. Bot. France 88: 291-356.

AMANN, J. 1918 - Flore des mousses de la Suisse. Deuxième partie: Bryogéographie de la Suisse. 414 p. Lausanne.

— 1922 - Les mousses du vignoble de Lavaux. Mem. Soc. Vaud. Sc. Nat. 1: 77.

— 1929 - L'Hygrothermie du climat, facteur déterminant la répartition des espèces atlantiques. Rev. Bryol. N. S. 2: 126-133.

— & Ch. MEYLAN. 1918 - Flore des mousses de la Suisse. Première partie: Tableaux synoptiques pour la détermination des mousses. 215 p. Lausanne.

ANDERSON, D. J. 1965 - Classification and ordination in vegetation science: controversy over a non-existent problem? Journ. Ecol. 53: 521-526.

ANZALONE, B. 1951 - Flora e vegetazione dei muri di Roma. Annali Bot. 23: 393-497.

ARÈNES, J. 1929 - Les associations végétales de la Basse-Provence. 284 p. Paris.

ARNOLD, F. 1891-1901 - Zur Lichenflora von München. 147 + 76 + 45 + 82 + 100 + 24 p. München.

AUGIER, J. 1966 - Flore des bryophytes. Morphologie, anatomie, biologie, écologie, distribution géographique. 702 p. Paris.

BAAS BECKING, L. G. M. 1934 - Geobiologie; of inleiding tot de milieukunde. 263 p. Den Haag.

BAKER, H. G. & G. L. STEBBINS. 1965 - The genetics of colonizing species. 588 p. London, New York.

BAKKER, D. 1962 - Migratie en vestiging. Jaarb. Kon. Ned. Bot. Ver. 1962: 50-51.

BARKMAN, J. J. 1947a - Een en ander over de mosflora rondom Leiden. Buxbaumia 1, no. 2: 2-12.

— 1947b - Mosgezelschappen aan meeroevers. De Lev. Natuur 50: 80-83.

— 1958 - On the ecology of cryptogamic epiphytes, with special reference to the Netherlands. 202 p. Assen.

— 1962 - Kartering van de epiphytenvegetatie van Belgisch Limburg in verband met de industrialisatie. Jaarb. Kon. Ned. Bot. Ver. 1962: 40-41.

— 1966 - De variëteiten van Hypnum cupressiforme Hedw. in Nederland. Buxbaumia 20: 1-6.

— 1968 - Das synsystematische Problem der Mikrogesellschaften innerhalb der Biozönosen. In: R. Tüxen (ed.), Pflanzensoziologische Systematik: 1-53. Den Haag.

—, H. DOING & S. SEGAL. 1964 - Kritische Bemerkungen und Vorschläge zur quantitativen Vegetationsanalyse. Acta Bot. Neerl. 13: 394-419.

Barnêwitz, A. 1898 - Die auf der Stadtmauer von Brandenburg a.H. wachsenden Pflanzen. Verh. Bot. Ver. Prov. Brandenb. 40, Abh.: 97-108.

Becking, R. W. 1957 - The Zürich-Montpellier School of phytosociology. Bot. Rev. 23: 411-488.

Beeftink, W. G. 1965 - De zoutvegetaties van Z.W.-Nederland beschouwd in Europees verband. Meded. Landbouwhogeschool Wageningen 65-1: 1-167.

Beger, H. 1930 - Praktische Richtlinien der strukturellen Assoziationsforschung im Sinne der von der Zürich-Montpellier-Schule geübten Methode. In: E. Abderhalden (ed.), Handbuch der biologischen Arbeitsmethoden, Abt. 11, Tl. 5: 481-526. Berlin, Wien.

Béguinot, A. 1911-1915 - La flora delle mure e delle vie di Padova. Malpighia 24: 413-428; 25: 61-84; 27: 244-259, 439-454, 547-582.

Beylsmit, L. & E. Maten. 1965 - Grachtkantenonderzoek. 17 p. Ned. Jeugdb. Natuurst. afd. Amersfoort. Mimeographed.

Bing, K. 1947 - Efflorescence on masonry. Trans. Chalmers Univ. Technol. Gothenburg 58: 1-150.

Birse, E. L. & C. H. Gimingham. 1954 - Changes in the structure of Bryophytic communities with the progress of succession on sand-dunes. Trans. Brit. Bryol. Soc. 2: 523-531.

Bleasdale, J. K. A. 1959 - The effects of air pollution on plant growth. In: W.B. Yapp (ed.), The effect of pollution on living material: 81-87. Oxford.

Blondel, R. 1941 - La végétation forestière de la région de Saint-Paul près de Montpellier. SIGMA-Comm. 79: 307-383. Lausanne.

Blum, A. 1925 - Beiträge zur Kenntnis der annuellen Pflanzen. Bot. Archiv 9: 3-36.

Boerboom, J. H. A. 1960 - De plantengemeenschappen van de Wassenaarse duinen. Meded. Landbouwhogeschool Wageningen 60-10: 1-135.

Bonner, J. T. 1965 - Size and cycles: An essay on the structure of biology. 219 p. Princeton.

Boycott, A. E. 1934 - The habitat of land Mollusca in Britain. Journ. Ecol. 22: 1-38.

Braun-Blanquet, J. 1915 - Les Cévennes méridionales (Massif de l'Aigoual). 207 p. Genève.

— 1928 - Pflanzensoziologie. 330 p. Berlin.

— 1931 - Aperçu des groupements végétaux du Bas-Languedoc. SIGMA-Comm. 9: 35-40. Montpellier.

— 1949 - Uebersicht der Pflanzengesellschaften Rätiens (II). Vegetatio 1: 129-146.

— 1959 - Grundfragen und Aufgaben der Pflanzensoziologie. In: Vistas in Botany: 145-171. London.

— 1964 - Pflanzensoziologie; Grundzüge der Vegetationskunde. 865 p., 3rd ed. Wien, New York.

— 1966 - Vegetationsskizzen aus dem Baskenland mit Ausblicken auf das weitere Ibero-Atlantikum. I. Teil. Vegetatio 13: 117-147.

— 1967 - Idem. II. Teil. Vegetatio 14: 1-126.

— , N. Roussine & R. Nègre. 1952 - Les groupements végétaux de la France méditerranéenne. 297 p. Vaison-la-Romaine.

— & R. Tüxen. 1952 - Irische Pflanzengesellschaften. Veröff. Geobot. Inst. Rübel Zürich 25: 224-415.

Briquet, J. 1910 - Prodrome de la Flore Corse. I. Genève, Bâle, Lyon. 656 p.

Büker, R. 1939 - Die Pflanzengesellschaften des Messtischblattes Lengerich in Westfalen. Abh. Landesmus. Prov. Westfalen, Mus. Naturk. 10, no. 1: 1-108.

— 1942 - Beiträge zur Vegetationskunde des südwestfälischen Berglandes. Beih. Bot. Centralbl. 61, Abt. B: 452-558.

CAIN, S. A. 1944 - Foundations of plant geography. 556 p. New York, London.
— 1947 - Characteristics of natural areas and factors in their development. Ecol. Monogr. 17: 185-200.
— & G. M. DE OLIVEIRA CASTRO. 1959 - Manual of vegetation analysis. 325 p. New York.
CALLÉJA, M. 1962 - Etude de la courbe aire espèce et de l'aire minimale. Bull. Serv. Carte Phytogéogr., Sér. B, 7: 161-179.
CARLES, J. 1948 - Le spectre biologique réel. Bull. Soc. Bot. France 95: 340-343.
CHATIN, A. 1861 - Sur les plantes des vieux châteaux. Bull. Soc. Bot. France 8: 359-365.
C. J. N. 1959 - Plantengroei op de Utrechtse grachtmuren. Correspondentiebl. Flor. 14: 145-147.
CLAPHAM, A. R., T. G. TUTIN & E. F. WARBURG. 1962 - Flora of the British Isles. 1269 p., 2nd ed. Cambridge.
CLAUSEN, J. 1961 - Introgression facilitated by apomixis in polyploid Poas. Euphytica 10: 87-94.
CLEMENTS, F. E. 1928 - Plant succession and indicators. 453 p. New York.
CLIFFORD, H. T. 1956 - Sead dispersal on footwear. Proc. Bot. Soc. Brit. Isles 2: 129-131.
COESEL, P. F. M. 1962 - De associatie van Didymodon recurvirostris en Tortella flavovirens Boerboom 1960. 22 p. RIVON, Bilthoven. Mimeographed.
CONTANDRIOPOULOS, J. 1957 - Caryologie et localisation des espèces végétales endémiques de la Corse. Bull. Soc. Bot. France 104: 53-55.
CRAWFORD, R. M. M. & D. WISHART. 1966 - A multivariate analysis of the development of dune slack vegetation in relation to coastal accretion at Tentsmuir, Fife. Journ. Ecol. 54: 729-743.
— & D. WISHART. 1967 - A rapid multivariate method for the detection and classification of groups of ecologically related species. Journ. Ecol. 55: 505-524.
CUFODONTIS, G. 1947 - Die Gattung Cymbalaria Hill.; Nachträge und Zusammenfassung. Botaniska Notiser 107: 135-156.

DAGNELIE, P. 1960 - Contribution à l'étude des communautés végétales par l'analyse factorielle. Bull. Serv. Carte Phytogéogr., Ser. B, 5: 7-71.
* DAMANTI, P. 1903 - Proemio ad una flora murale dei dintorni di Palermo. Rend. Congr. Bot. Naz. Palermo, Maggio 1902: 190.
DANSEREAU, P. 1951 - Description and recording of vegetation upon a structural basis. Ecology 32: 172-229.
DARLINGTON, C. D. & A. P. WYLIE. 1955 - Chromosome atlas of flowering plants. 519 p. London.
DAVEY, J. 1961 - Biological flora of the British isles: Epilobium nerterioides A. Cunn. Journ. Ecol. 49: 753-759.
* DEAKIN, M. 1855 - Flora of the Colosseum of Rome. London.
DELVOSALLE, L. 1968 - Observations botaniques faites à la Toussaint 1966 dans le Kent et le Sussex. Les Natural. Belg. 49: 161-165.
DEPASSE, S. 1957 - Fougères de la région Senne-Sennette-Samme. Bull. Soc. Roy. Bot. Belg. 90: 49-62.
* DE ROSA, F. 1905 - Contributo alla flora murale e ruderale di Napoli. Boll. Soc. Nat. Napoli 19: 219.
DHIEN, R. 1964 - Cyrtomium falcatum en France. Bull. Mens. Soc. Linn. Lyon 33: 307-309.
DIEMONT, W. H., G. SISSINGH & V. WESTHOFF. 1940 - Het Dwergbiezenverbond (Nanocyperion flavescentis) in Nederland. Ned. Kruidk. Arch. 50: 215-271.

DIXON, H. N. 1924 - The student's handbook of British mosses. 582 p., 3rd ed. London.

DOING KRAFT, H. 1956 - De tegenwoordige opvattingen omtrent het associatiebegrip en de systematiek van de plantengemeenschappen volgens de methode van Braun-Blanquet. Vakbl. Biol. 36: 222-234.

DOING (KRAFT), H. 1962 - Systematische Ordnung und floristische Zusammensetzung niederländischer Wald- und Gebüschgesellschaften. Wentia 8: 1-85.

DUDGEON, W. 1923 - Succession of epiphytes in the Quercus incana-forest at Lanour, Western Himalayas. J. Indian Bot. Soc. 3: 270-272.

DÜGGELI, M. 1930 - Die Mitwirkung von Bakterien bei der Gesteinsverwitterung. Verh. Schweiz. Naturf. Ges. 111: 307-308.

DU RIETZ, G. E. 1932 - Vegetationsforschung auf soziationsanalytischer Grundlage. In: E. Abderhalden (ed.), Handbuch der biologischen Arbeitsmethoden, Abt. 11, Tl. 5, 1. Hälfte: 293-480.

— 1936 - Classification and nomenclature units 1930-1935. Svensk Bot. Tidskr. 30: 580-589.

DURIN, L. 1955 - Les Filicariae du département du Nord, répartition et écologie. Ann. Sc. Nat., 2e Sér.: 481-492.

DUVIGNEAUD, P. 1946 - La variabilité des associations végétales. Bull. Soc. Roy. Bot. Belg. 78: 107-134.

EBERHARDT, P. 1903 - Influence de l'air sec et de l'air humide sur la forme et sur la structure des végétaux. Ann. Sci. Nat. Bot. VIII, 18: 61-153.

EEDEN, F. W. VAN. 1867 - De bosschen van Kennemerland. Album der Natuur 1867: 193-219.

EGLER, F. E. 1961 - The nature of naturalization. In: Recent advances in botany. II: 1340-1345. Toronto.

ELLENBERG, H. 1956 - Aufgaben und Methoden der Vegetationskunde. 136 p. Stuttgart.

— 1963 - Vegetation Mitteleuropas mit den Alpen. 943 p. Stuttgart.

— & M.-L. SNOY. 1957 - Physiologisches und ökologisches Verhalten von Ackerunkräutern gegenüber der Bodenfeuchtigkeit. Mitt. Staatsinst. Allg. Bot. Hamburg 11: 47-87.

EVANS, F. C. 1956 - Ecosystem as the basic unit in ecology. Science 123: 1127-1128.

FABER, A. 1933 - Pflanzensoziologische Untersuchungen im Süddeutschland; Ueber Waldgesellschaften in Württemberg. Bibl. Bot. 108: 1-68.

FITTER, R. S. R. 1945 - London's natural history. 282 p. London.

Flora Europaea. Vol. 1. Lycopodiaceae to Platanaceae (ed. T. G. Tutin, et al.). 464 p. Cambridge.

FOERSTER, E. 1968 - Zur systematischen Stellung artenarmer Lolium-Weiden. In: R. Tüxen (ed.). Pflanzensoziologische Systematik: 183-190. Den Haag.

FOURNIER, P. 1946 - Les quatre flores de la France. 2nd ed., 1091 p. Paris.

FREY, E. 1924 - Die Berücksichtigung der Lichenen in der soziologischen Pflanzengeographie, speziell in den Alpen. Verh. Naturf. Ges. Basel 35: 303-320.

FRIEDRICH, M. 1954 - Beitrag zur Pflanzengeographie des Gardasee-Gebietes. Prešt. God. Biol. Inst. Sarajevu 7: 3-24.

FRÖDE, E. 1958 - Die Pflanzengesellschaften der Insel Hiddensee. Wiss. Z. Univ. Greifswald, Math.-Nat. 7: 277-305.

GABELLI, L. 1915 - Contributo alla flora murale e ruderale del Senese. Atti Pont. Accad. Nuovi Lincei 68: 137-146.

* Gagnepain, F. 1897 - Végétation calamicole et murale des environs de Cercy-la-Tour (Nièvre). Bull. Soc. Hist. Nat. Autun 9: 230.

Gams, H. 1918 - Prinzipienfragen der Vegetationsforschung. Vierteljahrschr. Naturf. Ges. Zürich 63: 293-493.

— 1938 - Oekologie der extratropischen Pteridophyten. In: F. Verdoorn, Manuel of Pteridology: 382-419. Den Haag.

— 1939 - Die Hauptrichtungen der heutigen Biozönotik. Chron. Bot. 5: 133-140.

— 1941 - Pflanzengesellschaften der Alpen. II. Die Vegetation der Felsen. Jahrb. Ver. Schutze Alpenpfl. Tiere 13: 12-26.

— 1942 - Idem. III. Die Besiedlung des Felsschutz. Jahrb. Ver. Schutze Alpenpfl. Tiere 14: 16-44.

— 1954 - Vegetationssystematik als Endziel oder Verständigungsmittel? Veröff. Geobot. Inst. Rübel Zürich 29: 35-40.

— 1957 - Die Moos- und Farnpflanzen. 4th ed., 240 p. Stuttgart.

Gaschott, O. 1925 - Malakologisches aus Süddeutschland. Archiv Molluskenk. 57: 269-273.

Gaussen, H. 1933 - Géographie des plantes. 222 p. Paris.

Géhu, J. M. 1961 - Les groupements végétaux du bassin de la Sambre Française. III. Vegetatio 10: 257-372.

— & J. Géhu-Franck. 1961 - Recherches sur la végétation et le sol de la réserve de l'Ile des Landes (Ill-et-Vilaine) et de quelques ilots de la Côte Nord-Bretagne. Bull. Lab. Maritime Dinard 47: 19-57.

Geiger, R. 1961 - Das Klima der bodennahen Luftschicht. 4th ed., 646 p. Braunschweig.

Giesenhagen, K. 1910 - Die Moostypen der Regenwälder. Ann. Jard. Bot. Buitenzorg, Suppl. 3, no. 2: 711-791.

Gillham, M. E. 1956a - Ecology of the Pembrokeshire islands. IV. Effect of treading and burrowing by birds and mammals. Journ. Ecol. 44: 51-82.

— 1956b - Idem. V. Manuring by the seabirds and mammals, with a note on seed distribution by gulls. Journ. Ecol. 44: 429-454.

Gilbert, O. L. 1968 - Bryophytes as indicators of air pollution in the Tyne Valley. New Phytol. 67: 15-30.

Gimingham, C. H. & E. M. Birse. 1957 - Ecological studies on growth-forms in bryophytes. I. Correlations between growth-form and habitat. Journ. Ecol. 45: 533-545.

— & E. T. Robertson. 1950 - Preliminary investigations on the structure of bryophytic communities. Trans. Brit. Bryol. Soc. 1: 330-348.

Gleason, H. A. 1922 - On the relation between species and area. Ecology 3: 158-162.

— 1925 - Species and area. Ecology 6: 66-74.

— 1926 - The individualistic concept of the plant association. Bull. Torrey Bot. Cl. 53: 7-26.

Goodall, D. W. 1952 - Quantitative aspects of plant distribution. Biol. Rev. 27: 194-245.

— 1953a - Objective methods for the classification of vegetation. I. The use of positive interspecific correlation. Austr. Journ. Bot. 1: 39-63.

— 1953b - Idem. II. Fidelity and indicator value. Austr. Journ. Bot. 1: 434-456.

Goslin, C. R. 1958 - The Holly-fern, Cyrtomium falcatum, outdoors in Ohio. Am. Fern Journ. 48: 84.

Grabandt, R. A. J. 1952 - Floristisch onderzoek van oude muren. 15 p. Hugo de Vries-laboratorium, Amsterdam. Manuscript.

Gräfe, K. 1956 - Strahlungsempfang vertikaler ebener Flächen; Globalstrahlung von Hamburg. Ber. Deuts. Wetter D. 5: 1-15.

GREBE, C. 1918 - Studien zur Biologie und Geographie der Laubmoose. Hedwigia 59: 1-208.

GREEN, H. E. 1961 - Some notes on seed dispersal. Proc. Bot. Soc. Brit. Isles 4: 153-154.

GREIG-SMITH, P. 1961 - The use of pattern analysis in ecological investigations. In: Recent advances in botany. II: 1354-1358. Toronto.

— 1964 - Quantitative plant ecology. 2nd ed., 256 p. London.

GUITTART, J. 1957 - De flora van de waterkant in en bij Dordrecht. Correspondentiebl. Flor. 6: 65-67.

HAGERUP, O. 1931 - Ueber Polyploidie in Beziehung zu Klima, Oekologie und Phylogenie. Hereditas 16: 19-40.

HARTOG, C. DEN & S. SEGAL. 1964 - A new classification of the waterplant communities. Acta Bot. Neerl. 13: 367-393.

HEGI, G. 1931-1963 - Illustrierte Flora von Mitteleuropa. I. 2nd ed., 528 p. II. 2nd ed., 532 p. III/1, 2nd ed., 452 p. III. 607 p. IV/1, 2nd ed., 547 p. IV/2+3, 1251 p. V/1-4, 2631 p., VI/1-2, 1386 p., VII, 562 p.

HEIMANS, J. 1954 - L'accenssibilité, terme nouveau en phytogéographie. Vegetatio 5-6: 142-146.

— 1960 - De flora als inleiding tot de bijzondere plantkunde. In: E. Heimans, H. W. Heinsius & J. P. Thijsse, Geïllustreerde flora van Nederland: 1089-1110.

HEINTZE, A. 1932-1935 - Handbuch der Verbreitungsökologie der Pflanzen. I. II. 266 p. Stockholm.

HEPPER, F. N. 1954 - Flora of Caldy Island, Pembrokeshire. Proc. Bot. Soc. Brit. Isles 1: 21-36.

HEUKELS, H. & S. J. VAN OOSTSTROOM. 1962 - Flora van Nederland. 15th ed., 892 p. Groningen.

HILITZER, A. 1925 - La végétation épiphyta de la Bohème. Publ. Fac. Sc. Univ. Charles Prague 41: 1-200.

HILLMANN, J. & V. GRUMMANN. 1957 - Kryptogamenflora Mark Brandenburg VIII. Flechten. 898 p. Berlin-Nikolassee.

HOPKINS, B. 1955 - The species-area relations of plant communities. Journ. Ecol. 43: 409-426.

— 1957 - The concept of minimal area. Journ. Ecol. 45: 441-449.

HORA, F. B. 1947 - The pH-ranges of some cliff plants on rock of different geological origin in the Cader Idris area of North Wales. Journ. Ecol. 35: 158-165.

HORVATIĆ, S. 1934 - Flora i vegetacija otoka Paga (Flora und Vegetation der nordadriatischen Insel Pag). Bull. Int. Acad. Yougosl. 28: 86-157.

— 1963 - Vegetacijska karta otoka Paga s opcím pregledom vegetacijskih jedinica hrvatskog primorja. Acta Biol. 4: 1-187. Zagreb.

HÜBSCHMANN, A. VON, 1950 - Die Grimmia pulvinata-Tortula muralis-Ass. im nordwestdeutschen Flachlande. Mitt. Flor.-soz. Arbeitsgem. N.F. 2: 6-11.

— 1967 - Ueber die Moosgesellschaften und das Vorkommen der Moose in den übrigen Pflanzengesellschaften des Moseltales. Schriftenreihe Veg. kunde 2: 63-121.

HUMBOLDT, A. VON & A. BONPLAND. 1807 - De distributione geographica plantarum. 254 p. Montmartre.

IVERSEN, J. 1936 - Biologische Pflanzentypen als Hilfsmittel in der Vegetationsforschung. 224 p. København.

IVIMEY-COOK, R. B. & M. C. F. PROCTOR. 1966 - The plant communities of the Burren, Co. Clare. Proc. Roy. Irish Acad. 64, Sect. B, 14: 211-301.

JACCARD, P. 1902 - Gesetze der Pflanzenverteilung der alpinen Region. Flora 1902: 349-377.

JACKSON, F. 1958 - Asplenium viride. Brit. Fern Gaz. 8: 234-236.

JAECKEL, S. 1943 - Eine Molluskenausbeute aus Belgien und Frankreich während des Westfeldzuges 1940. Archiv Molluskenk. 75: 285-306.

* JOURDAN, P. 1867 - Flore murale de la ville de Tlemcen, province d'Oran (Algérie). Gaz. Med. Alg., and Bull. Soc. Climat Alg.

* — 1872 - Flore murale de la ville d'Alger. Bull. Soc. Climat Alg.

JOVET, P. 1941 - La végétation anthropophile du Pays basque français. Bull. Soc. Bot. France 88: 254-269.

— 1957 - Compte rendu des excursions. Bull. Soc. Bot. France 104; Sess. 83: 9-37.

JURKO, A. & V. PECIAR. 1963 - Pflanzengesellschaften an schattigen Felsen in den Westkarpaten. Vegetatio 11: 199-209.

KAISER, E. 1926 - Die Pflanzenwelt des Hennebergisch-Fränkischen Muschelkalkgebietes. Rep. Spec. Nov. Regni Veg. Beih. 44: 1-280.

KATZ, N. 1933 - Die Grundprobleme und die neue Richtung der Phytosoziologie. Beitr. Biol. Pfl. 21: 133-166.

KENT, D. H. 1960 - Senecio squalidus L. in the British Isles. 2. The spread from Oxford (1879-1939). Proc. Bot. Soc. Brit. Isles 3: 375-379.

— 1961 - The flora of Middlesex walls. Lond. Nat. 40: 29-43.

KETTENACKER, L. 1930 - Ueber die Feuchtigkeit von Mauern. Gesundheits Ing. 53: 721-728.

KILBURN, P. D. 1963 - Exponential values for the species-area relation. Science 141: 1276.

KIRSCHLEGER, M. F. 1862 - Sur les plantes des vieux châteaux, dans la région alsatovosgienne. Bull. Soc. Bot. France 9: 15-18.

KNAPP, R. 1948 - Einführung in die Pflanzensoziologie. 2. Die Pflanzengesellschaften Mitteleuropas. 94 p. Stuttgart.

— 1960 - Kennzeichnung der sozialen Beziehungen der gegenseitigen Beeinflussung und der Konkurrenzkraft der Pflanzen bei Vegetationsanalysen. Ber. Deuts. Bot. Ges. 73: 418-428.

— 1961 - Vegetations-Einheiten der Wegränder und der Eisenbahn-Anlagen in Hessen und im Bereich des unteren Neckar. Ber. Oberhess. Ges. Natur-Heilk. Giessen, N.F. Naturw. Abt. 31: 122-154.

KNOERR, A. 1960 - Le milieu, la flore, la végétation, la biologie des halophytes dans l'archipel de Riou et sur la côte sud de Marseille. Bull. Mus. Hist. Nat. Marseille 20: 89-173.

KONINGSDAAL, C. J. VAN & J. REYNDERS, et. al. 1956 - De begroeiing van de Amsterdamse grachten. 36 p. Ned. Jeugdb. Natuurstudie. Amsterdam. Mimeographed.

KOTLABA, F. 1962 - Nálezy fosilního jelení̇ho jazyku - Phyllitis scolopendrium (L.) Newm. - v Ceskoslovensku a poznámky k jeho recentnímu rozšíření. Preslia 34: 255-267.

KRATZER, A. 1956 - Das Stadtklima. 2nd ed., 184 p. Braunschweig.

KRAUSCH, D. 1955 - Neufunde von Farnen der Felsspaltengesellschaften. Wiss. Z. Pädag. Hochsch. Potsdam, Math.-Naturw. R. 2: 228-229.

KRAUSP, C. 1962 - Zur malakologischen Faunistik SO-Thüringens. Mitt. Berliner Malak. 17: 14-20.

KRUSENSTJERNA, E. VON. 1945 - Bladmossvegetation och bladmossflora i Uppsalatrakten. Acta Phytogeogr. Suec. 19: 1-250.

KUHN, K. 1937 - Die Pflanzengesellschaften in Neckargebiet der Schwäbischen Alb. 340 p. Oehringen.

KUIPER, J. G. J. 1953 - Een verzamelreis in de Pyreneeën. II. III. Correspondentiebl. Ned. Molac. Ver. 49: 454-459; 51: 475-480.
KUŠAN, F. 1933 - Die Flechtenflora und die Flechtenvegetation des nord-westlichen Gebirgzuges von Crna Gora (Montenegro). Bull. Int. Acad. Yougosl., Cl. Sc. Math. Nat. 27: 142-172.

LAMBINON, J., J. DUVIGNEAUD (& A. LAWALRÉE). 1965 - Un événément marquant dans l'histoire de la floristique européenne: La publication du Volume 1 de "Flora Europaea"; Ses incidences sur la floristique Belge et Luxembourgeoise. Natura Mosana 18: 1-17.
LANDWEHR, J. (& J. J. BARKMAN). 1966 - Atlas van de Nederlandse bladmossen. 504 p. Hoogwoud.
LAWALRÉE, A. 1950 - Flore générale de Belgique: Pteridophytes. 195 p. Bruxelles.
LEA, F. M. & C. H. DESCH. 1935 - The chemistry of cement and concrete. 429 p. London.
LEBRUN, J., A. NOIRFALISE, P. HEINEMANN & C. VANDEN BERGHEN. 1949 - Les associations végétales de Belgique. Bull. Soc. Roy. Bot. Belg. 82: 105-207.
LECOMTE DU NOÜY, P. 1936 - Le temps et la vie. 268 p. Paris.
LEEUWEN, C. G. VAN & E. VAN DER MAAREL. 1966 - Plantenoecologie en natuurbehoud in Groot-Brittannië; verslag van een studiereis naar Engeland en Wales, september 1966. 26 p. Rapp. RIVON, Zeist. Mimeographed.
LEFORT, F. L. 1950 - Les Corydalis de la flore Luxembourgeoise. Bull. Soc. Roy. Bot. Belg. 83: 257-258.
* LEPAGE. 1861 - Des plantes du vieux châteaux et des environs de Gisors. Mémoire Acad. Médec.
LEWIN, R. A. 1948 - Biological flora of the British isles: Sonchus L. (S. oleraceus L. and S. asper (L.) Hill). Journ. Ecol. 36: 203-223.
LIETH, H. 1954 - Die Porenvolumina der Grünlandböden und ihre Beziehungen zur Bewirtschaftung und zum Pflanzenbestand. Z. Acker-Pflanzenbau 98: 453-460.
LINDMAN, A. M. 1912 - Wie ist die Kollektivart Polygonum aviculare zu spalten? Svensk Bot. Tidskr. 6: 672-694.
LIPPERT, W. 1966 - Die Pflanzengesellschaften des Naturschutzgebietes Berchtesgaden. Ber. Bayer. Bot. Ges. 39: 67-122.
LIPPMAA, T. 1933 - Grundzüge der Pflanzensoziologischen Methodik nebst einer Klassifikation der Pflanzenassoziationen Estlands. Acta Inst. Horti Bot. Univ. Tartuensis 3, no. 3: 1-169 (In Esthonian with German summary).
— 1934. - La méthode des association unistrates et le système écologique des associations. Acta Inst. Horti Bot. Univ. Tartuensis 4: 1-7.
LITARDIÈRE, R. DE. 1928 - Contributions à l'étude phytosociologique de la Corse. Archives Bot. 2, Mém. 4: 1-184.
— 1933 - Un nouvel exemple de station d'Eupteris aquilina en terrain alcalin. Bull. Soc. Bot. France 80: 230-233.
LITZELMANN, E. & M. 1959 - Physiologie und Oekologie der Felsflurvegetation auf dem Isteiner-Klotz. Beitr. Naturk. Forsch. Südwestdeutschl. 18: 144-174.
LÖVE, A. 1965-1967 - IOPB chromosome number reports. V. Taxon 14: 191-196. VI. Taxon 15: 117-128. VII. Taxon 15: 155-163. IX. Taxon 16: 62-66. XI. Taxon 16: 215-222.
— & D. 1949 - The geobotanical significance of polyploidy. I. Portugal. Acta Biol., Ser. A.: 273-352.
— & D. 1961 - Chromosome numbers of central and Northwest European plant species. Opera Bot. 5: 1-581.
— & O. T. SOLBRIG. 1965-1965 - IOPB chromosome number reports. I. Taxon 13: 99-110. III. Taxon 14: 50-57. IV. Taxon 14: 86-92.

LORENZONI, G. G. 1961 - Ricerche sulle stazione a Ceterach officinarum Lam. et DC. delle Valli del Natisone. Accad. Scienze Lettere Arti Udine, Ser. 8, no. 1: 1-29.
LOUSLEY, J. E. 1949 - Botanical records for 1948. London Naturalist 28: 26-30.
LOVIS, J. D. & T. REICHSTEIN. 1964 - A diploid form of Asplenium rutamuraria. Brit. Fern Gaz. 9: 141-146.

MAAREL, E. VAN DER. 1966 - Over vegetatiestructuren, -relaties en -systemen, in het bijzonder in de duingraslanden van Voorne. 170 p. Zeist. Mimeographed.
MAGROU, J. 1950 - Transformation du Poa annua L. en plante vivace à rhizomes. Bull. Soc. Bot. France 97: 9-11.
MALCUIT, G. 1928 - Les associations végétales de la vallée de la Lanterne. Arch. Bot. 2, Mém. 6: 1-211.
MANTON, I. 1951 - Problems of cytology and evolution in the Pteridophyta. 316 p. Cambridge.
MARGADANT, W. D. 1959 - Mossentabel. 3rd ed., 155 p. Amsterdam.
MARTÍNEZ, S. R. 1960 - Roca, clima y comunidades rupicolas; sinopsis de las allianzas hispanas de "Asplenietea rupestris". Annales Real Acad. Farm. 1960: 153-168.
MATTICK, F. 1951 - Wuchs- und Lebensformen, Bestand- und Gesellschaftsbildung der Flechten. Engl. Bot. Jahrb. 75: 378-423.
MAYER, H. 1966 - Zur Waldbaulichen Beurteilung anthropogen beeinflusster Fichten-Tannen-Buchenwälder (Abieti-Fagetum) in den Chiemgauer Alpen. In: R. Tüxen (ed.), Anthropogene Vegetation: 347 (discussion S. Segal). Den Haag.
MAZZANTI, E. F. 1875-1877 - Florula del Colosseo. Comun. Atti Acad. Pontif. Nuovi Lincei 28: 8-13; 127-133; 254-261; 305-307; 397-400; 29: 8-15; 54-58; 122-125; 236-239; 457-462; 30: 97-105; 31: 155-167.
McVEAN, D. N. & D. A. RATCLIFFE. 1962 - Plant communities of the Scottish Highlands. 445 p. London.
MEDWECKA-KORNAŚ, A. 1950 - Biologie de la dissémination des associations végétales des rochers du Jura Cracovien. Bull. Acad. Polon. Sci. Lettre, Sér. B. Sc. Nat. 1949: 151-173.
MEIER, H. & J. BRAUN-BLANQUET. 1934 - Prodrome des groupements végétaux. 2. Classe des Asplenietales rupestres - Groupements rupicoles. 47 p. Montpellier.
MEINDERS-GROENEVELD, J. & S. SEGAL. 1967 - Pteridologische aantekeningen. 3. De eikvaren (Polypodium vulgare L. sensu lato) in Nederland. Gorteria 3: 183-199.
MELZER, H. 1963 - Neues zur Flora von Steiermark (VI). Mitt. Naturwiss. Ver. Steiermark 93: 274-290.
MENNEMA, J. 1968 - Floristische en vegetatiekundige aspecten van Noord-Wales. Natura 65: 73-79.
— & S. SEGAL. 1967. Het geslacht Parietaria L. in Nederland. 1. Gorteria 3: 96-102. 2. Gorteria 3: 109-118.
MEUSEL, H. 1935 - Wuchsformen, uns Wuchstypen der europäischen Laubmoose. Nova Acta Leopoldina N.F. 3: 123-277.
— 1939 - Die Vegetationsverhältnisse der Gipsberge im Kyffhäuser und im südlichen Harzvorland. Hercynia 2: 1-372.
— , E. JÄGER & E. WEINERT. 1965 - Vergleichende Chorologie der zentraleuropäischen Flora. Textband. 583 p. Kartenband. 258 p. Jena.
MEIJER, W. 1943 - Langs de Amsterdamse grachten. De Lev. Natuur 47: 156-157.

MEYER, D. E. 1957 - Zur Zytologie der Asplenien Mitteleuropas (I-XV). Ber. Deuts. Bot. Ges. 70: 57-66.

— 1958 - Idem (XVI-XX). Ber. Deuts. Bot. Ges. 71: 11-20.

MEIJER DREES, E. 1954 - The minimum area in tropical rain forest with special reference to some types in Bangka (Indonesia). Vegetatio 5-6: 517-523.

MIYAWAKI, A. 1964 - Trittgesellschaften auf den Japanischen Inseln. Bot. Mag. Tokyo 77: 365-374.

MÖNKEMEYER, W. 1927 - Die Laubmoose Europas; Andreaeales-Bryales. L. Rabenhorst Kryptogamenflora IV, Ergänzungsband. 960 p. Leipzig.

MOLINIER, R. 1937 - Les Iles d'Hyères; Etude phytosociologique. Ann. Soc. Hist. Nat. Toulon 21.

— 1959 - Etude des groupements végétaux terrestre du Cap Corse. Bull. Mus. Hist. Nat. Marseille 19: 5-75.

MOORE, J. J. 1962 - The Braun-Blanquet system. A reassessment. Journ. Ecol. 50: 761-769.

MORTON, F. & H. GAMS. 1925 - Höhlenpflanzen. Speläologische Monographien, Vol. 5. 227 p. Wien.

MORTON, C. V. 1965 - Report on botanical excursion to the Boreal forest region in northern Quebec and Ontario. Am. Fern Journ. 55: 85-89.

— 1967 - New data on North American oak ferns, Gymnocarpium. Am. Fern Journ. 57: 141-142.

MORTON, F. 1955 - Absolute Lichtmessungen im Dachsteinhöhlenparke und in der Koppenbrüllerhöhle. Mitt. Höhlenkommission Wien 1955: 41-53.

MROSE, H. 1941 - Die Verbreitung baumbewohnender Flechten in Abhängigkeit vom Sulfatgehalt der Niederschlagswässer. Biokl. Beibl. Meteor. Z. 8: 38-60.

MÜLLER, Th. 1966 - Die Wald-, Gebüsch-, Saum- und Trocken- und Halbtrockenrasengesellschaften des Spitzbergs. In: Der Spitzberg bei Tübingen; Die Natur- und Landschaftsschutzgebiete Baden-Württembergs 3: 278-475.

MUSSACK, A. 1933 - Untersuchungen über Cystopteris fragilis. Beih. Bot. Centralbl. 51, Abt. 1: 204-254.

NÈGRE, R. 1966 - Les thérophytes. Bull. Soc. Bot. France, Mém. 1966: 92-108.

NICKL-NAVRÁTIL, H. 1960 - Mooskleingesellschaften der Städte. Nova Hedw. 2: 425-462.

NICOL, A. 1950 - Aperçu bibliographique sur les agents et les causes de la corrosion du béton. Centre Et. Rech. Ind. Liants Hydraul. Note Inf. 10: 1-92.

NICOT, J. 1951 - Dégradation des murs de plâtre par les moisissures. Rev. Mycol. 16: 168-172.

NICKFELD, H. & H. MEIER. 1962 - Ueber die Pflanzengesellschaften der Fels- und Mauerspalten Südfrankreichs. S.B. Oest. Akad. Wiss., Math.-Nat. Kl., Abt. I, 171: 389-411

NORDHAGEN, R. 1936 - Versuch einer neuen Einteilung der subalpinen-alpinen Vegetation Norwegens. Bergens Mus. Aarb. 1936, Naturvid. Rekke 7, no. 1: 1-88.

— 1954 - Vegetation units in the mountain areas of Scandinavia. Veröff. Geobot. Inst. Rübel 29: 81-95.

OBERDORFER, E. 1938 - Ein Beitrag zur Vegetationskunde des Nordschwarzwaldes. Beitr. Naturk. Forsch. Südwestdeutschl. 3: 150-270.

— 1949 - Die Pflanzengesellschaften der Wutachschlucht. Beitr. Naturk. Forsch. Südwestdeutschl. 8: 22-60.

— 1954 - Ueber Unkrautgesellschaften der Balkanhalbinsel. Vegetatio 4: 379-411.

— 1956 - Uebersicht der Süddeutschen Pflanzengesellschaften. Beitr. Naturk. Forsch. Südwestdeutschl. 15: 11-29.
— 1957 - Süddeutsche Pflanzengesellschaften. Pflanzensoziologie 10: 1-564.
— 1962 - Pflanzensoziologische Exkursionsflora für Süddeutschland und die angrenzenden Gebiete. 2nd ed., 987 p. Stuttgart.
— et al. 1967 - Systematische Uebersicht der westdeutschen Phanerorgamen- und Gefässkryptogamen-Gesellschaften; Ein Diskussionsentwurf. Schriftenreihe Veg.kunde 2: 7-62.
Ochsner, F. 1928 - Studien über die Epiphytenvegetation der Schweiz. Jahrb. St. Gall. Naturwiss. Ges. 63: 1-106.
Odum, E. P. 1959 - Fundamentals of ecology. 2nd ed., 546 p. Philadelphia, London.
Omura, M. 1950 - Life-forms of epiphitic lichens. Bot. Mag. Tokyo 63: 155-160. (Japanese, with an English summary).
Oppenheimer, H. R. & J. Halevy. 1962 - Anabiosis of Ceterach officinarum Lam. et DC. Bull. Res. Counc. Israel, Sect. D 11: 127-147.
Ordnuff, R. 1965 - Index to plant chromosome numbers for 1965. Regnum Veg. 50: 1-128.
Orloci, L. 1966 - Geometric models in ecology. I. The theory and application of some ordination methods. Journ. Ecol. 54: 193-215.
Ouden, P. L. Den. 1958 - Het meten van luchtsnelheden. VI. Rapp. Inst. Gezondheidstechniek TNO 27: 232-240.

Paclt, J. 1959 - Ueber die Variabilität von Parietaria officinalis. Phyton 8: 171-174.
Paine, S. G. 1936 - The occurrence of autotrophic bacteria in the stones of buildings and their relationship to the decay of stone. Rep. Proc. 2nd Intern. Congress Microbiol. London 1936: 277-278; discussion 279 (H. G. Thornton).
Pankow, H. & P. Fischer. 1965 - Beiträge zur Moosflora Mecklenburgs. V. Die Lewitz. Wiss. Z. Univ. Rostock, Math.-Naturw. R. 14: 511-532.
Parker, C. D. 1947 - Species of Sulphur bacteria associated with the corrosion of concrete. Nature 159: 439-440.
Passarge, H. 1963 - Wege zur planmässigen Vegetationstypenforschung, dargestellt an Hand des Beispiels von Trittpflanzengesellschaften. Feddes Rep. Beih. 140: 7-18.
— 1964 - Pflanzengesellschaften des nordostdeutschen Flachlandes. I. Pflanzensoziologie 13: 1-324.
— 1965 - Ueber einige interessante Stromtalgesellschaften der Elbe unterhalb von Magdeburg. Abh. Ber. Naturk. Vorgesch. Magdeburg 11: 83-93.
Paton, J. A. 1956 - Bryophyte succession on the Wealden sandstone rocks. Trans. Brit. Bryol. Soc. 3: 103-114.
Petterson, B. 1962 - Om Ceterach officinarum och dess ekologi. Botaniska Not. 115: 237-240.
Pfeiffer, H. 1937 - Eine die Bürgersteige grosstädtischer Vororte begleitende Pflanzengesellschaft. Beih. Bot. Centralbl. 57B: 599-606.
— 1943 - Ueber örtliche Feinheiten der Assoziationsverteilung. Biologia Gen. 17: 147-163.
Pichi-Sermolli, R. E. G. 1958 - The higher taxa of the Pteridophyta and their clasification. Uppsala Univ. Aarsskr. 6: 70-90.
Pignatti, S. 1953 - Introduzione allo studio fitosociologico della pianura Veneta orientale con particolare riguardo alla vegetazione litoranea. Atti Ser. 5, no. 11: 92-258.

— 1960 - Il significato delle specie poliploidi nelle associazioni vegetali. Atti Ist. Ven. Sci. Lett. Acad. 118: 75-98.

PLATT, R. B. & J. F. GRIFFITHS. 1964 - Environmental measurement and interpretation. 235 p. New York, London.

PLITT, C. 1927 - Succession in lichens. Bryologist 30: 1-4.

POELT, J. 1954 - Moosgesellschaften im Alpenvorland. I. II. Sitzungsber. Oest. Acad., Math.-Nat. Kl. 163, Abt. 1: 141-174; 495-539.

— 1961 - Systematik der Farnpflanzen. Fortschr. Bot. 23: 57-64.

POORE, M. E. D. 1955 - The use of phytosociological methods in ecological investigations. I. The Braun-Blanquet-system. Journ. Ecol. 43: 226-244. II. Practical issues involved in an attempt to apply the Braun-Blanquet-system. Journ. Ecol. 43: 245-269. III. Practical application. Journ. Ecol. 43: 606-651.

— 1956 - Idem. IV. General discussion of phytosociological problems. Journ. Ecol. 44: 28-50.

— 1962 - The method of successive approximation in descriptive ecology. Adv. Ecol. Res. 1: 35-68.

POTTS, R. & W. T. PENFOUND. 1948 - Water relations of the polypody fern Polypodium polypodioides. Ecology 29: 43-53.

PRAEGER, R. L. 1901 - Irish topographical botany; compiled largely from original material. Proc. Roy. Irish Acad., 3rd Ser. 7: CLXXXVIII + 410 p.

PREIS, K. 1937 - Die Besiedlung der Blockhalden in der Biberklamm. Vegetationsstudien im böhmischen Mittelgebirge. Beih. Bot. Centralbl. 57B: 521-576.

PRESTON, F. W. 1962 - The canonical distribution of commonness and rarity. Ecology 43: 185-215; 410-432.

PROCTOR, M. C. F. 1960 - Mosses and liverworts of the Malham district. Field Studies 1, no. 2: 1-25.

PULLÈ, A. A. 1952 - Compendium van de terminologie, nomenclatuur en systematiek der zaadplanten. 3rd ed., 376 p. Utrecht.

RAMENSKY, L. G. 1925 - Die Grundgesetzmässigkeiten im Aufbau der Vegetationsdecke. 37 p. Woronesh. (In Russian; abstract in Bot. Centralbl. N.F. 49: 453-455 (1926).

— 1930 - Zur Methodik der vergleichenden Bearbeitung und Ordnung von Pflanzenlisten und anderen Objekten, die durch mehrere, verschiedenartig wirkende Faktoren bestimmt werden. Beitr. Biol. Pfl. 18: 269-304.

RAUNKIAER, C. 1918 - On the biological normal spectrum. Kungl. Danske Vidensk. Selsk., Biol. Medd. 1, no. 4: 1-80 and in C. Raunkiaer - 1934: The life forms of plants and statistical plant geography. 632 p. Oxford.

REICHLING, L. 1951 - Nouvelles observations du Gastéropode, Helix aspera Müller au Grand-Duché de Luxembourg. Bull. Soc. Natur. Luxembourg. N.S. 45: 24-25.

REY, P. 1956 - Aires ombrothermiques et milieux naturels. Bull. Serv. Carte Phytogéogr., Sér. A 1: 33-40.

REYNDERS, J. & S. SEGAL. 1963 - De flora van grachtmuren. De Lev. Natuur 66: 49-53.

REYNDERS, W. J. 1956 - De paddestoelen. In: C. van Koningsdaal & J. Reynders (ed.), De begroeiing van de Amsterdamse grachten: 17-21. Amsterdam. Mimeographed.

RICHARD, O. J. 1888 - Florule des clochers et des toitures des églises de Poitiers (Vienne). Paris.

RICHARDS, P. W. 1888 - Bryophyta. In: Vistas in Botany: 387-420. London.

RIKLI, M. 1903 - Die Anthropochoren und der Formenkreis des Nasturtium palustre DC. Ber. Zürich. Bot. Ges. 8: 71-82.
RIOUX, J. A. & P. QUÉZEL. 1951 - Note floristique sur la "Région" de Montpellier. Le Monde des Plantes 280-281: 41-43.
RISHBETH, J. 1948 - The flora of Cambridge walls. Journ. Ecol. 36: 136-148.
RODE, H. 1962 - Solitärbienen als Schädlinge an Mauerwerk. Anzeiger Schädlingsk. 35: 72-73.
ROTHMALER, W. 1956 - Taxonomische Monographie der Gattung Antirrhinum. Feddes Rep. Beih. 136: 1-124.
— 1962 - Exkursionsflora von Deutschland. II. Gefässpflanzen. 503 p. Berlin.
— et. al. 1963 - Exkursionsflora von Deutschland. IV. Kritischer Ergänzungsband Gefässpflanzen. 622 p. Berlin.
ROUY, G. & J. FOUCAULD. 1893 - Flore de France. I. 264 p. Asnières, Rochefort.
ROUX, J. & C. LAHONDERE. 1961 - A propos de la végétation chasmophytique des falaises maritimes en Bretagne septentrionale et occidentale. Nat. Monspeliensia Sér. Bot. 12: 53-80.
RÜBEL, E. 1933 - Versuch einer Uebersicht über die Pflanzengesellschaften der Schweiz. Ber. Geobot. Forschungsinst. Rübel Zürich 1932: 19-30.
RYDZAK, J. 1958 - Influence of small towns on the lichen vegetation. VII. Discussion and general conclusions. Ann. Univ. Mariae Curie-Sklodowska, Sect. C 13: 275-323.

SAUVAGE, C. 1966 - Remarques sur la classification des types biologiques. Bull. Soc. Bot. France, Mém. 1966: 5-13.
SCAMONI, A., H. PASSARGE & G. HOFMANN. 1965 - Grundlagen zu einer objektiven Systematik der Pflanzengesellschaften. Feddes Rep. Beih. 142: 117-132.
SCHEUERMANN, R. H. 1948 - Zur Einteilung der Adventiv- und Ruderalflora. Ber. Schweiz. Bot. Ges. 58: 268-276.
SCHMITT, C. 1950 - Die alte Mauer und ihr Leben. 2nd ed., 32 p. Berlin-Kleinmachnow.
SCHNELL, F. 1939 - Die Pflanzenwelt der Umgebung von Lauterbach (Hessen). Feddes Rep. Beih. 112: 1-106.
SCHOLZ, H. 1960 - Bestimmungsschlüssel für die Sammelart Polygonum aviculare L. Verh. Bot. Ver. Prov. Brandenburg 98/100: 180-182.
— & H. SUKOPP. 1965 - Drittes Verzeichnis von Neufunden höherer Pflanzen aus der Mark Brandenburg und angrenzenden Gebieten. Verh. Bot. Ver. Prov. Brandenburg 102: 3-40.
SCHORLER, B. 1914 - Die Algenvegetation an den Felswänden des Elbsandstein Gebirges. Abh. Naturw. Ges. Isis Dresden 1914: 3-27.
SCHRÖTER, C. 1932 - Kleiner Führer durch die Pflanzenwelt der Alpen. 80 p. Zürich.
SCHWICKERATH, M. 1931 - Die Gruppenabundanz, ein Beitrag zur Begriffsbildung der Pflanzensoziologie. Engl. Bot. Jahrb. 64: 1-16.
SEBASTIANI, A. 1815 - Romanarum plantarum fasciculus alter, accedit enumeratio plantarum sponte nascentium in ruderibus in Amphitheatri Flavii. 81 p. Roma.
SEGAL, S. 1954 - Onderzoek naar de muurbegroeiingen in Nederland. Kruipnieuws 17, no. 1: 1-9.
— 1960 - Een vooronderzoek naar de systematische positie van Juncus ambiguus Guss. Correpondentiebl. Flor. 16: 169-172.
— 1961 - Vegetaties op oude muren. Jaarb. Kon. Ned. Bot. Ver. 1961: 59-61.
— 1962a - Pteridologische aantekeningen. 1. Asplenium adiantum-nigrum × A. ruta-muraria in Nederland. Gorteria 1: 56-59; 128.

— 1962b - De floristiek van oude muren. Gorteria 1: 71-74 and appendix 20 p.

— 1963a - Over het Sagino- en het Filici-Bryetum argentei. Jaarb. Kon. Ned. Bot. Ver. 1963: 41-42.

— 1963b - De biologie van Parietaria officinalis en P. judaica. Summary lecture Comm. Florist. Onderz. Kon. Ned. Bot. Ver. 2 p. Mimeographed.

— 1963c - Floristische aantekeningen. In: Verslag botanische excursie Lautaret (Hautes Alpes) van 16 tot 31 juli 1963: 10-14. Hugo de Vries-laboratorium, Amsterdam. Mimeographed.

— 1963d - Vegetatiekunde van de Lautaret. In: verslag botanische excursie Lautaret (Hautes Alpes) van 16 tot 31 juli 1963: 15-23. Hugo de Vries-laboratorium, Amsterdam. Mimeographed.

— 1964 - On some basic terms in ecology. Abstr. 10th Intern. Bot. Congr. Edinburgh 1964: 465.

— 1965 - Een vegetatieonderzoek van de hogere waterplanten in Nederland. Wet. Meded. Kon. Ned. Natuurhist. Ver. 57: 1-80.

— 1967 - Some notes on the ecology of Ranunculus hederaceus L. Vegetatio 15: 1-26.

— 1968a - Over Rhynchostegiella compacta in Nederland. Buxbaumia 22: 26-31.

— 1968b - Ramalina intermedia in Nederland. Gorteria 4: 118-119.

— 1969 - Some general trends in the succession of biotic communities, with special reference to plant ecology. In preparation.

— & J. J. Barkman. 1960 - Enige opmerkingen over abundantie en dominantie bij het opnemen van kwadraten. Jaarb. Kon. Ned. Bot. Ver. 1960: 39-40.

— & V. Westhoff. 1959 - Die vegetationskundliche Stellung von Carex buxbaumii Wahlenb. in Europa, besonders in den Niederlanden. Acta Bot. Neerl. 8: 304-329.

Senay, P. 1952 - Découverte de l'Asplenium viride Huds. dans la Seine-Inférieure. Bull. Soc. Bot. France 99: 306-308.

Sernander, R. 1906 - Entwurf einer Monographie der europäischen Myrmekochoren. 410 p. Uppsala, Stockholm.

— 1927 - Zur Morphologie und Biologie der Diasporen. 104 p. Uppsala.

Shaw, H. K. A. 1960 - On the identity and nomenclature of Galium erectum Huds. Kew Bull. 14: 63-65.

Sissingh, G. 1950 - Onkruid-associaties in Nederland. 224 p. 's-Gravenhage.

Sjögren, E. 1964 - Epilitische und epigäische Moosvegetation in Laubwäldern der Insel Oeland. Acta Phytogeogr. Suec. 48: 1-184.

Sjörs, H. 1955 - Remarks on ecosystems. Svensk Bot. Tidskr. 49: 155-169.

Skottsberg, C. 1954 - Antarctic flowering plants. Bot. Tidskr. 51: 330-338.

Smarda, J. 1947 - Mechová a lišejnihova společenstva CSR (The moss and lichen communities in Czechoslovakia). Cas. Zemsh. Mus. Brne 31: 39-88.

Snel, M. 1964 - Chemisch onderzoek van gewas en grond ten behoeve van oecologische probleemstellingen. 56 p. Hugo de Vries-laboratorium, Amsterdam. Mimeographed.

Sørensen, T. 1948 - A method of establishing groups of equal amplitude in plant sociology based on similarity of species content and its application to analyses of the vegetation on Danish commons. Kongl. Danske Vid. Selsk., Biol. Skr. 5, no. 4: 1-34.

Soest, J. L. van. 1934 - Aantekeningen over Hieracium. Ned. Kruidk. Arch. 44: 296-303.

Soper, J. H. 1963 - Ferns of Manitoulin Island, Ontario. Am. Fern Journ. 53: 71-81.

Stebbins, G. L. 1942 - Polyploid complexes in relation to ecology and the history of floras. Am. Nat. 76: 36-45.

STERNER, G. 1922 - Några floristika nyheter från Oeland. Svensk Bot. Tidskr. 16: 117-123.

STODIEK, E. 1937 - Soziologische und ökologische Untersuchungen an den xerotopen Moosen und Flechten des Muschelkalkes in der Umgebung Jenas. Feddes Rep. Beih. 99: 1-46.

STOUTJESDIJK, P. 1961 - Micrometeorological measurements in vegetations of various structure. Proc. Kon. Ned. Acad. Wet., Ser. C. 64: 171-207.

STYLES, B. T. 1962 - The taxonomy of Polygonum aviculare and its allies in Britain. Watsonia 5: 177-214.

SUKATSCHEW, W. 1929 - Ueber einige Grundbegriffe in der Phytosoziologie. Ber. Deuts. Bot. Ges. 47: 269-312.

SUKOPP, H. 1962 - Neophyten in natürlichen Pflanzengesellschaften Mitteleuropas. Ber. Deuts. Bot. Ges. 75: 193-205.

— 1966 - Neophyten in natürlichen Pflanzengesellschaften Mitteleuropas. In: R. Tüxen (ed.), Anthropogene Vegetation: 275-291. Den Haag.

SUTTER, R. 1962 - Beiträge zur Flora des Grignamassives (Comersee). Bauhinia 2: 50-54.

SZAFER, W. & B. PAWLOWSKI. 1927 - Die Pflanzenassoziationen des Tatra-Gebirges. A. Bemerkungen über die angewandte Arbeitsmethodik (zu den Teilen III, IV und V). Bull. Int. Acad. Polon. Sci. Lettr., Cl. Sci. Math. Natur., B. Sci. Natur. 1926 Suppl. 2: 1-12.

TANSLEY, A. G. 1939 - The British Islands and their vegetation. 930 p. Cambridge.

TERRETAZ, J.-L. 1964 - Anogramma leptophylla en Valais et en Ossola. Trav. Soc. Bot. Genève 7: 20-30.

THELLUNG, A. 1911-1912 - La flore adventice de Montpellier. Mém. Soc. Nat. Sci. Nat. Math. Cherb. 38: 57-728.

THOMAS, M. D. 1964 - The effect of air pollution on plants and animals. In: G. T. Goodman, R. W. Edwards & J. M. Lambert (ed.), Ecology and the industrial society: 11-33. Oxford.

TÜXEN, R. 1937 - Die Pflanzengesellschaften Nordwestdeutschlands. Mitt. Flor.-soz. Arbeitsgem. Niedersachsen 3: 1-170.

— 1947 - Der Pflanzensoziologische Garten in Hannover und seine bisherige Entwicklung. Jber. Naturhist. Ges. Hannover 94/98: 113-288.

— 1950 - Grundriss einer Systematik der nitrophilen Unkrautgesellschaften in der Eurosibirischen Region Europas. Mitt. Flor.-soz. Arbeitsgem. N.F. 2: 94-177.

— 1955 - Das System der nordwestdeutschen Pflanzengesellschaften. Mitt. Flor.-soz. Arbeitsgem. N.F. 5: 155-176.

— 1957 - Zur systematischen Stellung des Sagino-Bryetum argentei. Mitt. Flor.-soz. Arbeitsgem. N.F. 6/7: 170-171.

— & H. ELLENBERG. 1937 - Der systematische und der ökologische Gruppenwert. Mitt. Flor.-soz. Arbeitsgem. Niedersachsen 3: 171-184.

— & E. OBERDORFER. 1958 - Die Pflanzenwelt Spaniens. II. Teil. Eurosibirische Phanerogamen-gesellschaften Spaniens. Veröff. Geobot. Inst. Rübel Zürich 32: 1-328.

TUOMIKOSKI, R. 1942 - Untersuchungen über die Untervegetation der Bruchmoore in Ost-Finland. I. Zur Methodik der pflanzensoziologischen Systematik. Ann. Bot. Soc. Zool.-bot. Fenn. Vanamo 17: 1-203.

TWENHOFEL, W. H. 1950 - Principles of sedimentation. 2nd ed., 673 p. New York, Toronto, London.

VAARAMA, A. 1956 - A contribution to the cytology of some mosses of the British Isles. Irish Nat. Journ. 12: 30-40.

* VALLOT, M. J. 1884 - Essai sur la flore du pavé de Paris suivi d'une florule des ruines du Conseil de l'Etat. 122 p. Paris.
— 1887 - Florule de Panthéon. Journal Bot. 1: 52-55.
VARESCHI, V. 1953 - La influencia de los bosques y parques sobre el aire de la ciudad de Caracas. Acta Cient. Venezolana 4: 89-95.
VASILEVICH, V. I. 1962 - Association between species and the structure of a phytocoenosis. Kokl. Bot. Sci. Sect. Translation 139: 133-135.
VELDHUIZEN, A. VAN. 1912 - Varenrijkdom. De Lev. Natuur 17: 143-144.
VELTMAN, A. L. M. 1962 - Bryophytenvegetaties op muren in de omgeving van Amsterdam. 29 p. Hugo de Vries-laboratorium, Amsterdam. Manuscript.
VOGLER, P. 1904 - Die Eibe (Taxus baccata L.) in der Schweiz. Jahrb. St. Gallischen Naturw. Ges. 1903: 436-491.

WALDHEIM, S. 1944 - Moosvegetation i Dalby-Söderskogs Nationalpark. Kungl. Sv. Vet. Akad. Avh. Natursk. 4: 1-142.
— 1947 - Kleinmoosgesellschaften und Bodenverhältnisse in Schonen. Bot. Not. Suppl. 1: 1-203.
WALLACE, E. C. 1949 - Plant records. Watsonia 1: 37-61.
— 1968 - Plant records. Proc. Bot. Soc. Brit. Isles 7: 195-199.
WALTER, H. & H. LIETH. 1960 - Klimadiagramm-Weltatlas. 1. Lieferung. Jena.
WEBB, D. A. 1947 - The vegetation of Carrowkeel, a limestone hill in North-West Ireland. Journ. Ecol. 35: 105-129.
— & E. V. GLANVILLE. 1962 - The vegetation and flora of some islands in the Connemara Lakes. Proc. Roy. Irish Acad. 62, Sect. B: 31-54.
WEERS, D. J. VAN. 1965 - Microklimatologisch onderzoek van muurvegetaties. 39 p. Hugo de Vries-laboratorium, Amsterdam. Manuscript.
WESTHOFF, V. 1947 - The vegetation of dunes and salt-marshes on the Dutch islands of Terschelling, Vlieland and Texel. 131 p. Utrecht.
— 1951 - An analysis of some concepts and terms in vegetation study or phytocenology. Synthese 8: 194-206.
— 1965 - Plantengemeenschappen. In: Uit de plantenwereld: 288-349. Zeist, Arnhem.
— 1966 - Systeem der in Nederland voorkomende plantenassociaties. 13 p. RIVON, Zeist. Mimeographed.
— 1967 - Problems and use of structure in the classification of vegetation. The diagnostic evaluation of structure in the Braun-Blanquet system. Acta Bot. Neerl. 15: 495-511.
— , J. W. DIJK, H. PASSCHIER & G. SISSINGH. 1946 - Overzicht der plantengemeenschappen in Nederland. 2nd ed., 118 p. Amsterdam.
WEVER, A. DE. 1942 - Wat groeit er op rotsen, muren en daken? Maandbl. Natuurhist. Gen. Limb. 31: 117-122.
WHITTAKER, R. H. 1957 - Recent evolution of ecological concepts in relation to the eastern forests of North America. Am. Journ. Bot. 44: 197-206.
— 1967 - Gradient analysis of vegetation. Biol. Rev. 42: 207-264.
WILLIAMS, W. T. & J. M. LAMBERT. 1959 - Multivariate methods in plant ecology. I. Association-analysis in plant communities. Journ. Ecol. 47: 83-101.
— & J. M. LAMBERT. 1960 Idem. II. The use of an electronic digital computer for association-analysis. Journ. Ecol. 48: 689-710.
— & J. M. LAMBERT. 1961 - Idem. III. Inverse association-analysis. Journ. Ecol. 49: 717-729.
— , J. M. LAMBERT & G. N. LANCE. 1966 - Idem. V. Similarity-analysis and information-analysis. Journ. Ecol. 54: 427-445.
WOODELL, S. R. J. & J. ROSSITER. 1959 - The flora of Durham walls. Proc. Bot. Soc. Brit. Isles 3: 257-273.

Wijk, R. van der. 1962 - Lijst van de in Nederland voorkomende Bryophyta. Buxbaumia 16: 50-67.
— , W. D. Margadant & P. A. Florschütz. 1959-1967 - Index Muscorum. I (A-C). 548 p. II (D-H). 535 p. III (H-O). 529 p. IV (P-S). 604 p. Utrecht, and in Regnum Veg. 17; 26; 33; 48.

Zuure, M. J. A. B. 1962 - Een vegetatiekundig onderzoek van de Bryophyten-vegetaties op muren in Nederland. 22 p. Hugo de Vries-laboratorium, Amsterdam. Manuscript.

reprinted by
PASMANS - The Hague
The Netherlands

Fig. 1. Map of geographical regions

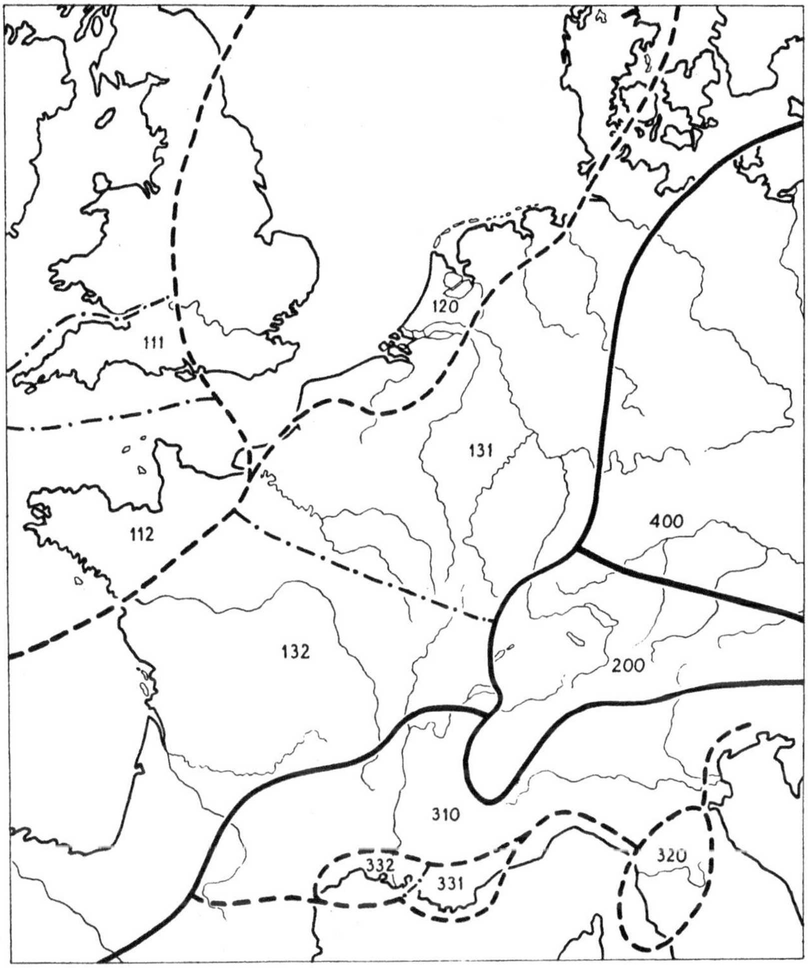

Explanation of abbreviations and symbols used in tables and text

CH	coverage of herb (shrub) layer (%)
CM	coverage of bryophyte layer (%)
DS	degree of sociability
Ex	exposure
GeoR	geographical region
Year	(after 1900)
MGP	mean group abundance in stands where species is present (%)

MGT	mean group abundance in all stands together (%)
MNS-MCH-MCM	mean number of species-mean coverage of herb (shrub) layer (%)-mean coverage of bryophyte layer (%) In brackets: MCM in stands where bryophytes are present
Nr C	number of column
Nr R	number of relevé (in year in which the analysis was made)
P	presence (%)
Pabs.	presence (absolute)
RH	relative humidity
U	(from) unweighted numbers
UW	ultimate weights
W	(from) weighted numbers
c	coastal region
s	scarce
st	stands
t	total
w	in immediate vicinity of water

Exposure

1	N	(NNW—ENE)
2	E	(ENE—SSE)
3	S	(SSE—WSW)
4	W	(WSW—NNW)
5	overshadowed	
6	unknown	

Weighted numbers (abundance and coverage percentage)

symbol	meaning		readjusted values (%) for group abundance
r	of sporadic occurrence in the stand (or ecosystem) as a whole		0
+r	sporadically (1 to 2 individual specimens)		0
+p	not frequent (about 3 to about 20 indiv. spec.), coverage	<1%	0.1
+a	,, ,, (,, 3 ,, ,, 20 ,, ,,), ,,	1-2%	1.5
+b	,, ,, (,, 3 ,, ,, 20 ,, ,,), ,,	2-5%	3.5
1p	frequent (,, 21 ,, ,, 100 ,, ,,), ,,	<1%	0.4
1a	,, (,, 21 ,, ,, 100 ,, ,,), ,,	1-2%	1.5
1b	,, (,, 21 ,, ,, 100 ,, ,,), ,,	2-5%	3.5
2p	very frequent (>100 indiv. spec.), ,,	<1%	0.4
2m	,, ,, (>100 ,, ,,), ,,	<5%	2.5
2a	number of individuals various, but coverage 5- 12%		8
2b	,, ,, ,, ,, , ,, ,, 12- 25%		18
3a	,, ,, ,, ,, , ,, ,, 25- 37%		32
3b	,, ,, ,, ,, , ,, ,, 37- 50%		44
4a	,, ,, ,, ,, , ,, ,, 50- 62%		56
4b	,, ,, ,, ,, , ,, ,, 62- 75%		69
5a	,, ,, ,, ,, , ,, ,, 75- 87%		81
5b	,, ,, ,, ,, , ,, ,, 87-100%		94
5c	,, ,, ,, ,, , ,, ,, >100%		105

Symbols

used in the species list and spectra

a. **Taxonomical spectrum**

00	Lichenes
01-22	Bryophyta
01-03	Hepaticae
01	Marchantiales
02	Metzgeriales
03	Jungermanniales
11-12	Musci
11	Polytrichales
12	Fissidentales
13	Dicranales
14	Grimmiales
15	Pottiales
16	Encalyptales
17	Funariales
18	Orthotrichales
19	Bryales
21	Leucodontales
22	Hypnales
31-39	Pteridophyta
31	Lycopsida
31	Selaginellales
33	Sphenopsida
33	Equisetales
36-39	Filicopsida
36	Aspidiales
37	Pteridales
38	Dicksoniales
39	Polypodiales
41-88	Spermatophyta
41-43	Coniferopsida
41	Taxales
42	Pinales
43	Cupressales
51-55	Monocotyledones
51	Alismatales
52	Arales
53	Poales
54	Liliales
55	Cyperales
61-88	Dicotyledones
61	Ranales
62	Clusiales
63	Cistales
64	Urticales
65	Fagales
66	Caryophyllales
67	Polygonales
68	Plumbaginales
69	Primulales
70	Geraniales
71	Malvales
72	Euphorbiales
73	Sapindales
74	Balsaminales
75	Apiales
76	Rubiales
77	Rosales
78	Santalales
79	Myrtales
80	Ericales
81	Brassicales
82	Oleales
83	Apocynales
84	Solanales
85	Plantaginales
86	Asterales
87	Salicales
88	Rhamnales

b. **Spectrum of degrees of ploidy**

2	diploid
3	di- and tetraploid
4	tetraploid
5	di-, (tetra-) and at least hexaploid
6	at least hexaploid
7	tetra- and at least hexaploid
8	hybrids
0	the remainder, including, e.g. unknown numbers, uncertain basic numbers (i.a. aneuploid series, apogamous or apomictic forms with 3n, 5n, etc.)

c. **Life form spectrum according to Raunkiaer**

1 therophytes
2 geophytes
3 hemicryptophytes
4 chamaephytes
5 nanophanerophytes
6 macrophanerophytes

d. **Life form spectrum according to Iversen**

Vascular plants

1-5 terriphytes
1 seasonal xerophytes
2 euxerophytes
3 hemixerophytes
4 mesophytes
5 hygrophytes
6 telmatophytes

Bryophytes

7 xerophytes
8 mesophytes
9 hygrophytes
0 poikilohygrophytes

e. **Growth form spectrum**

Vascular plants

11-14 trees and shrubs
11 pinids
12 fagids
13 ericids
14 vacciniids
21-29 herbs and suffrutices
21-22 Sociability 1
21 liliids
22 echiids
23-26 Sociability 2
23 poids
24 Asplenium-type
25 fumariids
26 fragariids
27-28 Sociability 3
27 corynephorids
28 androsacids
29 Sociability 4 or 5
29 soleiroliids

Bryophytes

32-35 orthotropes
32 bryids
33 polytrichids
34 grimmiids
35 leucobryids
41-46 plagiotropes
41 neckerids
42 leucodonids
43 isotheciids
44 hypnids
45 radulids
46 metzgeriids

Lichens

51 gelatinous types
51 collemids
52-53 crustaceous types
52 leprariids
53 caloplacids
54-55 foliaceous types
54 parmeliids
55 lobariids
57-59 fruticose types
57 cetrariids
58 usneids
59 cladoniids

f. **Sociability spectrum**

1 1 - 3 shoots
2 4 - 33 „
3 34 - 3×10^2 „
4 3×10^2 - 3×10^3 „
5 $> 3 \times 10^3$ „
or surface (almost) wholly covered

g. **Flower colour spectrum**

00 no flowers (Pteridophytes)
01 scarious
02 white
03 violet
04 blue
05 green
06 yellow
07 pink or lilac
08 red
09 others
26 white and yellow
36 pink and yellow
99 unknown

h. **Dissemination spectrum**

10 vegetative propagation
20 autochory
30 barochory
50 anemochory

51 dissemination potential high (diaspores as fine as dust)
52 dissemination potential medium (diaspores with floss)
53 dissemination potential low (diaspores - winged or relative large)
60 hydrochory
70 zoochory
71 endozoochory
72 epizoochory
73 myrmecochory
80 anthropochory
99 polychory
00 unknown
01 sterile

i. **Formation spectrum**

Vascular plants

10 rock surfaces, walls and gravelly or stony slopes
20 pioneer vegetation of dry sites including "open" dune country
30 shoals and mud flats (halophytes)
40 arable land including vineyards
50 ruderal places including fields lying fallow
60 banks, reed vegetation, broads and peat bogs
61 riparian vegetation
62 peat bogs and marshes
70 grassy sites including meadows, hay-fields, roadsides, dikes, etc., and Hochstauden vegetation
71 grassy sites
72 Hochstauden vegetation (tall herbs; generally on peaty soils)
80 dwarf shrub vegetation, scrubs and forests
81 heaths, moors, garigues and macchia
82 scrubs
83 forests
90 vegetation of contact zones and various other cases
91 contact zones of terrestrial formations and fresh water, or vegetation of habitats with a strongly fluctuating water tabel, which periodically run dry
92 contact zones of terrestrial formations and seawater (the foot of dunes along beaches, shoals and mud flats)
99 various, not restricted to one of the above-mentioned categories
00 unknown or doubtful
01 cultivated plants
02 adventitious forms

Bryophytes and fruticose Lichens

terrestrial forms
06 general and of constant occurrence, not only in pioneer vegetation
26 in open, and generally dry, habitats
96 in moist sites, more particularly along the sides of ditches and streams and in marshy habitats

epilithic forms
16 rocks, stones, boulders and walls

epiphytic forms
07 tree boles

combinations
17 terrestrial and epilithic
18 epilithic and epiphytic
19 epilithic and lignicole
98 indifferent or ubiquitous

j and k. **Distributional area**

10 Cosmopolitan
20 Supraholarctic
30 Holarctic
40 Eurasiatic
50 European
51 Alpine and montane European
52 W. European
53 S. European
54 Central European
55 S. and W. European
56 W. and C. European
57 S. and C. European
58 N. and C. European
59 W., S. and C. European
70 neophytes
80 adventitious forms
90 species that have run wild (cultivated plants)

Annotated species list

The symbols correspond with the numbers in appendix II

VASCULAR PLANTS	a	b	c	d	e	f	g	h	i	j	k
001 Acacia	77	0	6	4	14	1	06	80	01	90	90
002 Acanthus mollis L.	84	8	3	3	22	1	09	00	99	53	53
003 Acer campestre L.	73	2	6	4	12	1	05	53	83	59	59
004 Acer monspessulanum L.	73	0	6	4	12	1	05	53	82	53	53
005 Acer platanoides L.	73	2	6	4	12	1	05	53	83	50	50
006 Acer pseudoplatanus L.	73	4	6	4	12	1	05	53	83	59	59
007 Achillea millefolium L.	86	5	4	4	22	1	02	80	71	40	50
008 Adiantum capillus-veneris L.	37	2	3	5	24	2	00	51	10	20	53
009 Aethionema saxatile (L.) R.Br.	81	5	3	4	23	2	03	53	10	53	53
010 Agrostis gigantea Roth	53	6	3	4	23	2	05	99	99	30	59
011 Agrostis stolonifera L.	53	7	3	4	26	2	05	99	99	30	50
012 Agrostis tenuis Sibth.	53	4	3	4	23	2	05	99	20	30	50
013 Aira praecox L.	53	2	1	3	23	2	05	80	20	55	55
014 Ajuga reptans L.	84	4	3	5	23	2	04	73	99	59	59
015 Alliaria petiolata (Bieb.) Cavara et Grande	81	6	3	5	21	1	02	53	82	50	50
016 Alchemilla	77	0	3	4	26	2	05	00	00	00	00
018 Allium roseum L.	54	7	2	4	21	1	07	10	40	53	53
019 Allium paniculatum L.	54	2	2	4	21	1	07	10	99	57	57
020 Allium polyanthum Boreau	54	0	2	4	21	1	07	00	40	53	53
021 Allium spec.	54	0	2	4	21	1	99	00	00	00	00
023 Alnus glutinosa (L.) Vill.	65	3	6	4	12	1	05	53	83	50	50
024 Alopecurus pratensis L.	53	7	3	4	23	2	05	80	71	40	50
025 Alyssum alyssoides (L.) L.	81	4	1	3	21	1	06	53	20	50	50
026 Amaranthus deflexus L.	66	2	1	4	21	1	01	80	50	70	70
027 Amaranthus retroflexus L.	66	4	1	4	21	1	01	80	50	70	70
028 Anagallis arvensis L. ssp. arvensis	69	4	1	5	23	2	08	80	40	20	50
029 Anchusa officinalis L.	77	2	3	3	22	2	03	72	20	56	56
030 Anogramma leptophylla (L.) Link	37	5	1	5	23	2	00	51	10	10	53
031 Angelica archangelica L.	75	2	3	6	21	1	02	60	99	40	58
032 Angelica sylvestris L.	75	2	3	5	22	1	02	60	99	40	50
033 Antennaria dioica (L.) Gaertn.	86	4	4	3	23	2	02	53	81	30	58
034 Anthemis cotula L.	86	2	1	4	21	1	26	60	50	20	50
035 Anthoxanthum odoratum L.	53	3	3	4	23	2	05	53	71	20	50
037 Anthriscus sylvestris (L.) Hoffm.	75	2	3	5	21	1	02	99	99	58	58
038 Antirrhinum latifolium DC.	77	2	3	4	21	1	06	00	10	53	53
039 Antirrhinum majus L.	77	2	3	4	21	1	08	00	10	55	55
081 Apera spica-venti (L.) P.B.	53	2	1	4	21	1	05	80	40	40	50
040 Apium graveolens L.	75	2	3	5	21	1	02	80	99	20	50
041 Apium inundatum (L.) Rchb.f.	75	2	3	6	21	1	02	60	91	59	59
042 Apium nodiflorum (L.) Lag.	75	2	3	6	22	1	02	60	61	57	57
043 Aquilegia vulgaris L.	61	3	3	5	21	1	03	00	83	40	57
044 Arabidopsis thaliana (L.) Heynh.	81	2	1	4	22	1	02	53	20	40	50
045 Arabis alpina L.	81	3	3	3	22	1	02	53	10	40	50
046 Arabis hirsuta (L.) Scop.	81	3	3	3	22	1	02	53	20	30	50
047 Arctium pubens Bab.	86	3	3	3	21	1	09	72	50	50	50
048 Arenaria serpyllifolia L. (s.l.)	66	3	1	4	21	1	02	99	20	30	50
050 Armeria maritima (Mill.) Willd.	68	2	3	3	22	1	09	99	99	59	59
051 Armoracia rusticana G.,M.et Sch.	81	4	3	4	22	1	02	80	01	90	90
052 Arrhenatherum elatius (L.) P.B.ex J.et C.Presl	53	4	3	4	23	2	05	80	71	40	50

053	Artemisia absinthium L.	86	2	4	3	23	2	06	80	50	40	50
054	Artemisia alba Turra	86	0	1	4	21	1	02	53	10	53	53
055	Artemisia annua L.	86	2	1	4	21	1	02	80	50	40	53
056	Artemisia campestris L.	86	2	4	2	23	2	05	53	20	30	50
057	Artemisia vulgaris L.	86	2	3	4	23	2	06	80	50	30	50
058	Arum italicum Mill	52	7	2	5	21	1	02	71	82	53	53
059	Asparagus acutifolius L.	54	0	2	3	23	2	02	71	99	53	53
080	Asparagus officinalis L.	54	2	2	3	25	2	02	71	99	40	59
060	Asplenium adiantum-nigrum L.	36	3	3	4	24	2	00	51	10	40	55
061	Asplenium marinum L.	36	2	3	4	24	2	00	51	10	55	55
062	Asplenium onopteris L.	36	2	3	4	24	2	00	51	99	40	53
063	Asplenium ruta-muraria L.	36	3	3	3	24	2	00	51	10	30	50
064	Asplenium trichomanes L. ssp. quadrivalens D.E.Meyer	36	4	3	3	24	2	00	51	10	10	50
065	Asplenium viride Huds.	36	2	3	4	24	2	00	51	10	30	51
067	Asplenium ad-ni.x r.-m. A.x perardi Litard.	36	8	3	4	24	2	00	01	10	40	55
069	Asplenium r.-m.x viride = A.x meyeri Rothm.	36	8	3	4	24	2	00	01	10	30	51
070	Asplenium r.-m.x Cet.offic. = x Asplenoceterach badense D.E.Meyer	36	8	3	3	24	2	00	01	10	40	55
071	Aster novi-belgii L.	86	5	3	5	23	2	36	80	01	90	90
072	Aster tripolium L.	86	2	1	6	21	1	36	52	30	40	50
073	Asterolinum stellatum (L.) Hg. et Lk.	70	0	1	4	21	1	02	53	20	53	53
074	Athyrium filix-femina (L.) Roth	36	2	3	5	23	2	00	51	83	20	50
075	Atriplex glabriuscula Edmonst.	66	2	1	3	21	1	01	99	92	52	52
076	Atriplex hastata L.	66	3	1	3	21	1	01	99	99	30	50
077	Atriplex littoralis L.	66	2	1	3	21	1	01	99	92	40	50
078	Avena barbata Pott ex Link	53	4	1	3	21	1	05	52	50	53	53
079	Avena sterilis L.	53	6	1	3	21	1	05	52	50	53	53

091	Ballota nigra L.	84	2	3	4	23	2	07	80	50	55	55
092	Barbarea vulgaris R.Br.	81	2	3	4	22	1	06	80	50	40	50
093	Bellis perennis L.	86	2	3	4	22	1	26	99	71	50	50
094	Beta vulgaris L. ssp. maritima (L.) Arcangeli = B.m.L.	66	2	3	4	22	1	01	53	92	40	55
095	Betula pendula Roth	65	4	6	4	12	1	01	53	83	40	50
096	Betula pubescens Ehrh. ssp. pubescens	65	6	6	4	12	1	01	71	83	40	50
097	Bidens connatus Muehlenb.	86	0	1	5	21	1	06	60	61	70	70
098	Bidens frondosus L.	86	4	1	5	21	1	06	60	61	70	70
099	Bidens tripartitus L.	86	7	1	5	21	1	06	60	91	40	50
100	Blackstonia perfoliata (L.) Huds.	69	4	1	5	22	1	06	00	91	55	55
101	Bonjeania hirsuta (L.) Rchb.	77	0	4	4	23	2	07	00	10	53	53
102	Borago officinalis L.	84	2	1	4	22	1	04	73	50	53	53
103	Brachypodium phoenicoides R. et S.	53	4	3	3	23	2	05	53	20	53	53
104	Brachypodium pinnatum (L.) P.B.	53	4	3	3	23	2	05	80	71	40	50
105	Brachypodium ramosum (L.) R. et S.	53	0	4	3	23	2	05	53	81	53	53
106	Brassica napus L.	81	2	4	4	22	1	06	80	01	90	90
107	Brassica nigra (L.) Koch	81	2	1	4	21	1	06	80	40	10	50
108	Bromus erectus Huds.	53	6	3	3	23	2	05	80	71	40	57

109	Bromus inermis Leyss.	53	6	3	3	23	2	05	80	71	30	58
110	Bromus madritensis L.	53	7	1	3	21	1	05	99	50	40	55
111	Bromus mollis L.	53	4	1	4	21	1	05	99	99	40	50
117	Bromus racemosus L.	53	4	1	3	21	1	05	80	71	50	50
112	Bromus rubens L.	53	4	1	3	21	1	05	00	20	20	53
113	Bromus sterilis L.	53	3	1	3	21	1	05	99	50	59	59
114	Bromus tectorum L.	53	2	1	3	21	1	05	80	20	50	50
115	Bromus spec.	53	0	1	3	21	1	05	00	00	00	00
116	Buddleja davidii Franch.	83	4	5	4	21	1	07	80	50	70	70
126	Calamagrostis epigeiois (L.) Roth	53	7	3	3	23	2	05	52	91	40	50
127	Calamintha nepeta (L.) Savi	84	0	3	3	21	1	07	73	82	55	55
128	Calendula arvensis L.	86	4	1	3	21	1	06	80	40	53	53
129	Calystegia sepium (L.) F.Br.	84	2	3	5	21	1	02	80	99	20	50
130	Calystegia soldanella (L.) R.et Sch.	84	2	3	4	21	1	07	53	20	20	55
131	Campanula erinus L.	86	2	1	4	22	1	04	53	20	53	53
132	Campanula rapunculoides L.	86	6	3	4	21	1	04	53	99	50	50
133	Campanula rotundifolia L.	86	3	3	4	21	1	04	73	99	30	50
134	Campanula spec.	86	0	3	3	21	1	04	00	00	00	00
135	Capparis spinosa L.	81	2	5	3	21	1	02	71	10	53	53
136	Capsella bursa-pastoris (L.) Med.	81	4	1	4	22	1	02	80	40	10	50
137	Cardamine amara L.	81	3	3	5	22	1	03	60	61	40	50
138	Cardamine flexuosa With.	81	4	1	5	22	1	02	53	83	30	50
139	Cardamine hirsuta L.	81	2	1	5	22	1	02	53	20	40	50
140	Cardamine pratensis L. ssp. pratensis	81	5	3	5	22	1	07	60	71	30	50
142	Cardamine spec.	81	0	3	5	22	1	99	00	00	00	00
143	Cardaminopsis arenosa (L.) Hayek	81	3	1	5	22	1	02	53	10	54	54
144	Cardaria draba (L.) Desv.	81	6	3	4	22	1	02	99	20	40	50
145	Carduus crispus L.	86	2	3	3	21	1	07	52	50	40	50
146	Carduus pycnocephalus L.	86	0	1	3	21	1	07	52	50	53	53
147	Carex acutiformis Ehrh.	55	0	3	4	23	2	05	60	61	20	50
148	Carex arenaria L.	55	0	3	2	22	1	05	80	20	30	50
149	Carex digitata L.	55	0	3	5	23	2	05	73	83	50	50
150	Carex divisa L.	55	0	3	4	23	2	05	00	99	20	55
151	(incl. var. chaetophylla Steud.)	55	0	3	4	23	2	05	00	99	53	53
152	Carex extensa Good.	55	0	3	3	22	1	05	60	30	20	50
153	Carex flacca Schreb.	55	0	3	3	23	2	05	00	71	40	50
154	Carex hirta L.	55	0	3	4	23	2	05	80	99	20	50
155	Carex mairei Coss. et G.	55	0	3	4	23	2	05	00	62	53	53
156	Carex otrubae Podp.	55	0	3	4	27	3	05	60	91	40	50
157	Carex remota L.	55	0	3	4	23	2	05	60	83	30	50
158	Carex scandinavica E.W.Davies	55	0	3	4	23	2	05	80	91	30	50
159	Carpinus betulus L.	65	5	6	4	12	1	01	53	83	50	50
160	Catapodium marinum (L.) Hubb.	53	2	1	3	21	1	05	80	92	55	55
161	Catapodium rigidum (L.) Hubb.	53	2	1	3	21	1	05	53	20	40	55
162	Centaurea aspera L.	86	2	3	4	21	1	07	53	99	53	53
163	Centaurea nigra L.	86	4	3	4	21	1	07	53	50	52	52
164	Centaurea pratensis Thuill.	86	2	3	4	21	1	07	53	99	59	59
165	Centaurea solstitialis L.	86	2	1	4	21	1	07	53	50	40	53
213	Centaurea stoebe L.	86	2	1	4	21	1	07	52	99	54	54
166	Centranthus ruber (L.) DC.	76	3	3	4	23	2	08	00	10	55	55
167	Cerastium arvense L. ssp. arvense	66	5	4	4	23	2	02	53	99	30	50
168	Cerastium diffusum Pers.ssp. diffusum Curtis	66	7	3	4	23	2	02	53	20	55	55

'- 4 -'

169	Cerastium fontanum Baumg. ssp. triviale (Link) Jalas	66	6	4	4	23	2	02	99	99	10	50
170	Cerastium glomeratum Thuill.	66	7	1	4	22	1	02	80	99	10	50
171	Cerastium pumilum Curtis ssp. pallens (F.W.Schultz) Schinz et Thell.	66	6	1	4	22	1	02	53	20	40	57
172	Cerastium semidecandrum L. ssp. semidecandrum	66	3	1	4	22	1	02	53	20	59	59
174	Ceterach officinarum Lamk. et DC.	36	4	3	3	24	2	00	51	10	40	55
175	Chamaecyparis	43	0	6	3	11	1	01	80	01	90	90
176	Chaenorhinum origanifolium (L.) Fourr.	84	2	3	4	26	2	03	53	10	53	51
177	Chamaenerion angustifolium (L.) Scop.	79	6	3	4	21	1	07	52	99	30	50
178	Cheilanthes fragrans (L.f.) Swartz	37	0	2	4	24	2	00	51	10	20	53
179	Cheiranthus cheiri L.	81	2	4	4	24	2	06	53	10	55	55
180	Chelidonium majus L.	81	2	3	4	22	1	06	73	50	40	50
181	Chenopodium album L.ssp. album	66	7	1	4	21	1	01	80	50	10	50
182	Chenopodium bonus-henricus L.	66	4	3	4	23	2	01	80	50	30	50
183	Chenopodium ficifolium Sm.	66	2	1	4	21	1	01	80	50	40	59
184	Chenopodium murale L.	66	2	1	4	21	1	01	99	50	10	50
185	Chenopodium rubrum L.	66	4	1	4	21	1	01	99	50	40	59
186	Chrysanthemum leucanthemum L.	86	3	3	4	22	1	26	71	71	40	50
187	Chrysanthemum parthenium (L.) Bernh.	86	2	3	4	22	1	26	80	50	59	59
188	Chrysanthemum segetum L.	86	2	1	4	22	1	06	80	40	40	50
189	Cicuta virosa L.	75	2	3	6	21	1	02	60	61	40	58
190	Circaea lutetiana L.	79	2	2	5	21	1	02	72	83	30	50
191	Cirsium arvense (L.) Scop.	86	2	2	3	22	1	07	52	50	40	50
192	Cirsium vulgare (Savi) Ten.	86	4	3	3	22	1	07	52	99	40	50
193	Clematis flammula L.	61	2	5	4	12	2	02	52	82	53	53
194	Clematis vitalba L.	61	2	5	4	12	2	02	52	82	40	57
195	Cochlearia danica L.	81	6	1	3	22	1	02	99	92	52	52
196	Convolvulus arvensis L.	84	6	3	4	22	1	07	80	20	20	50
197	Coronilla minima L.	77	0	3	4	25	2	06	00	20	53	53
198	Coronopus didymus (L.) Sm.	81	4	1	4	22	1	02	80	99	10	50
199	Coronopus squamatus (Forsk.)Aschrs.	81	4	1	4	22	1	02	80	50	50	50
200	Corrigiola litoralis L.	66	3	1	4	26	2	01	99	99	55	55
201	Corydalis lutea (L.) DC.	81	5	3	4	25	2	06	73	10	59	59
202	Corynephorus canescens (L.) P.B.	53	2	3	3	27	2	05	80	20	59	59
203	Corylus avellana L.	65	2	6	4	12	1	05	53	83	50	50
204	Cotoneaster integerrima Med.	77	4	5	3	14	1	07	71	10	40	50
205	Crepis capillaris (L.) Wallr.	86	2	1	4	11	1	06	52	71	50	50
206	Crepis vesicaria L.ssp.taraxacifolia (Thuill.) Thell.	86	3	1	4	11	1	06	52	99	55	55
207	Crithmum maritimum L.	75	2	3	3	25	2	02	00	10	55	55
208	Cyclamen repandum Sibth. et Sm.	69	4	2	4	22	1	08	00	99	53	53
209	Cynodon dactylon (L.) Pers.	53	4	2	3	21	1	05	80	20	10	50
210	Cynoglossum officinale L.	84	3	1	3	22	1	04	72	20	40	50
211	Cynosurus echinatus L.	53	2	1	3	21	1	05	80	20	53	53
212	Cystopteris fragilis (L.) Bernh.	36	5	3	5	24	2	00	51	10	10	50

221	Dactylis glomerata L.	53	7	3	4	23	2	05	99	71	30	50
222	Daucus carota L. ssp. carota	75	2	1	5	22	1	02	99	99	40	50
223	Daucus carota L.ssp. gummifer Hook f.	75	2	1	5	22	1	02	53	10	55	55
224	Deschampsia cespitosa (L.) P.B.	53	3	3	3	27	3	05	53	71	20	50
225	Dianthus arenarius L.	66	4	3	3	23	2	08	00	20	54	54
226	Dianthus caryophyllus L.	66	5	3	3	23	2	08	00	10	53	53
227	Digitaria ischaemum (Schreb.) Muehlenb.	53	4	3	3	21	1	05	80	20	20	59
228	Diplotaxis erucoides (L.) DC.	81	2	1	4	23	2	06	80	02	80	80
229	Diplotaxis muralis (L.) DC.	81	3	1	4	23	2	06	80	99	50	50
230	Diplotaxis tenuifolia (L.) DC.	81	2	2	4	23	2	06	80	20	50	50
231	Dipsacus pilosus L.	76	2	1	3	22	1	07	00	82	40	56
232	Draba muralis L.	81	4	1	4	22	1	02	53	20	59	59
233	Dryopteris abbreviata (DC.) Newm.	36	2	3	5	23	2	00	51	10	56	56
234	Dryopteris carthusiana (Villar) H.P.Fuchs	36	4	3	5	23	2	00	51	10	30	58
235	Dryopteris cristata (L.) A.Gray	36	4	3	5	23	2	00	51	62	30	58
236	Dryopteris dilatata (Hoffm.) A.Gray	36	4	3	5	23	2	00	51	83	30	58
237	Dryopteris filix-mas (L.) Schott	36	4	3	5	23	2	00	51	83	20	50
241	Echium italicum L.	84	2	3	3	22	1	04	72	20	53	53
242	Echium vulgare L.	84	3	3	3	22	1	04	72	20	40	50
243	Elymus arenarius L.	53	6	3	2	23	2	05	53	20	40	58
244	Elytrigia pungens (Pers.) Tutin	53	6	3	2	23	2	05	53	20	55	55
245	Elytrigia repens (L.) Nevski	53	6	3	3	23	2	05	99	50	30	50
246	Elytrigia pu.x re.=E.x oliveri (Druce) Ooststr.	53	8	3	4	23	2	05	53	20	55	55
247	Empetrum nigrum L.	80	2	4	2	23	2	07	71	81	30	52
248	Epilobium adnatum Griseb.	79	6	3	5	21	1	07	52	99	40	50
249	Epilobium collinum Gmel.	79	6	3	4	21	1	07	52	10	51	51
250	Epilobium hirsutum L.	79	5	3	6	21	1	07	52	61	40	50
251	Epilobium lanceolatum Seb. et Mauri	79	0	3	4	21	1	07	52	10	55	55
252	Epilobium montanum L.	79	6	3	5	21	1	07	52	99	40	50
253	Epilobium nerterioides A.Cunn.	79	6	3	5	21	1	07	52	99	70	70
254	Epilobium obscurum Schrb.	79	6	3	5	21	1	07	52	99	50	50
255	Epilobium parviflorum Schreb.	79	6	3	5	21	1	07	52	99	40	50
256	Epilobium roseum Schreb.	79	6	3	5	21	1	07	52	91	40	50
257	Epilobium lanc.x mont.=E. x neogradiense Borbas	79	8	3	5	21	1	07	01	10	55	55
259	Equisetum arvense L.	33	0	2	2	21	1	00	51	99	20	50
260	Equisetum ramosissimum Desf.	33	0	2	2	21	1	00	51	10	20	57
261	Eragrostis poaeoides P.B.	53	4	1	4	23	2	05	80	50	30	53
262	Erica cinerea L.	80	2	3	3	23	2	08	53	81	52	52
263	Erigeron acer L.	86	2	1	4	22	1	02	52	20	30	50
264	Erigeron canadensis L.	86	2	1	4	22	1	02	52	20	10	50
265	Erigeron mucronatus DC.	86	4	4	4	25	2	26	52	10	70	70
266	Erigeron naudini (Bonnet) G.Bonnier	86	0	1	4	22	1	02	52	00	70	70
267	Erigeron spec.	86	0	1	4	22	1	99	00	00	00	00
269	Erinus alpinus L.	84	2	4	3	25	2	03	00	10	53	53
270	Erodium cicutarium (L.) LHerit ex Ait.	70	4	1	4	22	1	09	20	20	50	50

271	Erophila verna (L.) Chevall.	81	3	1	4	22	1	02	53	20	40	59
272	Erysimum cheiranthoides L.	81	2	1	4	22	1	06	53	50	30	50
273	Eupatorium cannabinum L.	86	2	3	5	23	2	09	60	72	50	50
274	Euphorbia characias L.	72	2	3	4	21	1	06	73	20	53	53
275	Euphorbia cyparissias L.	72	3	3	4	23	2	06	73	20	57	57
276	Euphorbia exigua L.	72	5	1	4	21	1	06	73	40	50	50
277	Euphorbia helioscopia L.	72	6	1	4	21	1	06	73	40	40	50
278	Euphorbia peplis L.	72	0	1	4	26	2	06	80	92	53	53
279	Euphorbia peplus L.	72	2	1	5	21	1	06	99	40	40	50
280	Euphorbia segetalis L.	72	2	1	4	21	1	06	73	20	53	53
281	Euphorbia serrata L.	72	0	1	4	21	1	06	73	40	53	53
286	Fagus sylvatica L.	65	2	6	3	12	1	01	71	83	59	59
287	Ferula glauca (L.) Ry et C.	75	2	3	5	21	1	02	00	20	53	53
288	Festuca ovina L.	53	5	3	3	27	3	05	71	20	30	50
289	(incl. F. longifolia Thuill.)	53	6	3	3	27	3	05	71	20	30	50
290	Festuca pratensis Huds.	53	2	3	4	23	2	05	80	71	40	50
291	Festuca rubra L.	53	5	3	4	23	2	05	53	99	30	50
292	(incl. F. juncifolia St.Amans)	53	6	3	4	23	2	05	53	20	52	52
302	Ficus carica L.	64	2	5	3	12	1	05	71	10	40	53
293	Filago gallica L.	86	0	1	3	21	1	02	80	40	55	55
294	Filago germanica (L.) L.	86	4	1	3	21	1	02	80	40	40	57
295	Filipendula ulmaria (L.) Maxim.	77	2	3	4	21	1	02	60	72	40	50
296	Foeniculum vulgare Mill.	75	2	3	4	21	1	02	00	99	53	53
297	Fragaria vesca L.	77	2	3	4	26	2	02	71	83	40	50
298	Fraxinus excelsior L.	82	2	6	4	12	1	05	53	83	50	50
299	Fumaria capreolata L.	81	6	1	5	25	2	02	73	50	55	55
300	Fumaria officinalis L.	81	3	1	5	25	2	07	73	40	50	50
301	Fumaria parviflora Lam.	81	4	1	5	25	2	07	73	50	55	55
306	Galactites tomentosa Moench	86	0	3	3	22	1	06	52	20	53	53
307	Galeopsis ladanum L.	84	2	1	4	21	1	07	73	10	57	57
308	Galeopsis tetrahit L.	84	4	1	5	21	1	07	71	99	40	50
309	Galinsoga ciliata (Raf.) Blake	86	4	1	5	21	1	02	80	50	70	70
310	Galinsoga parviflora Cav.	86	2	1	5	21	1	02	80	50	70	70
311	Galium aparine L.	76	6	1	4	23	2	02	72	99	40	50
312	Galium mollugo L.	76	4	3	4	23	2	02	00	99	40	50
313	(incl.ssp.erectum Syme)	76	4	3	4	23	2	02	00	99	55	55
314	Galium murale All.	76	0	1	4	26	2	02	00	20	53	53
315	Galium palustre L.	76	2	3	5	23	2	02	60	62	50	50
316	Galium parisiense L.	76	0	1	4	26	2	02	00	20	59	59
317	Galium pumilum Murr.	76	6	3	4	23	2	02	00	10	51	51
318	Galium tricornutum Dandy	76	4	3	4	23	2	02	00	40	53	53
319	Galium verum L.	76	4	3	2	23	2	02	80	71	50	50
336	Galium spec.	76	0	3	4	26	2	02	00	00	00	00
321	Geranium lucidum L.	70	2	1	5	21	1	07	20	99	55	55
322	Geranium molle L.	70	2	1	4	22	1	07	20	20	50	50
323	Geranium pyrenaicum Burm.f.	70	3	3	4	21	1	03	20	82	50	50
324	Geranium robertianum L.	70	6	1	4	21	1	07	20	99	20	50
325	Geranium rotundifolium L.	70	2	1	4	21	1	07	20	40	40	55
326	Geranium sanguineum L.	70	6	3	4	21	1	08	20	20	50	50
327	Geum urbanum L.	77	6	3	4	22	1	06	72	82	40	50

328	Glaucium flavum Crantz	81	3	3	3	22	1	06	00	92	55	55
329	Glaux maritima L.	69	2	3	4	26	2	07	80	30	30	52
330	Glechoma hederacea L.	84	3	3	4	23	2	04	73	99	40	50
331	Glyceria fluitans (L.) R.Br.	53	7	3	5	23	2	05	60	61	20	50
332	Gnaphalium luteo-album L.	86	2	1	4	21	1	06	80	99	20	59
333	Gnaphalium uliginosum L.	86	2	1	4	21	1	02	80	91	40	50
334	Gymnocarpium dryopteris (L.) Newm.	36	3	2	5	21	1	00	51	83	30	58
335	Gymnocarpium robertianum (Hoffm.) Newm.	36	4	2	5	21	1	00	51	10	30	57
337	Gymnocarpium dr.x rob.= G.x hybridum	36	8	2	5	21	1	00	01	10	30	54
341	Halimione portulacoides (L.) Aellen	66	3	4	3	25	2	01	99	30	30	50
342	Hedera helix L.	75	4	5	3	26	2	02	99	99	59	59
343	Helianthus annuus L.	86	2	1	4	21	1	06	71	01	90	90
344	Helichrysum arenarium (L.) Moench	86	3	3	3	23	2	06	53	20	54	54
345	Helichrysum italicum (Roth) Guss.	86	0	3	3	23	2	06	53	81	53	53
346	Helictotrichon pubescens (Huds.) Pilger	53	2	3	4	23	2	05	80	71	40	58
347	Heracleum sphondylium L.	75	2	3	5	22	1	02	60	99	40	50
348	Herniaria glabra L.	66	2	1	4	26	2	01	80	20	40	50
349	Hieracium amplexicaule L.	86	4	3	3	22	1	06	52	10	53	53
350	(incl.ssp.speluncarum A.-T. Zahn)	86	4	3	3	22	1	06	52	10	53	53
351	Hieracium bauhinii Bess.	86	4	3	2	22	1	06	52	20	40	00
352	Hieracium lachenalii C.C.Gmel.	86	0	3	3	21	1	06	52	99	40	50
353	Hieracium laevigatum Willd.	86	0	3	3	21	1	06	52	82	30	50
354	Hieracium maculatum Sm.	86	0	3	4	22	1	06	52	83	56	56
355	Hieracium murorum L.	86	0	3	3	22	1	06	52	99	40	50
356	Hieracium pilosella L.	86	0	3	4	22	1	06	52	20	50	50
357	Hieracium praecox Schultz-Bip.	86	0	3	3	22	1	06	52	83	00	00
358	Hieracium umbellatum L.	86	0	3	2	21	1	06	52	99	30	50
372	Hieracium spec.	86	0	3	4	22	1	06	52	00	00	00
361	Holcus lanatus L.	53	2	3	4	23	2	05	53	99	40	50
362	Holcus mollis L.	53	5	3	4	23	2	05	53	82	50	50
363	Honckenya peploides (L.) Ehrh.	66	4	3	2	23	2	02	99	20	30	50
364	Hordeum marinum Huds.	53	3	1	3	21	1	05	60	92	30	55
365	Hordeum murinum L.	53	3	1	3	21	1	05	80	50	30	57
366	Hornungia petraea (L.) Rchb.	81	2	1	4	21	1	02	53	20	55	55
367	Hyoscyamus albus L.	84	4	3	3	21	1	06	00	10	53	53
373	Hyoseris radiata L.	86	0	3	4	22	1	06	52	20	61	61
368	Hypericum hircinum L.	62	4	4	4	26	2	06	80	10	70	70
369	Hypericum perforatum L.	62	4	3	4	21	1	06	53	99	40	50
370	Hypochaeris glabra L.	86	3	3	4	22	1	06	52	20	59	59
371	Hypochaeris radicata L.	86	2	3	4	22	1	06	52	20	50	50
376	Iberis amara L.	81	2	1	4	25	2	02	53	40	52	52
377	Impatiens glandulifera Royle	74	2	1	5	21	1	08	99	99	70	70
378	Impatiens parviflora DC.	74	3	1	5	21	1	06	99	50	70	70
379	Inula viscosa (L.) Ait.	86	2	4	3	23	2	06	53	82	53	53
380	Isatis tinctoria L.	81	4	3	4	21	1	06	00	50	80	80
381	Jasione montana L.	86	2	3	3	22	1	04	53	20	50	50
382	Juncus bufonius L.	54	7	1	4	26	2	01	99	91	10	50

383	Juncus hybridus Brot.	54	2	1	4	26	2	01	99	92	20	50
384	Juncus tenuis Willd.	54	3	3	4	23	2	01	80	91	70	70
386	Knautia arvensis (L.) Coulter	86	3	3	4	22	1	04	53	71	50	50
387	Koeleria albescens DC.	53	0	3	3	23	2	05	00	20	00	00
391	Lactuca serriola L.	86	2	1	3	21	1	06	52	20	40	53
392	Lagoseris sancta (L.) Maly	86	0	1	4	21	1	06	52	40	53	53
393	Lagurus ovatus L.	53	2	1	3	21	1	05	52	20	53	53
394	Lamium album L.	84	2	3	4	23	2	02	73	99	40	50
395	Lamium amplexicaule L.	84	2	1	4	21	1	07	73	40	40	50
396	Lamium maculatum L.	84	2	3	4	21	1	07	73	82	59	59
397	Lamium purpureum L.	84	2	1	4	23	2	03	99	40	50	50
398	Lapsana communis L.	86	2	1	4	21	1	06	52	83	40	50
399	Larix decidua Miller	42	3	6	3	11	1	01	53	83	51	51
400	Lathyrus sylvestris L.	77	2	3	4	21	1	07	00	82	57	57
401	Lavatera maritima Gouan	71	0	4	4	21	1	07	72	10	53	53
402	Leontodon autumnalis L.	86	5	3	4	22	1	06	52	71	40	50
403	Leontodon hispidus L.	86	2	3	4	22	1	06	52	71	50	50
404	Leontodon nudicaulis (L.) Banks ex Lowe	86	2	3	4	22	1	06	52	71	59	59
405	Lepidium graminifolium L.	81	2	3	4	21	1	02	99	50	53	53
406	Lepidium ruderale L.	81	4	1	4	21	1	00	80	50	40	50
407	Limonium binervosum (G.E.Sm.) C.E. Salmon	69	4	3	3	22	1	07	00	10	52	52
408	Limonium lychnidifolium (Girard) Kuntze	69	0	3	3	22	1	07	00	10	52	52
409	Limonium vulgare Mill.	69	4	3	3	22	1	07	99	30	30	52
410	Linaria arvensis (L.) Desf.	84	2	1	4	21	1	04	53	40	59	59
411	Linaria cymbalaria (L.) Mill.	84	2	4	4	25	2	03	20	10	40	59
412	Linaria purpurea (L.) Mill.	84	2	3	4	23	2	03	53	10	70	70
413	Linaria repens (L.) Mill.	84	2	3	4	23	2	03	53	20	52	52
414	Linaria vulgaris Mill.	84	2	3	4	23	2	06	53	20	40	50
415	Linaria spec.	84	0	3	4	23	2	99	00	00	00	00
416	Lithospermum arvense L.	84	3	1	4	21	1	06	71	40	40	59
417	Lobularia maritima (L.) Desv.	81	4	1	4	25	2	02	53	20	53	53
418	Lolium multiflorum Lamk.	53	2	3	4	23	2	05	80	71	55	55
419	Lolium perenne L.	53	2	3	4	23	2	05	80	71	40	50
420	Lolium rigidum Gaud.	53	2	1	4	21	1	05	80	50	40	53
421	Lonicera periclymenum L.	76	3	5	4	14	2	07	71	83	55	55
422	Lotus corniculatus L.	77	4	3	2	25	2	06	53	20	20	50
423	Luzula campestris (L.) DC. ssp. campestris	54	0	3	4	23	2	01	99	83	50	50
424	Luzula luzuloides (Lamk.) Dandy et Wilmott	54	0	3	4	23	2	02	53	99	10	50
425	Lycium halimifolium Mill.	84	4	5	3	14	2	07	71	82	57	57
426	Lycopsis arvensis L.	84	6	1	4	21	1	08	73	40	40	50
427	Lycopus europaeus L.	84	2	3	6	23	2	04	60	61	40	50
428	Lysimachia vulgaris L.	69	3	3	6	21	1	06	60	72	40	50
429	Lythrum salicaria L.	79	7	3	6	21	1	08	53	72	20	50

436	Malva neglecta Wallr.	71	6	1	4	23	2	07	99	50	50	50
437	Malva sylvestris L.	71	6	3	4	23	2	07	99	50	40	50
438	Marrubium vulgare L.	84	2	3	4	23	2	02	00	50	40	53
439	Matricaria chamomilla L.	86	2	1	4	22	1	26	80	40	40	50
440	Matricaria maritima L. ssp. inodora (L.) Clapham	86	3	3	4	22	1	26	73	50	40	56
441	Matricaria maritima L. ssp. maritima	86	2	3	4	22	1	26	60	30	40	56
442	Matricaria matricarioides (Less.) Porter	86	2	1	4	22	1	26	80	50	30	50
443	Matthiola incana (L.) R.Br.	81	2	4	2	21	1	03	53	10	53	53
444	Medicago lupulina L.	77	3	3	4	26	2	06	99	20	30	50
445	Melica bauhinii All.	53	0	3	4	23	2	05	73	10	53	53
446	Melica ciliata L.	53	2	3	3	23	2	05	73	10	30	50
447	Melica minuta L.	53	0	3	4	23	2	05	73	10	53	53
448	Melica nutans L.	53	2	3	4	23	2	05	73	83	50	50
449	Mentha aquatica L.	84	7	3	5	23	2	07	60	61	20	50
450	Mentha arvensis L.	84	5	3	4	23	2	07	99	91	40	50
451	Mentha pulegium L.	84	5	3	4	23	2	07	00	91	59	59
452	Mentha rotundifolia (L.) Huds.	84	5	3	4	23	2	07	73	99	55	55
453	Mercurialis annua L. (incl. M. hueti Hanry)	72	2	1	5	21	1	05	73	50	50	50
455	Minuartia mediterranea (Link) K. Maly	66	3	1	4	21	1	02	53	20	53	53
456	Minuartia mutabilis Schinz et Thell. ex Becherer	66	2	1	4	21	1	02	53	20	53	51
457	Muscari comosus (L.) Mill.	54	2	2	4	23	2	04	71	40	53	53
458	Muscari neglectus Guss.	54	6	2	4	23	2	04	71	20	53	53
459	Mycelis muralis (L.) Dum.	86	2	3	4	21	1	06	52	80	50	50
462	Myosotis arvensis (L.) Hill.	84	6	1	4	22	1	04	99	40	40	50
460	Myosotis ramosissima Rochel ex Schult.	84	6	1	3	22	1	04	99	20	40	50
461	Myosotis scorpioides L.	84	6	3	5	23	2	04	60	61	30	50
466	Oenanthe aquatica L.	75	2	3	6	21	1	02	60	91	40	50
467	Oenanthe crocata L.	75	2	3	6	21	1	02	60	61	55	55
468	Origanum vulgare L.	84	3	4	4	23	2	03	99	99	40	50
469	Orobanche hederae Duby	84	2	2	5	21	1	06	51	99	55	55
470	Oryzopsis miliacea (L.) Benth. et Hook. ex Aschrs. et Schweinf.	53	3	4	4	23	2	05	53	10	53	53
471	Ostrya carpinifolia Scop.	65	2	5	3	12	1	01	53	82	53	53
472	Osyris alba L.	78	4	5	3	13	2	02	00	20	53	53
473	Oxalis acetosella L.	70	2	3	5	25	2	02	20	82	30	50
474	Oxalis corniculata L.	70	7	1	5	25	2	06	20	90	53	53
475	Oxalis europaea Jord.	70	4	1	5	25	2	06	20	40	57	57
476	Oxalis pes-caprae L.	70	4	2	5	25	2	06	80	02	80	80

481	Panicum miliaceum L.	53	4	3	4	23	2	05	80	01	01	01
482	Papaver argemone L.	81	5	1	5	21	1	08	80	40	59	59
483	Papaver rhoeas L.	81	2	1	5	21	1	08	80	40	40	50
484	Parapholis strigosa (Dum.) Hubbard	53	2	1	4	21	1	05	53	92	52	52
485	Parietaria judaica L.	64	2	3	4	24	2	01	73	10	55	55
486	Parietaria lusitanica L.	64	0	3	3	24	2	01	73	10	53	53
487	Parthenocissus	88	0	6	4	12	2	05	71	01	90	90
488	Petrorhagia prolifera (L.) P.W.Ball et Heywood	66	2	1	4	23	2	08	53	10	40	59
489	Petroselinum crispum (Mill.) Nyman	75	2	3	5	21	1	02	80	01	90	90
490	Peucedanum palustre (L.) Moench	75	2	3	5	21	1	02	60	62	40	50
491	Phagnalon sordidum (L.) DC.	86	0	4	3	23	2	06	53	10	53	53
492	Phalaris arundinacea L.	53	5	3	3	21	1	05	60	72	30	50
493	Phleum arenarium L.	53	2	1	4	22	1	05	53	20	55	55
494	Phleum pratense L.	53	6	3	4	21	1	05	80	71	30	50
495	Phyllitis scolopendrium (L.) Newman	36	2	3	4	23	2	00	51	99	30	50
496	Picris echioides L.	86	2	1	4	22	1	06	52	99	40	50
497	Picris hieracioides L.	86	2	3	4	22	1	06	52	99	53	53
498	Pimpinella major (L.) Huds.	77	2	3	4	22	1	02	00	99	50	50
499	Pimpinella saxifraga L.	77	3	3	2	22	1	02	53	99	50	50
500	Pimpinella tragium Villars	77	2	3	2	22	1	02	00	20	53	53
501	Pinus halepensis Miller	42	2	6	2	11	1	01	53	83	53	53
550	Plantago alpina L.	85	3	3	3	22	1	01	80	99	51	51
502	Plantago coronopus L.	85	5	3	3	22	1	01	99	92	59	59
503	Plantago lanceolata L.	85	5	3	4	22	1	01	99	71	40	50
504	Plantago major L. ssp. intermedia (Godr.) Lange	85	3	3	4	22	1	01	00	80	91	00
505	Plantago major L. ssp. major	85	2	3	4	22	1	01	80	50	10	50
506	Plantago maritima L.	85	3	3	3	22	1	01	60	30	56	56
507	Plantago media L.	85	3	3	4	22	1	01	99	71	40	50
508	Plantago psyllium L.	85	2	3	4	21	1	01	80	20	53	53
552	Poa alpina L.	53	3	3	4	23	2	05	80	71	30	51
509	Poa angustifolia L.	53	6	3	4	23	2	05	99	20	40	59
510	Poa annua L.	53	4	1	4	23	2	05	99	99	10	50
511	Poa bulbosa L.	53	5	3	4	23	2	05	99	20	40	57
512	Poa chaixii Vill.	53	2	3	4	23	2	05	80	01	90	90
513	Poa compressa L.	53	6	3	4	23	2	05	53	20	10	50
514	Poa nemoralis L.	53	7	3	4	23	2	05	53	99	30	50
515	Poa palustris L.	53	7	3	4	23	2	05	60	61	30	50
516	Poa pratensis L.	53	7	3	4	23	2	05	53	71	20	50
517	Poa supina Schrad.	53	2	3	4	23	2	05	80	10	50	51
518	Poa trivialis L.	53	3	3	4	23	2	05	53	71	58	58
519	Poa cf.angustif. x compressa	53	8	3	4	23	2	05	53	10	40	59
520	Poa "pseudocompressa"	53	0	3	4	23	2	05	53	10	54	54
521	Podospermum laciniatum (L.) DC.	86	2	3	4	21	1	06	52	99	40	57
522	Polycarpon tetraphyllum (L.) L.	66	6	1	4	21	1	02	53	20	53	53
523	Polygonum aequale Lindman	67	4	1	4	26	2	01	80	50	10	50
524	Polygonum calcatum Lindman	67	4	1	4	23	2	01	80	50	54	54
525	Polygonum heterophyllum Lindman = P.aviculare L. s.s.	67	6	1	4	26	2	01	80	99	10	50
526	Polygonum hydropiper L.	67	2	1	5	21	1	01	60	91	30	50
527	Polygonum lapathifolium L.	67	2	1	4	21	11	01	99	99	30	56
528	Polygonum persicaria L.	67	4	1	4	21	1	01	99	91	40	50
529	Polypodium australe Fee	39	2	3	4	23	2	00	51	99	53	53
530	Polypodium interjectum Shivas	39	6	3	4	23	2	00	51	99	59	59

'- '1 -'

531	Polypodium vulgare L.	39	4	3	4	23	2	00	51	99	20	50
532	Polypodium interj.x vulg. = P. x mantoniae Rothm.	39	8	3	4	23	2	00	01	99	59	59
533	Polystichum aculeatum (L.) Roth	36	4	3	4	23	2	00	51	83	40	59
534	Polystichum falcatum (L.f.)Diels	36	0	3	4	23	2	00	80	01	90	90
535	Polystichum lonchitis (L.) Roth	36	2	3	5	23	2	00	51	10	30	50
536	Polystichum setiferum (Forsk.) Woynar	36	2	3	5	23	2	00	51	82	10	59
551	Potentilla arenaria Borkh.	77	4	3	4	21	1	06	99	20	54	54
537	Potentilla argentea L.	77	5	3	4	21	1	06	99	20	30	50
538	Potentilla caulescens L.	77	2	3	4	24	2	02	71	10	51	51
539	Potentilla micrantha Ramond	77	2	3	4	26	2	02	71	99	53	53
540	Potentilla recta L.	77	7	3	4	21	1	06	71	20	40	57
541	Potentilla reptans L.	77	4	3	4	26	2	06	99	99	40	50
542	Potentilla tabernaemontani Aschrs.	77	7	3	4	26	2	06	71	20	59	59
544	Prunella vulgaris L.	84	4	3	5	23	2	03	80	99	20	59
545	Psoralea bituminosa L.	77	2	3	4	21	1	03	00	99	20	53
546	Pteridium aquilinum (L.) Kuhn	38	6	2	4	21	1	00	51	99	10	50
547	Puccinellia distans (L.) Parl.	53	5	3	3	23	2	05	80	92	40	50
548	Puccinellia maritima (Huds.) Parl.	53	5	3	3	23	2	05	60	30	30	50
549	Pulmonaria spec.	84	0	3	5	22	1	08	00	00	00	00
566	Quercus ilex L.	65	2	6	2	12	1	01	71	83	53	53
567	Quercus petraea (Mattuschka)Liebl.	65	2	6	3	12	1	01	71	83	50	50
568	Quercus robur L.	65	2	6	3	12	1	01	71	83	50	50
571	Ranunculus acris L.	61	5	3	4	22	1	06	99	71	20	50
572	Ranunculus bulbosus L.	61	2	3	5	22	1	06	99	71	30	50
573	Ranunculus ficaria L.	61	3	3	5	22	1	06	10	82	50	50
574	Ranunculus repens L.	61	3	3	5	26	2	06	80	91	40	50
575	Ranunculus sceleratus L.	61	4	1	5	22	1	06	60	91	30	50
576	Ranunculus spec.	61	0	3	4	22	1	06	00	00	00	00
577	Reichardia picroides (L.) Roth	86	0	3	4	22	1	06	52	99	53	53
578	Reseda lutea L.	81	6	3	4	21	1	06	80	50	40	50
579	Reseda luteola L.	81	4	3	4	21	1	06	80	50	59	59
580	Reseda phyteuma L.	81	2	3	4	21	1	02	80	99	53	53
581	Ribes rubrum L.	77	2	5	4	12	2	08	71	83	52	52
582	Ribes uva-crispi L.	77	2	5	4	12	2	08	71	82	40	50
583	Rorippa amphibia (L.) Besser	81	3	3	6	21	1	06	60	61	40	50
584	Rorippa islandica (Oeder) Borbas	81	3	1	5	21	1	06	60	91	10	50
585	Rorippa sylvestris (L.) Besser	81	6	3	5	21	1	06	60	91	50	50
586	Rosmarinus officinalis L.	84	4	5	2	13	2	04	73	81	53	53
587	Rubia peregrina L.	76	6	4	3	23	2	06	71	82	55	55
588	Rubia tinctorum L.	76	3	3	3	23	2	06	71	82	53	53
589	Rubus fruticosus agg.	77	3	5	4	12	2	99	71	82	00	00
590	(incl. R.ulmifolius Schott)	77	2	5	4	12	2	07	71	82	55	55
591	Rubus idaeus L.	77	3	5	4	12	2	02	71	82	30	50
592	Rumex acetosa L.	66	2	3	5	22	1	01	53	71	30	50
593	Rumex acetosella L.	66	5	3	4	22	1	01	53	20	30	50

603	Rumex alpinus L.	66	2	3	5	22	1	01	99	50	40	51
594	Rumex bucephalophorus L.	66	2	1	4	22	1	01	53	20	53	53
595	Rumex conglomeratus Murray	66	2	3	5	22	1	01	99	91	59	59
596	Rumex crispus L.	66	6	3	5	22	1	01	99	91	10	50
597	Rumex hydrolapathum L.	66	6	3	6	22	1	01	60	61	59	59
598	Rumex maritimus L.	66	4	1	5	22	1	01	60	91	40	59
599	Rumex obtusifolius L.	66	4	3	5	22	1	01	99	91	50	50
600	Rumex palustris Sm.	66	4	3	5	21	1	01	60	91	40	50
601	Rumex sanguineus L.	66	2	3	4	22	1	01	53	83	59	59
602	Rumex scutatus L.	66	2	3	4	22	1	01	53	10	57	57
606	Sagina apetala Ard. ssp. apetala	66	2	1	5	23	2	05	99	99	20	57
607	Sagina maritima G. Don fil	66	2	1	5	23	2	05	99	92	30	50
608	Sagina nodosa (L.) Fenzl	66	5	3	5	23	2	02	99	91	50	50
609	Sagina procumbens L.	66	2	3	5	23	2	05	99	99	30	50
610	Salicornia europaea L.	66	3	1	3	21	1	01	60	30	20	50
611	Salix aurita L.	87	3	6	4	12	1	01	52	82	50	50
612	Salix caprea L.	87	3	6	4	12	1	01	52	82	40	50
613	Salvia pratensis L.	84	2	3	4	22	1	04	00	71	59	59
614	Salvia verticillata L.	84	2	3	4	22	1	04	00	71	54	54
615	Sambucus ebulus L.	76	3	3	4	12	1	02	71	99	57	57
616	Sambucus nigra L.	76	3	6	4	12	1	02	71	82	50	50
617	Sambucus racemosa L.	76	3	6	4	12	1	02	71	83	30	57
618	Samolus valerandi L.	69	2	3	5	22	1	02	00	91	10	50
619	Sanguisorba minor Scop.	77	4	3	4	22	1	05	71	20	40	59
620	Saponaria officinalis L.	66	3	3	4	22	1	02	00	99	40	57
621	Satureja acinos (L.) Scheele	84	2	1	4	21	1	07	00	20	50	50
622	Saxifraga granulata L.	77	4	3	5	22	1	02	00	99	52	52
623	Saxifraga hirculus L.	77	4	3	5	22	1	02	00	62	40	51
624	Saxifraga rosacea Moench ssp. sponhemica (C.C.Gmel.)D.A.Webb	77	6	3	4	23	2	02	00	10	52	52
625	Saxifraga tridactylites L.	77	2	1	4	22	1	02	53	20	50	50
626	Saxifraga umbrosa L.	77	4	3	5	22	1	02	80	01	90	90
627	Scabiosa maritima L.	86	2	3	4	21	1	07	00	20	53	53
628	Scabiosa spec.	86	0	3	4	21	1	07	00	00	00	00
629	Scandix pecten-veneris L.	75	2	1	4	22	1	02	99	40	40	57
630	Schoenus nigricans L.	55	2	3	4	27	3	01	00	62	20	50
631	Scleranthus annuus L.	66	3	1	3	25	2	01	80	20	40	50
632	Scleranthus perennis L.	66	3	4	3	25	2	01	80	20	40	50
633	Scorzonera humilis L.	86	2	3	4	21	1	06	52	71	50	50
634	Scrophularia balbisii Hornem.	84	0	3	4	22	1	09	00	99	40	59
635	Scrophularia nodosa L.	84	4	3	4	21	1	09	00	82	40	50
636	Scrophularia peregrina L.	84	0	3	4	21	1	09	00	99	53	53
637	Scrophularia scorodonia L.	84	0	3	4	21	1	09	00	82	52	52
638	Scrophularia umbrosa Dum.	84	4	3	4	22	1	09	60	61	00	00
697	Scrophularia spec.	84	0	3	4	21	1	99	00	00	00	00
639	Scutellaria galericulata L.	84	4	3	6	23	2	04	60	61	30	50
640	Sedum acre L.	77	7	4	1	26	2	06	99	20	50	50
641	Sedum album L.	77	7	4	1	26	2	02	53	10	57	57

642	Sedum cepaea L.	77	2	1	1	21	1	08	99	82	55	55
643	Sedum dasyphyllum L.	77	7	4	1	24	2	02	53	10	59	59
644	Sedum forsteranum Sm.	77	0	4	1	25	2	06	53	10	52	52
645	Sedum reflexum L.	77	3	4	1	25	2	06	53	10	58	58
646	Sedum rubens L.	77	0	1	1	26	2	07	99	20	53	53
647	Sedum sediforme (Jacq.) Pau, non Hamlet	77	4	4	1	25	2	06	53	10	53	53
648	Sedum sexangulare L.	77	6	4	1	26	2	06	53	10	57	57
649	Sedum spurium Bieb.	77	4	4	1	23	2	08	80	01	90	90
650	Sedum telephium L.	77	7	4	1	21	1	03	99	99	40	56
699	Sedum spec.	77	0	4	1	24	2	00	00	00	00	00
652	Selaginella denticulata (L.) Link	31	0	4	5	26	2	07	51	10	53	53
653	Sempervivum tectorum L.	77	7	3	1	22	1	06	99	10	54	54
700	Senecio congestus (R.Br.) DC.	86	7	1	5	21	1	06	52	91	54	54
654	Senecio jacobea L.	86	7	3	4	22	1	06	52	99	40	50
655	Senecio lividus L.	86	0	1	4	21	1	06	52	50	53	53
656	Senecio squalidus L.	86	3	1	4	21	1	06	52	50	70	70
657	Senecio viscosus L.	86	7	1	4	21	1	06	52	50	50	50
658	Senecio vulgaris L.	86	7	1	4	21	1	06	52	50	40	50
660	Setaria viridis (L.) P.B.	53	2	1	4	23	2	05	80	40	40	50
661	Sherardia arvensis L.	76	2	1	4	26	2	04	53	40	50	50
662	Silene alba (Mill.) E.H.L.Krause ssp. alba	66	2	3	5	22	1	02	20	99	40	50
663	Silene italica (L.) Pers.	66	2	3	4	22	1	02	20	20	53	53
664	Silene otites (L.) Wibel	66	2	3	3	22	1	06	20	20	40	57
698	Silene viscosa (L.) Pers.	66	0	3	3	22	1	08	20	20	54	54
665	Silene vulgaris (Moench) Garcke	66	2	3	3	22	1	02	99	99	40	50
666	Silene spec.	66	0	3	4	22	1	99	00	00	00	00
667	Sinapis arvenis L.	81	2	1	4	21	1	06	80	40	10	50
668	Sisymbrium austriacum Jacq.	81	2	1	4	21	1	06	80	02	80	80
669	Sisymbrium irio L.	81	5	1	4	21	1	06	80	50	40	53
670	Sisymbrium officinale (L.) Scop.	81	2	1	4	21	1	06	80	50	40	50
671	Sium erectum Huds.	75	2	3	6	21	1	02	60	61	30	59
672	Solanum dulcamara L.	84	4	5	4	12	1	07	71	82	50	50
673	Solanum nigrum L.	84	6	1	5	21	1	02	71	50	10	50
674	Soleirolia soleirolii (Req.) Dandy	64	2	4	5	29	4	01	73	10	70	70
675	Solidago gigantea Ait.	86	3	3	4	23	2	06	80	01	90	90
676	Sonchus arvensis L.	86	5	3	3	22	1	06	52	99	40	50
677	(incl. S. maritimus L.)	86	2	3	3	22	1	06	52	20	55	55
678	Sonchus asper (L.) Hill	86	2	1	4	22	1	06	52	50	40	50
679	Sonchus oleraceus L.	86	4	3	4	22	1	06	52	50	50	50
680	Sonchus palustris L.	86	2	3	5	22	1	06	52	72	59	59
681	Sorbus aucuparia L.	77	2	6	4	12	1	02	71	82	50	50
682	Spergula arvensis L.	66	2	1	4	26	2	02	80	40	10	50
683	Spergularia marina (L.) Griseb.	66	4	1	3	26	2	02	60	30	20	50
684	Spergularia media (L.) C.Presl	66	2	3	3	26	2	02	99	92	20	50
685	Spergularia rubra (L.) J. et C.Presl	66	4	1	4	26	2	07	99	99	30	50
686	Spergularia rupicola Lebel ex le Jolis	66	4	4	3	25	2	07	53	10	55	55
687	Stachys officinalis (L.) Trevisan	84	2	3	4	22	1	03	00	71	59	59
688	Stachys palustris L.	84	6	3	6	21	1	07	60	72	30	50
689	Stellaria alsine Grimm	66	2	3	5	21	1	02	00	91	30	50
690	Stellaria graminea L.	66	2	3	5	21	1	02	80	99	40	50
691	Stellaria media (L.) Vill.ssp.media	66	4	1	5	23	2	02	99	50	10	50
692	(incl.var.apetala Goudin)	66	2	1	5	23	2	05	99	50	10	50

693	Stenactis annua (L.) N.ab E.	86	0	1	4	21	1	07	52	99	70	70
694	Stenactis strigosa (Muehhlenb. ex Willd.) DC.	86	4	1	4	23	2	02	52	02	80	80
695	Symphytum officinale L.	84	4	3	4	21	1	03	73	99	50	50
696	Syringa vulgaris L.	82	3	6	4	12	1	02	80	01	90	90
701	Tanacetum vulgare L.	86	2	3	4	23	2	06	52	99	40	50
702	Taraxacum	86	0	3	4	22	1	06	52	00	00	00
703	Taxus baccata L.	41	2	6	3	11	1	01	71	83	40	99
704	Teucrium montanum L.	84	6	3	4	23	2	06	73	10	56	56
705	Teucrium scorodonia L.	84	3	3	4	23	2	06	73	83	56	56
706	Thalictrum flavum L.	61	7	3	5	21	1	06	60	72	40	50
707	Thelypteris palustris Schott	36	2	2	5	21	1	00	51	61	30	50
708	Thlaspi arvense L.	81	2	1	4	21	1	02	53	40	40	50
709	Thymus serpyllum L.	84	4	4	3	25	2	07	73	20	40	50
710	Thymus vulgaris L.	84	2	4	3	25	3	07	73	81	53	53
711	Thymus spec.	84	0	4	3	25	2	99	00	00	00	00
712	Tilia cordata Mill.	71	2	6	5	12	1	05	53	83	50	50
713	Torilis arvensis (Huds.) Link	75	2	1	5	21	1	02	99	40	20	57
714	Torilis japonica (Houtt.) DC.	75	2	1	5	21	1	02	99	82	40	50
715	Trifolium arvense L.	77	2	1	4	21	1	07	99	20	40	50
716	Trifolium dubium Sibth.	77	3	1	4	21	1	07	99	71	59	59
717	Trifolium pratense L.	77	2	3	4	21	1	08	99	71	40	50
718	Trifolium repens L.	77	7	3	4	26	2	02	99	71	40	50
719	Trifolium spec.	77	0	1	4	21	1	99	00	00	00	00
720	Triglochin maritima L.	51	5	3	4	21	1	05	60	30	30	50
721	Trisetum flavescens (L.)P.B.	53	3	3	4	23	2	05	80	71	30	50
722	Tussilago farfara L.	86	5	2	4	21	1	06	52	99	40	50
723	Ulmus (incl.U.x hollandica Mill.)	64	0	6	4	12	1	05	53	01	90	90
724	Umbilicus rupestris (Salisb.)Dandy	77	6	4	1	24	2	07	53	10	55	55
725	Urtica atrovirens Req.ex Loisel.	77	2	3	5	23	2	01	99	50	53	53
726	Urtica dioica L.	77	7	3	5	23	2	01	99	50	30	50
727	Urtica urens L.	77	3	1	5	23	2	01	99	50	30	50
731	Vaillantia muralis (L.) DC.	76	2	1	4	26	2	02	53	20	53	53
732	Valeriana officinalis L.	76	5	3	5	21	1	07	52	72	40	50
733	Valerianella carinata Loisl.	76	2	1	5	21	1	04	53	40	57	57
734	Valerianella eriocarpa Desv.	76	2	1	5	21	1	04	53	40	53	53
735	Verbascum lychnitis L.	84	3	3	3	22	1	06	99	20	59	59
736	Verbascum nigrum L.	84	2	3	3	22	1	06	99	20	50	50
737	Verbascum phlomoides L.	84	3	3	3	22	1	06	99	20	53	53
738	Verbascum sinuatum L.	84	2	3	3	22	1	06	99	20	53	53

739	Verbascum thapsus L.	84	3	3	3	22	1	06	99	20	59	59
740	Verbascum spec.	84	0	3	3	22	1	06	00	00	00	00
742	Verbena officinalis L.	84	2	3	3	22	1	07	80	50	20	59
743	Veronica agrestis L.	84	3	1	4	21	1	04	80	40	58	58
744	Veronica arvensis L.	84	2	1	4	23	2	04	73	40	40	50
745	Veronica chamaedrys L.	84	4	4	4	23	2	04	99	99	40	50
746	Veronica cymbalaria Bodard	84	0	1	5	26	2	04	73	99	53	53
747	Veronica hederifolia L.	84	6	1	5	26	2	04	73	99	40	50
748	Veronica officinalis L.	84	3	4	4	26	2	04	99	99	30	50
749	Veronica polita Fr.	84	2	1	4	21	1	04	73	40	40	50
750	Veronica serpyllifolia L.	84	2	3	5	21	1	04	73	71	30	50
751	Veronica teucrium L.	84	6	3	4	23	2	04	00	82	40	57
752	Veronica spec.	84	0	3	4	23	2	99	00	00	00	00
753	Vinca maior L.	83	4	4	4	26	2	04	00	82	53	53
754	Viola arvensis L.	63	2	1	5	21	1	06	20	40	10	50
755	Viola odorata L.	63	3	3	5	21	1	04	20	82	50	50
756	Viola tricolor L. ssp. curtisii (Forst.)Syme	63	2	1	5	21	1	06	20	20	40	50
757	Vulpia myuros (L.) C.C. Gmel.	53	5	1	3	23	2	05	52	20	20	57
761	Wisteria sinensis (Sims) Sweet	77	2	6	4	12	2	04	80	01	90	90

BRYOPHYTES

		a	d	e	f	i	j	k
801	Aloina aloides (Schultz) Kindb.	15	7	32	3	17	30	59
802	Aloina rigida (Hedw.) Limpr. (incl. var.ambigua (B.S.G.) Craig.)	15	7	32	3	17	30	59
804	Amblystegium juratzkanum Schimp.	22	8	44	2	98	30	50
805	Amblystegium serpens (Hedw.) B.S.G.	22	8	44	2	98	30	50
806	Amblystegium varium (Hedw.) Lindb.	22	9	44	2	98	30	58
807	Anisothecium varium (Hedw.) Mitt.	22	8	44	2	06	30	56
808	Anomodon viticulosus (Hedw.)Hook. et Tayl.	22	8	42	2	98	30	50
809	Atrichum undulatum (Hedw.) P.Beauv.	11	9	32	2	06	30	50
810	Barbula acuta (Brid.) Brid.	15	7	32	3	17	30	59
811	Barbula fallax Hedw.	15	8	32	3	17	30	50
812	Barbula hornschuchiana Schultz	15	8	32	2	26	59	59
813	Barbula revoluta Brid.	15	7	32	4	16	59	59
814	Barbula sinuosa (Mitt.) Grav.	15	8	32	2	16	54	54
815	Barbula unguiculata Hedw.	15	0	32	3	17	30	50
816	Barbula vinealis Brid. (incl. var. cylindrica (Tayl.) Boul.)	15	8	32	3	16	30	59
818	Bartramia stricta Brid.	19	8	34	3	17	30	53
819	Brachythecium albicans (Hedw.) B.S.G.	22	7	44	3	26	30	50
820	Brachythecium glareosum (Spruc.) B.S.G.	22	8	44	2	17	40	50
821	Brachythecium mildeanum (Schimp.) Schimp.	22	8	44	2	06	40	58
822	Brachythecium populeum (Hedw.) B.S.G.	22	8	44	2	98	30	50
823	Brachythecium rivulare B.S.G.	22	9	44	2	17	30	50
824	Brachythecium rutabulum (Hedw.) B.S.G.	22	0	44	2	98	20	50
825	Brachythecium salebrosum (Web. et Mohr) B.S.G.	22	8	44	2	98	30	50
826	Brachythecium velutinum (Hedw.) B.S.G.	22	8	44	3	98	30	50
827	Bryoerythrophyllum recurvirostre (Hedw.) Chen	15	0	32	3	17	30	50
828	Bryum argenteum Hedw.	19	0	32	3	17	10	50
829	Bryum atrovirens Brid.	19	7	32	2	06	30	50
830	Bryum bicolor Dicks.	19	7	32	3	17	20	59
831	Bryum caespiticium Hedw. (incl. ssp.kunzei Podp.)	19	0	32	3	17	30	50
833	Bryum capillare Hedw. (incl. ssp. torquescens (De Not.) Kindb.)	19	0	32	3	98	30	50

835	Bryum creberrimum Tayl.	19	9	32	3	17	30	58
836	Bryum inclinatum (Brid.) Bland.	19	7	32	2	17	30	50
837	Bryum intermedium (Brid.) Bland.	19	9	32	2	17	30	58
838	Bryum murorum (Schimp.) Berk.	19	8	32	3	16	55	55
839	Bryum pallescens Schwaegr.	19	9	32	2	16	30	58
840	Bryum pseudotriquetrum (Hedw.) Gaertn., Meyer et Scherb. (incl. var. bimum (Schreb.) Hartm.)	19	9	32	2	96	30	50
843	Bryum spec.	19	8	32	3	00	00	00
844	Calliergonella cuspidata (Hedw.) Loesk.	22	0	44	2	06	20	50
845	Camptothecium aureum (Lag.) B.S.G.	22	7	42	3	26	53	53
846	Camptothecium lutescens (Hedw.) B.S.G.	22	7	42	3	17	30	50
847	Campylium hispidulum (Brid.) Mitt.	22	9	44	2	17	30	50
848	Cephaloziella hampeana (Nees) Schiffn.	03	7	45	3	26	20	50
849	Ceratodon purpureus (Hedw.) Brid.	13	7	32	3	98	30	50
850	Conocephalum conicum (L.) Dum.	01	9	46	4	17	30	50
851	Cratoneuron commutatum (Hedw.) Roth	22	9	44	2	17	30	50
852	Cratoneuron filicinum (Hedw.) Spruc.	22	8	44	2	17	20	50
853	Ctenidium molluscum (Hedw.) Mitt.	22	8	44	2	06	30	50
854	Dicranella heteromalla (Hedw.) Schimp.	13	8	32	2	06	20	50
974	Dicranella subulata (Hedw.) Schimp.	13	8	32	2	06	30	50
855	Didymodon rigidulus Hedw.	15	8	32	3	17	30	50
856	Didymodon tophaceus (Brid.) Lisa	15	9	32	4	16	30	50
857	Didymodon trifarius (Hedw.) Roehl.	15	8	32	3	17	30	57
858	Distichium capillaceum (Hedw.) B.S.G.	13	8	32	3	17	30	51
859	Ditrichum flexicaule (Schwaegr.) Hamp.	13	7	32	3	17	30	59
860	Drepanocladus aduncus (Hedw.) Warnst.	22	9	44	2	96	20	50
861	Encalypta streptocarpa Hedw.	16	8	32	2	16	30	50
862	Encalypta vulgaris Hedw.	16	8	32	2	17	30	50
863	Eucladium verticillatum (Brid.) B.S.G.	15	9	35	4	16	30	59
864	Eurhynchium pulchellum (Hedw.) Jenn.	22	8	44	2	17	30	50
976	Eurhynchium striatum (Hedw.) Schimp.	22	8	44	2	17	50	50
865	Fissidens bryoides Hedw.	12	8	34	3	06	30	50
866	Fissidens cristatus Mitt.	12	8	34	2	17	30	50

867	Fissidens osmundoides Hedw.	12	9	34	3	96	30	50
868	Fissidens taxifolius Hedw.	12	8	34	2	06	30	59
869	Frullania dilatata (L.) Dum.	03	8	45	2	18	40	55
870	Frullania tamarisci (L.) Dum.	03	8	45	2	18	30	50
871	Funaria calcarea Wahlenb. ssp. mediterranea (Lindb.) Kindb.	17	7	32	4	16	30	53
872	Funaria hygrometrica Hedw.	17	0	32	2	17	10	50
873	Grimmia pulvinata (Hedw.) Sm. (incl. Grimmia pulvinata var.africana (Hedw.) Hook f. et Wils.)	14	7	34	4	16	20	50
875	Grimmia trichophylla Grev.	14	7	34	3	16	20	50
876	Grimmia unicolor Hook.	14	7	34	3	16	30	58
877	Gymnostomum aeruginosum Sm.	15	8	32	3	16	30	50
878	Gyroweisia tenuis (Hedw.) Schimp.	15	8	32	4	16	30	50
879	Homalothecium sericeum (Hedw.) B.S.G.	22	7	42	3	18	30	50
880	Hygroamblystegium tenax (Hedw.) Jenn.	22	9	44	2	19	30	50
881	Hygrohypnum luridum (Hedw.) Jenn.	22	9	44	2	19	30	50
882	Hymenostomum tortile (Schwaegr.) B.S.G.	15	8	32	2	17	59	59
883	Hypnum cupressiforme Hedw. (incl. Hypnum cupressiforme var. lacunosum Brid. Hypnum cupressiforme var. resupinatum (Tayl.) Schimp. Hypnum cupressiforme f. tectorum (Brid.) C.Jens.)	22	8	44	2	98	10	50
887	Isothecium myurum Brid.	21	8	43	2	18	50	50
888	Leiocolea muelleri (Nees.) Joerg.	03	9	45	2	17	30	58
889	Leptobryum pyriforme (Hedw.) Wils.	19	8	32	3	17	20	50
890	Leptodictyum riparium (Hedw.) Warnst.	22	9	44	4	19	20	50
891	Leptodon smithii (Hedw.) Web. et Mohr.	21	8	42	3	18	20	55
892	Leskea polycarpa Hedw.	22	9	42	3	18	30	50
893	Leucodon sciuroides (Hedw.) Schwaegr.	21	7	42	3	18	30	50
894	Lophocolea bidentata (L.) Dum.	03	9	45	2	17	30	50
895	Lunularia cruciata (L.) Dum.	01	9	46	4	17	20	55

896	Marchantia polymorpha L.	01	9	46	2	17	10	50
897	Metzgeria furcata (L.) Dum.	02	8	46	3	15	20	50
898	Mnium affine Funck.	19	8	32	2	06	30	50
899	Mnium hornum Hedw.	19	8	32	3	17	30	50
900	Mnium marginatum (With.) P. Beauv.	19	8	32	2	17	30	50
901	Mnium rostratum Schrad.	19	8	32	2	17	10	50
902	Mnium stellare Hedw.	19	8	32	2	17	30	50
903	Mnium undulatum Hedw.	19	9	32	2	06	30	50
904	Neckera complanata (Hedw.) Hueb.	21	8	41	2	18	30	50
905	Neckera crispa Hedw.	21	8	41	2	18	50	50
906	Greoweisia bruntonii (Sm.) Mild.	13	8	32	3	16	59	59
907	Orthotrichum affine Brid.	18	7	34	3	17	30	50
908	Orthotrichum anomalum Hedw.	18	7	34	3	16	30	50
909	Orthotrichum cupulatum Brid.	18	7	34	3	16	30	50
910	Orthotrichum diaphanum Brid.	18	7	34	3	17	30	50
911	Orthotrichum lyellii Hook. et Tayl.	18	7	42	3	17	30	56
912	Orthotrichum rivulare Turn.	18	9	34	2	18	56	56
913	Orthotrichum rupestre Schleich. ex Schwaegr.	18	7	34	3	16	20	50
914	Orthotrichum spec.	18	7	34	3	00	00	00
915	Oxyrrhynchium hians (Hedw.) Loesk.	22	8	44	2	17	30	50
916	Oxyrrhynchium praelongum (Hedw.) Warnst.	22	8	44	3	98	30	50
917	Oxyrrhynchium schleicheri (Hedw.f.)Roell	22	9	45	2	17	57	57
918	Oxyrrhynchium speciosum (Brid.) Warnst.	22	9	44	2	19	56	56
919	Oxyrrhynchium swartzii (Turn.) Warnst.	22	8	44	2	98	30	50
920	Oxystegus cylindricus (Brid.) Hilp.	15	8	32	3	17	30	50
921	Pellia endiviaefolia (Dicks.) Dum.	02	9	46	2	96	30	50
922	Phascum floerkeanum Web. et Mohr.	15	8	32	3	26	30	56
923	Plasteurhynchium meridionale (B.S.G.) Fleisch.	21	8	44	3	16	53	53
924	Plasteurhynchium striatulum (Spruc.) Fleisch.	21	8	44	2	16	57	57
925	Pleurochaete squarrosa (Brid.) Lindb.	15	7	32	3	06	30	59
926	Pleurozium schreberi (Brid.) Mitt.	22	7	44	1	06	30	50
927	Pohlia camptotrachela (Ren. et Card.) Broth.	19	8	32	3	96	30	50
928	Pohlia nutans (Hedw.) Lindb.	19	8	32	2	06	30	50
929	Polytrichum commune Hedw.	11	9	33	3	06	10	50
930	Polytrichum formosum Hedw.	11	8	33	3	06	30	59
931	Polytrichum piliferum Hedw.	11	7	33	2	06	10	50

932	Porella platyphylla (L.) Lindb.	03	8	45	2	17	30	55
933	Pottia heimii (Hedw.) Hamp.	15	0	32	2	17	30	52
934	Pottia intermedia (Turn.) Fuernr.	15	8	32	2	17	30	50
935	Pottia lanceolata (Hedw.) C.Muell.	15	8	32	2	26	40	59
936	Preissia quadrata (Scop.) Nees.	01	9	46	3	17	30	50
937	Pseudoscleropodium purum (Hedw.) Fleisch.	22	9	44	2	06	30	50
938	Pterogonium gracile (Hedw.) Sm.	21	8	43	2	18	20	57
939	Pylaisia polyantha (Hedw.) B.S.G.	22	8	44	2	18	30	50
940	Radula complanata (L.) Dum.	03	8	45	2	07	30	59
941	Reboulia haemisphaerica (L.) Raddi	01	9	46	2	16	10	50
942	Racomitrium canescens (Hedw.) Brid.	14	7	34	3	26	30	50
943	Racomitrium heterostichum (Hedw.) Brid.	14	7	34	4	16	20	50
944	Racomitrium lanuginosum (Hedw.) Brid.	14	7	34	3	17	20	50
945	Rhynchostegiella tenella (Dicks.) Limpr.	22	9	45	2	16	59	59
946	Rhynchostegium confertum (Dicks.) B.S.G.	22	8	44	2	18	57	57
947	Rhynchostegium megapolitanum (Web. et Mohr.) B.S.G.	22	8	44	2	06	59	59
948	Rhynchostegium murale (Hedw.) B.S.G.	22	8	44	2	16	59	59
949	Rhytidiadelphus squarrosus (Hedw.) Warnst.	22	8	44	2	06	30	50
950	Schistidium apocarpum (Hedw.) B.S.G.	14	8	32	3	16	20	50
951	Scleropodium cespitans (C.Muell.) L.Koch	22	8	43	3	16	30	55
952	Scorpiurium circinatum (Brid.) Fleisch. et Loesk.	21	8	43	3	17	55	55
953	Solenostoma cordifolium (Hook.) Steph.	03	9	45	2	19	30	51
954	Streblotrichum convolutum	15	8	32	2	17	30	50
955	Targionia hypophylla L.	01	9	46	2	16	10	59
956	Thuidium tamariscinum (Hedw.) B.S.G.	22	8	44	2	06	30	50
975	Timmiella anomala (B.S.G.) Limpr.	15	8	32	2	17	30	53
957	Tortella flavovirens (Bruch) Loesk.	15	7	32	3	06	30	55
958	Tortella nitida (Lindb.) Broth.	15	7	32	3	16	30	53
959	Tortella tortuosa (Hedw.) Limpr.	15	8	32	2	17	30	50
960	Tortula intermedia (Brid.) De Not.	15	7	34	4	16	59	59
961	Tortula laevipila (Brid.) Schwaegr.	15	7	32	2	07	30	59
962	Tortula marginata (B.S.G.) Spruc.	15	8	32	2	16	30	55
963	Tortula muralis Hedw.	15	7	32	3	16	10	50
964	Tortula ruralis (Hedw.) Gaertn., Meyer et Scherb. (incl. Tortula ruralis var.arenicola (Besch.) Wildeman)	15	7	34	4	17	30	50
966	Tortula subulata Hedw.	15	8	32	2	17	30	50

967	Tortula virescens (De Not.) De Not.	15	7	34	4	18	54	54
968	Trichostomum brachydontium Bruch (incl. Trichostomum brachydontium ssp. cuspidatum (Braithw.) Glac.)	15	7	32	3	16	40	59
971	Trichostomum crispulum Bruch (incl. Trichostomum crispulum var. brevifolium B.S.G.	15	7	32	3	17	30	50
972	Weisia fallax Sehlm.	15	8	32	3	16	30	53
973	Zygodon viridissimus (Dicks.) Brid.	18	8	32	2	07	30	50

<u>LICHENES</u>

981	Cladonia coniocraea (Flk.) Sandst. = C.fimbriata var. c. (Flk.) Zahlbr.	00	7	59	2	17	00	00
982	Cladonia fimbriata (L.) Fr.	00	7	59	2	17	00	00
983	Cladonia mitis Sandst.	00	7	59	2	26	20	50
984	Cladonia pyxidata (L.) Fr.	00	7	59	2	17	30	50
985	Cladonia spec.	00	7	59	2	00	00	00
991	Collema	00	7	51	2	00	00	00
995	Lepraria	00	9	52	3	00	00	00
996	Peltigera canina (L.) Willd.	00	8	55	1	00	00	00
999	Xanthoria parietina (L.) Th. Fr.	00	7	54	2	00	00	00

GPSR Compliance

The European Union's (EU) General Product Safety Regulation (GPSR) is a set of rules that requires consumer products to be safe and our obligations to ensure this.

If you have any concerns about our products, you can contact us on ProductSafety@springernature.com

In case Publisher is established outside the EU, the EU authorized representative is:

Springer Nature Customer Service Center GmbH
Europaplatz 3
69115 Heidelberg, Germany

Batch number: 10824476

Printed by Printforce, the Netherlands